516.15 Spi

Spiral symmetry.

**The Lorette Wilmot Library
Nazareth College of Rochester**

SPIRAL SYMMETRY

"Spiral" This curious lava form, beaten and shaped by constant pounding from ocean waves, is the home of various limu (seaweeds). While on an early morning seashore walk on the North Shore of Oahu, I was drawn to this composition by the spiraling combination of opposing characteristics: dark and light, static and dynamic, solid and liquid.

© James Kastner, 1990.
(Kastner Design Associates — 707 Queen Street, Honolulu HI 96813–5109, USA).

SPIRAL SYMMETRY

editors

István Hargittai
*Technical University of Budapest
and Hungarian Academy of Sciences
Budapest, Hungary*

Clifford A. Pickover
*IBM T. J. Watson Research Center
Yorktown Heights, New York*

World Scientific
Singapore • New Jersey • London • Hong Kong

Published by

World Scientific Publishing Co. Pte. Ltd.
P O Box 128, Farrer Road, Singapore 9128
USA office: Suite 1B, 1060 Main Street, River Edge, NJ 07661
UK office: 73 Lynton Mead, Totteridge, London N20 8DH

Library of Congress Cataloging-in-Publication data is available.

SPIRAL SYMMETRY

Copyright © 1992 by World Scientific Publishing Co. Pte. Ltd.

All rights reserved. This book, or parts thereof, may not be reproduced in any form or by any means, electronic or mechanical, including photocopying, recording or any information storage and retrieval system now known or to be invented, without written permission from the Publisher.

ISBN 981-02-0615-1

Cover illustration: Computer Graphics, by C. A. Pickover.

Printed in Singapore by General Printing Services Pte. Ltd.

PREFACE

> If the cosmos were suddenly frozen, and all movement ceased, a survey of its structure would not reveal a random distribution of parts. Simple geometrical patterns, for example, would be found in profusion—from the spirals of galaxies to the hexagonal shapes of snow crystals. Set the clockwork going, and its parts move rhythmically to laws that often can be expressed by equations of surprising simplicity. And there is no logical or a priori reason why these things should be so.
>
> Martin Gardner, *Order and Surprise* (1985)

"Whatever can be done once can always be repeated," begins Louise B. Young in *The Mystery of Matter* when describing the shapes and structures of nature. From the tiny twisted DNA molecules in all living cells to the gargantuan curling arms of many galaxies, the physical world contains a startling repetition of spiral patterns. The fundamental rules underlying the apparent general repetition of nature has led to the search to identify, measure, and define these patterns in precise scientific terms (see bibliography). A seminal early book concerned with spirals in nature is *On Growth and Form* by D'Arcy Thompson. In his book, Thompson defines and illustrates the remarkable natural occurrences of spirals (he capitalizes this word) with pictures and words:

> ...a Spiral is a curve which, starting from a point of origin, continually diminishes in curvature as it recedes from that point; or, in other words, whose *radius of curvature* continually increases.... Of true organic spirals we have no lack. We think at once of horns of ruminants, and of still more exquisitely beautiful molluscan shells—in which (as Pliny says) *magna ludentis Naturae varietas*.

The present collection of papers presents aesthetically appealing and scientifically interesting patterns from a range of scientific, historical, and artistic realms. The resulting pictures should be of interest to a range of scientists as well as home-computer artists. The term "spiral" is used generically to describe any geometrical smooth curve, or collection of discrete points, that winds about a central point or axis while also receding from it. When thinking of examples of spirals, both the mundane and exotic easily come to mind, for example: the gentle curl of a fern tendril, the shape of an octopus's retracted arm, the death-form assumed by a centipede, the spiral intestine of a giraffe, the shape of a butterfly's tongue, the spiral cross-section of a scroll, the shape of the Yellow Brick Road in Munchkinland in the film classic "The Wizard of Oz", and even the characters of several written languages. For additional pictorial examples, see Figs. 1–3.

One goal of this book is to emphasize the important and conspicuous role that spirals play in science and, in some cases, to show the reader how to create such spirals using a computer. Another goal is to demonstrate how research in simple mathematical formulas can reveal an inexhaustible new reservoir of magnificent shapes and images. Indeed, structures produced by these equations include shapes of startling intricacy. The graphics experiments presented, with the variety of accompanying parameters, are good ways of characterizing the complexity of such behavior.

From an artistic standpoint, spiral equations provide a vast and deep reservoir from which artists can draw. The computer is a machine which, when guided by an artist, can render images of captivating power and beauty. New "recipes", such as those outlined here, interact with such traditional elements as form, shading, and color to produce futuristic images and effects. The recipes function as the artist's helper, quickly taking care of much of the repetitive and sometimes tedious detail. By creating an environment of advanced computer graphics, artists with access to computers will gradually change our perception of art.

As far as the physical universe is concerned, spiral shapes are one of nature's most fundamental forms. As science writer Kathleen Stein once pointed out, spirals appear early in the chain of animal evolution: cilia, worms' gills, fly larvae, and some shark-egg capsules have them. The cochlea in the ear of every mammal is screw-shaped, and so are many corals. Many spiral shapes correspond to magnificently complicated structures which no one could fully have appreciated or suspected before

Fig. 1. Fractal spiral produced by a "mathematical feedback loop" in the complex plane. For more information, see Pickover (1990).

the age of the computer. The richness of resultant forms often contrasts with the simplicity of the generating formula.

Precisely *why* the spiral is ubiquitous in nature and in civilization is a deep question. Whether one is considering the movement of stars, the development of an embryo, the motion of a pencil on a page, or much of the phenomena which makes up the fabric of our universe, it is clear that symmetry operations are often nature's guiding hand. Spiral patterns

Fig. 2. 18th Century engraving depicting a spiral musical instrument. This "voice amplifier" appeared in Jesuit Filippo Bonanni's book *Gabinetto Armonico*.

Fig. 3. Spirals in an Art Nouveau decorative stair railing.

often occur spontaneously in matter which is organized through symmetry transformations: change of size (growth) and rotation. Form follows function, and the spiral form can allow for the compaction of a relatively long length. Long-yet-compact tubes are useful in spiral molds, brass instruments, mollusks, and cochleas for obvious reasons including physical strength and increased surface area. For some phenomena, such as in doodles, written languages, and spiral galaxies, the precise "reason" for spiral forms is less clear.

The mathematical concept of similarity holds one of the keys to understanding the processes of growth in the natural world. As a member of a species grows to maturity it generally transforms in such a way that its parts maintain approximately the same proportion with respect to each other, and this is probably a reason why nature is often constrained to exhibit self-similar spiral growth. Through time, humans have imitated the spiral motifs around them in their art forms, and occasionally they invent new spirals which presently are not known to have specific counterparts in the natural world. It is probable that the multitude of "traditional" and "nontraditional" spiral formulas will help scientists better understand the fundamental rules underlying the apparent spiral repetition of nature since they can now generate, define, and predict these patterns using simple algorithms.

January 1991   István Hargittai and Clifford A. Pickover

Bibliography

The following books and papers contain interesting examples of spirals in a variety of scientific and historical disciplines.

M. Gardner, *Spirals, In The Unexpected Hanging.* Simon and Schuster, New York (1969).
I. Hargittai, ed. *Symmetry: Unifying Human Understanding.* Pergamon Press, New York (1986).
I. Hargittai, ed. *Symmetry 2: Unifying Human Understanding.* Pergamon Press, Oxford (1989).

Y. Kawaguchi, *A Morphological Study of the Form of Nature*. Computer Graphics (ACM-SIGGRAPH), Boston (1982).
T. Cook, *The Curves of Life*. Dover, New York (1979).
R. Dixon, The mathematics and computer graphics of spirals in plants, *Leonardo* **16** (1983) 86–90.
R. V. Jean, *Mathematical Approach to Pattern and Form in Plant Growth*. Wiley, New York (1984).
C. Pickover, The world of chaos, *Computers in Physics* **4**(5) (1990) 460–470.
C. Pickover, Mathematics and beauty II: A sampling of spirals and "strange" spirals in nature, science, and art, *Leonardo* **21**(2) (1988) 173–181. Reprinted in *Computers, Pattern, Chaos and Beauty*. St. Martin's Press, New York (1990).
C. Pickover, *Computers and the Imagination*. St. Martin's Press, New York (1991).
T. Schwenk, *Sensitive Chaos*. Schocken Books, New York (1976).
D. Thompson, *On Growth and Form*. Cambridge University Press, Cambridge (1961).
L. Young, *The Mystery of Matter*. Oxford University Press, New York (1965).

CONTENTS

Preface	v
Contributors	xv
The Spiral in Nature, Myth, and Mathematics J. Kappraff	1
Does the Golden Spiral Exist, and if not, Where is its Center? A. L. Loeb and W. Varney	47
Pythagorean Spirals E. J. Eckert	63
Dynamical Spirals A. V. Holden	73
Random Spirals W. A. Seitz and D. J. Klein	83
Spiral Galaxies B. G. Elmegreen	95
Spiral-Based Self-Similar Sets K. Wicks	107
Symmetry and Spirals: An Artist's Personal Statement R. Newman	123
Spiral Structures in Julia Sets and Related Sets M. Michelitsch and O. E. Rössler	129

The Evolution of a Three-Armed Spiral in the Julia Set, and
Higher Order Spirals 135
 A. G. Davis Philip

Autonomous Organization of a Chaotic Medium into Spirals:
Simulations and Experiments 165
 M. Markus

Broken Symmetry and the Formation of Spiral Patterns in
Fluids 187
 I. Stewart

Oscillations, Waves, and Spirals in Chemical Systems 221
 E. Körös

Determination of Spiral Symmetry in Plants and Polymers 251
 D. Friedman

Electromagnetic Theory for Chiral Media 281
 A. Lakhtakia

Sunflower Quasicrystallography 295
 L. A. Bursill, J. L. Rouse, and A. Needham

On the Origins of Spiral Symmetry in Plants 323
 R. V. Jean

Green Spirals 353
 R. Dixon

The Form, Function, and Synthesis of the Molluscan Shell 369
 M. Cortie

Isometric Systems in Isotropic Space: An Artist's Personal
Statement on Spiral and Other Map Projections 389
 A. Denes

Helmet 395
 E. Kent

Spinning Descartes into Blake: Spirals, Vortices, and the
Dynamics of Deviation 399
 K. L. Cope

CONTRIBUTORS

István Hargittai
Technical University
 of Budapest and Hungarian
 Academy of Sciences
H-1431 Budapest
Hungary

Clifford A. Pickover
IBM T.J. Watson Research Center
Yorktown Heights,
 New York 10598
U.S.A.

Jay Kappraff
Dept. of Mathematics
New Jersey Institute of
 Technology
University Heights
Newark, New Jersey 07102
U.S.A.

Arthur L. Loeb and
 William Varney
Dept. Visual and Environmental
 Sciences
Carpenter Center for the Visual
 Arts
Harvard Univ.
Cambridge, Massachusetts 02138
U.S.A.

Ernest J. Eckert
Univ. of South Carolina at Aiken
171 University Parkway
Aiken, South Carolina 29801
U.S.A.

Arun V. Holden
Dept. of Physiology and Centre
 for Nonlinear Studies
The University, Leeds LS2 9JT
United Kingdom

W.A. Seitz and D.J. Klein
Marine Sciences
Texas A&M Univ. at Galveston
Galveston, Texas 77553-1675
U.S.A.

Bruce G. Elmegreen
IBM Research Division
T.J. Watson Research Center
Yorktown Heights,
 New York 10598
U.S.A.

Keith Wicks
Dept. of Mathematics and
 Computer Science
Univ. College of Swansea
Singleton Park, Swansea SA2 8PP
United Kingdom

Rochelle Newman
Pythagorean Education Project
P.O. Box 162
Bradford, Massachusetts 01835
U.S.A.

Michael Michelitsch and
 Otto E Rössler
Institute for Physical and
 Theoretical Chemistry
Univ. of Tübingen
W-7400 Tübingen
Germany

A.G. Davis Philip
Physics Dept.
Union College
Schenectady, New York 12308
U.S.A.

Mario Markus
Max Planck Institut für
 Ernährungsphysiologie
Rheinlanddamm 201
W-4600 Dortmund 1
Germany

Ian Stewart
Nonlinear Systems Laboratory
Mathematics Institute
Univ. of Warwick
Coventry CV4 7AL
United Kingdom

Endre Körös
Dept. of Inorganic and
 Analytical Chemistry
Eötvös Univ. H-1518
Budapest-112, Pf. 32
Hungary

Dawn Friedman
Dept. of Chemistry
Harvard Univ.
Cambridge, Massachusetts 02138
U.S.A.

Akhlesh Lakhtakia
Dept. of Engineering Science and
 Mechanics
Pennsylvania State Univ.
University Park,
 Pennsylvania 16802
U.S.A.

L.A. Bursill J.L. Rouse, and
  Alun Needham
School of Physics
Univ. of Melbourne
Parkville, Victoria 3052
Australia

Roger V. Jean
Univ. of Quebec
300 avenue des Ursulines
Rimouski, Quebec G5L 3A1
Canada

Robert Dixon
Mathographics
125 Cricklade Avenue
London SW2
United Kingdom

Michael Cortie
28 Francois Avenue
Bordeaux
Randburg 2194
South Africa

Agnes Denes
595 Broadway
New York, New York 10012
U.S.A.

Eleanor Kent
544 Hill Street
San Francisco, California 94114
U.S.A.

Kevin L. Cope
Dept. of English
Louisiana State Univ.
Baton Rouge, Louisiana 70803
U.S.A.

THE SPIRAL IN NATURE, MYTH, AND MATHEMATICS

Jay Kappraff

> *To see a World in a Grain of Sand*
> *And Heaven in a Wild Flower*
> *Hold Infinity in the palm of your hand*
> *And Eternity in an hour.*
>
> William Blake
> "Auguries of Innocence"

1. Introduction

Modern science has been remarkably successful in explaining and gaining control over nature. However, this knowledge has led us to substitute a synthetic world for the real one. Many people have lost touch with the natural world around them. It is this lack of awareness that enables us to do things that threaten the well-being of our environment. There appears to be some fundamental missing elements in our scientific model of the world. To some degree, this incompleteness is reflected in the difficulty science has when explaining the behaviour of living systems in contrast with its ability to explain the non-living and mechanical worlds. Science can describe the atom in isolation from its context better than it can explain how complex organic systems function.

The purpose of this paper is to show that our technological culture may have much to learn from looking at alternative descriptions of nature found in primitive societies. These societies were in tune with their environments and encoded their understanding of nature in myth and ritual. The spiral played an important symbolic role in these myths.

In Sec. 2, we shall examine the way in which myth and ritual of the Australian aborigines and the Fali of the Cameroons provide a framework for relating to nature, and we shall examine the symbolic role that the spiral plays in this, drawing on the work of Enrico Guidoni.[1] In Sec. 3, we shall present an outline of the observations and ideas of Theodor Schwenk on the creative force of water in the genesis of organic forms.[2]

In any attempt to build new theories that encompass the images portrayed by Guidoni and Schwenk, mathematics is bound to play an important role. After all, mathematics is the study of symbolic relationships. Euclidean geometry (the geometry that we learned in school) has been the primary tool in formulating mathematical models of the physical world. However, there is a more general geometry, namely projective geometry, of which Euclidean geometry is a special case. Projective geometry may be more successful in describing natural processes. After briefly introducing in Sec. 4 some of the central ideas of projective geometry, we shall show how Lawrence Edwards[3] used projective geometry to describe the shapes of plants and other biological forms, as well as the creation of the watery vortex. Again, the spiral plays a key role in these descriptions.

Section 5 is devoted to a brief discussion of how these three parts relate to each other. These connections are my own, and I invite the reader to make his or her own connections. I feel that a better understanding of the extraordinary patterns of interconnectedness and genesis of forms exhibited by the natural world, and echoed in the patterns of primitive thought, could be used to guide our modern world along less self-destructive paths.

2. Myth and Ritual

We will examine the myths based on the ancestral Gods of two primitive cultures, the Australian Aborigines and the Fali from the Cameroons.[1] The myths of these cultures are encoded in their social relationships, architecture, art, and survival mechanisms.[4]

2.1 *The Australian Aborigines*

The Aborigines are a nomadic people with a strong relationship to the territory through which they roam. In an ancient time, known as the "epoch of the sky," gods inhabited the territory. At a later time, known as "the epoch of the dream," these gods were replaced by legendary heroes and relegated to a mythical time past and to eternal idleness. The heroes, being less removed from men's experience than the Gods, present men with a model that they can emulate. Through the example of these heroes, men became capable of molding nature, of controlling what it has to give, and of introducing aspects of the natural world into their social systems, architecture, etc. In this way, all species and phenomena of nature are reduced by a tribe member to parts of his own organization of kinship class, clan, etc.

The landscape was said to be formed by the mythical ancestor-snake when he emerged from the sea and crawled across the dry land leaving his sinuous track imprinted there forever (see Fig. 1). This relationship

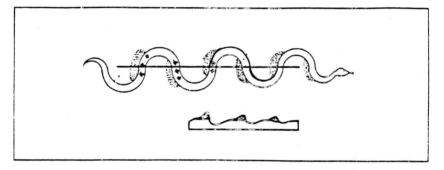

Fig. 1. Australian aborigines: Mythical formation of the dunes along a river. Schematic representation of the river-as-serpent (Australia). By courtesy of Elemond Milano.

can be understood by looking at the undulating pattern of a river as it traverses the landscape and comparing it with the meandering course of a snake slithering along the earth (meanders are related to spiral movements, as we shall see in Sec. 3). There are many different tribal groups, each with its own interlocking myths, which direct them along paths that often cross. The underpinning of tribal unity is, however, that the territory is conceived as a network of sacred centers that represented the campsites along the wandering snakelike paths. These centers were located at water holes and the path takes the nomads to places of abundant food and water as the seasons change. These centers are thought to have been used by their mythic progenitors to issue from and reenter the earth during their wanderings. These centers or water holes are symbolized by spirals.

More than one group can occupy a single campsite at the same time, but only according to specific rules laid down in the myths. Each tribe has a coat of arms consisting of connected bands cut with diagonal parallel lines inclined alternately to the left and the right to represent a particular stretch of watercourse inhabited by the tribe. The whole course of the river is thereby synthesized into a zigzag pattern that shows the position of the various tribes as they disperse themselves along the river relative to each other and as they follow the itinerary of the "culture hero" in their wanderings.

During their wanderings, the chief of the group carries a sacred pole with him to symbolize the sacred center, and he helps orient the tribe to the space and point to the directions of the path through it. Legend has said that the Achilpa tribe perished when this pole was accidently broken. The group became utterly disoriented and could not proceed. The Achilpa hero, Numbakulla, in a myth, is said to have climbed to the heavens on the pole, again emphasizing the pole's connection of heaven with earth.

Large numbers of "vital spirits" are always present at the sacred places. Conception is portrayed by the passing into the uterus of one of these "spirit babies." Since every individual existed prior to his life in a specific territorial "center," he considers himself more intimately linked to the place of his conception than of his birth.

Dances are used to reenact the myth. In the Bamba ceremony of the Walbirti, a pole about three feet high is erected and, like the dance that follows, is intended to increase the ant species. First a hole is dug and water poured on the ground. The moist earth is then made sacred with blood by soaking it with red ocher. The pole is decorated with white spots representing ants and topped with a tuft of leaves of the bloodrot plant considered to be the source of life, "containing a baby." The pole is placed in the hole, which symbolizes an anthill, while the circle and hole represent the encampment of ants. Dancers crawl across the symbolic campsite in imitation of the insect, coming closer to the center and finally symbolically entering the hole. This end to the dance represents the act of procreation that concludes every act of "entering" the sacred center. Viewing the ceremony in terms of two-dimensional design, the center represents the hole, the sacred source, and the concentric circles stand for the degrees of distance from the center with vertical movement back to the surface of the earth and to the present. The combination of inwardly moving circular motion results in a spiral path.

The representation of the sacred hole and the path leading to it carries a great many associations. The center is the point of contact with dreamtime; the concentric circle represents the primordial campsite; the path leading to it signifies the present time; the direction of ascent and the descent the sexual act and the male organ. The complex geometrical symbolism also gives a summary picture of the territory. It represents the routes the tribal groups take in movement through the territory and the seasons, giving primary attention to the need for water and relationships between the other tribal groups.

2.2 *The Fali*

According to Guidoni, every interpretation of the Fali culture can be traced back to the mythic creation of the universe through the balanced correspondence between two cosmic eggs: one of the tortoise and the other of the toad. This subdivision between two unequal parts corresponding to the tortoise and the toad is also reflected in the organization of the society, the territory, and the architecture of the Fali. Every subsequent differentiation of elements within the society came about through a series of alternate and opposing movements or "vibrations," which guaranteed the maintenance of equilibrium between opposites. Every region, every group, or architectural element either participates in one of those opposing movements or is a fixed point that acts as a pivot for the motion of the parts around it.

The form of the Fali's dwellings is an example of how this mythical organizing principle works. The huts are constructed with a "feminine" cylindrical part made of masonry and a "masculine" conical part made up of rafters and straw, as shown in Fig. 2. Although they are stationary, these parts can be imagined to circle in opposite directions to each other. All points participate, then, in a kind of virtual motion except the vertex of the cone that is fixed. "It was the tortoise that gave man the model for his house: Under its tutelage the first couple built the primordial house, whose constructive and decorative detail was established for all time and which must be faithfully imitated in all dwellings of the Fali."

The relationship between the Fali myths and the spiral is not obvious at this point in the essay, but will be made explicit in Sec. 5.

Fig. 2. Fali: Section of a granary of *bal do* type (Cameroon). By courtesy of Elemond Milano.

3. Flowing Forms in Water

3.1 *Introduction*

The book, *Sensitive Chaos* by Theodor Schwenk[2] is concerned with the creation of flowing forms in water and air. As Schwenk states:

> In the olden days, religious homage was paid to water, for men felt it to be filled with divine beings whom they could only approach with the greatest reverence. Divinities of the water often appear at the beginning of a mythology (for example, the Australian aborigines). Men gradually lost the knowledge and experience of the spiritual nature of water, until at last they came to treat it merely as a substance and a means of transmitting energy.

Water expresses itself in a vocabulary of spiral forms. Schwenk feels that these forms are the progenitors of life. The meandering stream, the breaking wave, the train of vortices created by a branch or other obstructions hanging in the water, and the watery vortex extending from the water's surface into its depths are the raw materials of living forms. According to Schwenk,

> Every living creature in the act of bringing forth its visible form passes through a liquid phase. Some creatures remain in this liquid state or solidify only slightly, others leave the world of water, densify and fall under the dominion of the earthly element. All reveal in their forms that at one time they passed through a liquid phase.

Let's see how spiral forms arise in water and manifest in living forms.

3.2 *Meanders*

A naturally flowing stream always takes a winding course. The rhythm of these meanders is a part of the nature of a river. A stream that has been artificially straightened looks lifeless and dreary.

A closer look at the flow patterns in a meandering stream shows that in addition to the forward motion of the stream, the flow of water revolves around the axis of the river in two contrary directions. The water flows predominantly from the inside of a bend to the outside, as shown in Fig. 3. The movement downstream combined with the

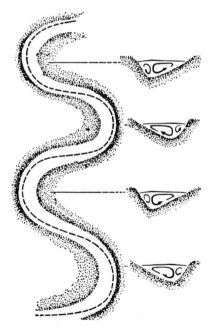

Fig. 3. Representation of the meander of a river showing the revolving secondary currents in the bed of the stream.

revolving circulation results in a spiraling motion. Actually, two spiralling streams lie next to each other along the river bed and form a kind of twisted rope of watery strands. To be more exact, rather than being strands, the water forms entire surfaces that twist together.

Fig. 4 illustrates two unicellular water animals that have incorporated the spiralling movement of water in their shapes. They usually propel themselves along like a meandering stream with a screw-like movement.

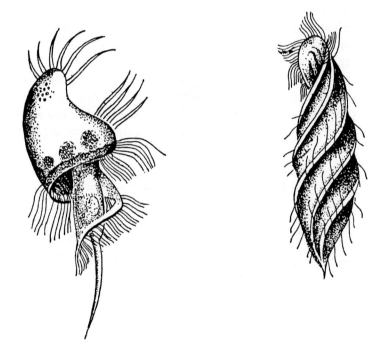

Fig. 4. Many unicellular water animals have incorporated the spiralling movement of water in their shapes (from Ludwig, after Kahl).

3.3 *Wave Movement*

A stone in the stream or a gentle breeze on the ocean will cause the water to respond immediately with rhythmical movement. The patterns that arise from these external influences are characteristic of the particular body of water, be it a lake, a stream, or an ocean.

In spite of the ceaseless flow of the stream and its swirling nature as it moves around stones and boulders, the flow pattern is stationary. The same wave forms remain behind the same rocks. On the other hand, on the open sea, the wave form wanders across the surface, the water remaining in the same place. Throw a piece of cork in the water and watch it bob up and down while the waves sweep over the surface. Schwenk notes that the wave is a newly formed third element at the surface of contact between water and wind. The wave is a form created simply out of movement. In this sense it is like all organic forms, which, in spite of chemical changes, remain intact as entities.

Wave forms are replete with complex movements of different types. For example, waves sort themselves out in different wavelengths, with the longer wave lengths moving faster than the shorter. Also, as the wave passes an element of water, the element moves vertically in circles. Flowing movements can also be superimposed on a wave, so that as the wave moves forward, a strong wind can cause the moving current on the back of the wave to move faster than the wave and overshoot the crest and break. In this process, all that is rhythmic in a wave becomes altered. It takes on a spiral form interspersed with hollow spaces in which air is trapped. Whenever hollow spaces are formed, water is drawn in to the hollows in a circular motion, and eddies and vortices arise. This presents us with a new formative principle:

The wave folds over and finally curls under to form a circling vortex.

This is illustrated in Fig. 5 and also by the famous painting "The Great Wave" by Katsushika Hokusai, shown in Fig. 6. Elements that until now were separate unite in turbulence and foam.

The different speeds of growth or development in organic forms also show evidence of the same folding, in the process in which organs are developed. Schwenk offers the development of the pupa of a butterfly as an example. As shown in Fig. 7, the organs, at first curled up, are pushed out when fully developed and appear as feelers, limbs, or the like.

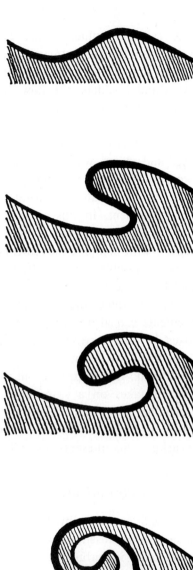

Fig. 5. A wave curls over to form a vortex.

Fig. 6. "The Great Wave" by Katsushika Hokusai. Courtesy of the Granger Collection.

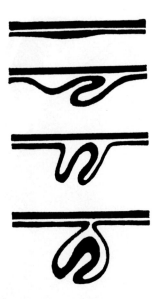

Fig. 7. In the rigid chrysalis of a butterfly, growth takes place at varying speeds. This leads to folding processes in preparation for the forming of the organs (from Eidmann, after Weber).

3.4 *Vortices and Vortex Trains*

This wave-forming motion not only occurs between air and water but also in the midst of water, as when two streams of water flow past each other at different speeds. For example, in a naturally flowing stream we can observe the patterns formed in the water at a place where a twig from a bush hangs into the water. The flowing water is parted by the obstructions and reunites when it has passed. But at the same time, a series of small vortex pairs, spiraling in alternate directions, arise as shown in Fig. 8 and travel downstream with the current. The vortices in

Fig. 8. A distinct train of vortices (after Homann).

this train of vortices are evenly spaced in a rhythm determined by the thickness of the obstruction. These vortices have the same effect as the breaking waves. The boundary of the vortex train entraps the stagnant fluid on the inside of the boundary and mixes it with the water of the swiftly moving stream exterior to the boundary. In this way, fluids of different states of the motion on either side of the boundary are gradually combined and made one.

A particularly clear picture of a train of vortices is exhibited by the bony structure in the nose of a deer, shown in Fig. 9. Large surfaces

Fig. 9. Enlarged detail of the bony structure in the nose of the deer.

are thus created, past which air can stream, giving the animal its acute sense of smell. In Fig. 10, the spirals are seen as a kind of "joint" where "ball and socket" lie opposite one another with flow lines passing straight across this "joint." The spongy structure that makes up the joints of humans and animals also closely follows the form of a single link within a vortex train, as Fig. 11 illustrates, with its stress lines running directly across the gap. Striking images of a vortex train have also been found in primitive designs, such as the one shown in Fig. 12.

14 *Jay Kappraff*

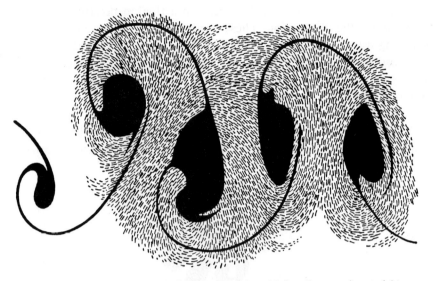

Fig 10. A vortex train seen as a ball and socket joint with flow lines passing straight across joint.

Fig. 11. Spongy bone structure in the human hip joint.

Fig. 12. Design on a palm leaf (May River, New Guinea), Volkerkundliches Museum, Basel.

3.5 *Vortex Rings*

Aside from the rhythmic processes of vortex formation at the surface of the water, there is also the three-dimensional nature of the vortex to consider. Every vortex, as shown in Fig. 13, is a funnel of downward suction. All flowing water, though it may seem to be entirely uniform, is really divided into extensive inner surfaces, each

Fig. 13. Vortex funnel.

rotating at a different speed. In the formation of vortices, these surfaces are drawn into the whirlpool. The inside of the vortex turns faster than the outside, and corkscrew-like surfaces appear on the surface of the vortex as the result of the disparity of the motion. The vortex is a figure complete in itself with its own forms, rhythms, and movements.

The vortex is like an isolated system closed off from the body of the water around it. For the most part, the vortex with its different speeds follows Kepler's Third Law, in which a planet moves fast when near the sun. In fact, Schwenk suggests that the vortex is a miniature planetary system with the sun corresponding to the center of the vortex, with the exception being that the planets move in slightly eccentric orbits in contrast to the circular vortex motion. It is interesting that Descartes' model of the universe consisted of vortices made up of the fine matter of the "ether" with the stars at the center.[5] The planets are carried about in the sun's vortex, and the moon is carried around the earth in the same way as shown in Fig. 14.

The vortex has another quality that suggests cosmic connections. If a small floating object with a fixed pointer is allowed to circulate in a vortex, the pointer always points in the direction that it was originally placed, that is, it always remains parallel to itself. In other words, it is always directed to the same point at infinity, just as the axis of the Earth points in the same direction as it revolves around the sun. The center of the vortex would rotate at infinite speed if this were possible. Since it is not, it instead creates a kind of negative pressure, which is experienced as suction.

Many forms in the organic world manifest themselves in the form of a vortex. For example, the twisting antlers of a horned animal or the multitude of snails and shells shown in Fig. 15, some spiral formations in the plant world, and, most strikingly, the human cochlea. Fig. 16 shows the fibers in the auditory nerve, arranged spirally just like a liquid vortex. However, it is literature and art that has captured the essence of the vortex. Edgar Allen Poe's classic tale, *A Descent into the Maelström,* presents a palpable description of the abyss at the base of the vortex and chaotic description of the multitude of inner surfaces of the vortex. Van Gogh's art portrays nature in a perpetual state of movement. Perhaps his masterpiece "Starry

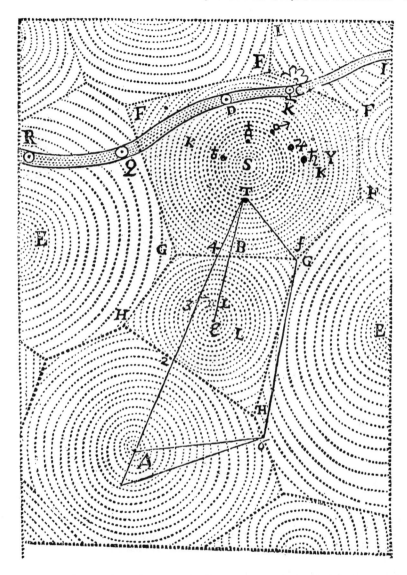

Fig. 14. System of vortices with which Descartes sought to account for the motion of the heavenly bodies consisted of whirlpools of ether. In the case of the solar system the vortex carried the planets around the sun (S). Irregular path across the top of the illustration is a comet, the motions of which Descartes believed could not be reduced to a uniform law. Courtesy of the Bodleian Library, Oxford, England.

Fig. 15. Spiral formation in snails and shells.

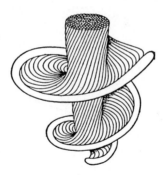

Fig. 16. Fibers in the auditory nerve, arranged spirally just like a liquid vortex, as though picturing an invisible vortex of forces (after De Burlet).

Night," shown in Fig. 17, best dramatizes the preeminence of the vortex in nature. This painting is an affirmation of the swirling harmony between the forces of nature. As Jill Purce observes,[6] "Not only do the clouds spiral into a Yin Yang formation (a pair of alternately spiralling vortices), but the opposing forces of sun and

The Spiral in Nature, Myth, and Mathematics 19

Fig. 17. "Starry Night" by Vincent Van Gogh. About this affirmation of the swirling harmony between the forces of nature, Van Gogh wrote: "First of all the twinkling stars vibrated, but remained motionless in space, then all the celestial globes were united into one series of movements ... Firmament and planets both disappeared, but the mighty breath which gives life to all things and in which all is bound up remain."[6] Reprinted by permission from The Museum of Modern Art, New York.

moon are unified. For Van Gogh, this was a decisive moment of union between inner self and outside world."

3.6 *Three Characteristic Features of Water*

The above discussion illustrates three characteristic features of water. The first is the activity of water in all metabolic processes. The second is the close connection with all rhythmic processes. The third, lesser known characteristic, is the sensitivity of water's boundary surfaces, which Schwenk sees as an indication that water is a cosmic sense organ of the earth.

We have seen evidence of the metabolic function where water churns up silt from a river bed and redeposits it, or where a breaking wave incorporates the air at its boundary into itself. The rhythmic patterns were evidenced in the meanders of the river and the moving vortex trains. The sensitivity at boundary surfaces was illustrated by the influence of the smallest environmental factors on the formation of waves or creation of vortex trains. This sensitivity is also passed on to the organic world in such structures as the deer's nose structure or the antelope's horns. The vortex itself is a mechanism to open up the inner surfaces of water to the influences of the moon and the stars, due to the disparities of fluid velocity from the center to exterior of the vortex.

All of these functions are well known to us in the world of living organisms. In humans, the intestines best represent an organ of metabolism; the heart, the center of rhythmic organization; the ear, a sense organ. These three organs are shaped by patterns similar to those found in flowing water. But, as Schwenk shows, just as the three characteristics do not specialize to any of the fundamental patterns of movement of water, they also do not specialize to any particular organ. In each movement of water and in each organ, all three are in evidence.

3.7 *Flowforms*

Theodor Schwenk founded the Institute for Flow Research at Herrischried, Germany. Here he studied the formative processes of water in great depth and has come to profound spiritual conclusions about the place of water and its resulting living forms as a mediator between the earthly and cosmic realms. His work has been continued by his son Wolfram Schwenk[7] and others, most notably a mathematician and sculptor, John Wilkes at the Flow Design Research Association at Emerson College in Sussex, England, and Jennifer Green, founder of the Flow Research Laboratory in Blue Hill, Maine. These individuals have made careful observations of the rhythms and forms that are assumed by water in a healthy state and have created sets of sculptured vessels that they call "Flowforms" based on these natural flow patterns in order to "enliven" the flow of fluids in artificially created water cascades.[8] As Green states, "Flowforms can stand alone as sculpture in order to demonstrate the wonder of water as Sculptor. However, preliminary research has shown that water enhanced rhythmically via Flowforms increases the efficiency of oxygen-transfer, augments biological activity in wastewater treatment, and improves effluent quality. Many applications are being investigated."[9]

Fig. 18 shows a sevenfold cascade created by Wilkes and a diagram of water moving from an entry vessel through the cascade. As in all of Wilkes' flowforms, the water moves in a kind of vortical meander forming figure-eight, lemniscate curves. The flowforms have three main functions: oxygenation; thorough mixing; and maintaining a rhythmically pulsing flow pattern (similar to the fluid movement that occurs in the heart). In an effort to create a theoretical foundation for Wilkes's

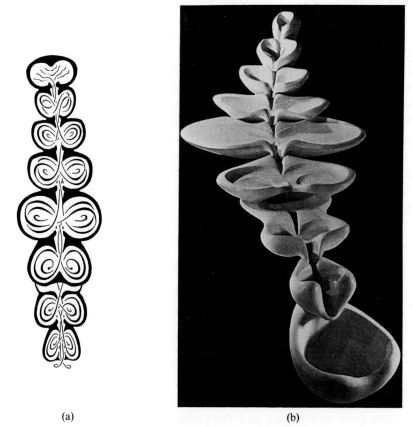

(a)   (b)

Fig. 18. (a) A sevenfold cascade design by John Wilkes, (b) diagram of water moving from an entry vessel through the cascade. Photo by John Wilkes.

empirical studies, a set of curves known as "path curves" and related to a group of projective transformations are being studied. The next section gives a brief discussion of projective geometry in order to describe these path curves.

4. Field of Form

4.1 *Introduction to Projective Geometry*

Whenever a two-dimensional object is projected by a point source of light from one plane to another, a projective image results. The object and image are considered to be projectively equivalent. Artists of the fifteenth century such as Brunelleschi, Albrecht Dürer, and Leonardo da Vinci developed the art and mathematics of projections and understood that it is connected with vision. The eye projects a scene from the horizontal plane to an imaginary screen in front of the body. Rays of light can be thought to connect points on the scene to points on this imaginary plane, in which case the scene is viewed in perspective. When a canvas replaces the imaginary screen, the scene may be rendered by recreating it at the points where the rays pass through the canvas, as shown in Fig. 19. This projective mapping of scene to canvas maps the infinitely distant line or horizon onto a real line on the canvas. Also, parallel lines receding from the viewer towards the horizon appear on the canvas as oblique lines meeting at some point on the horizon line.

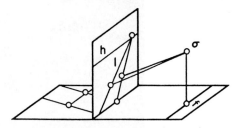

Fig. 19. A road receding to infinity depicted as converging to a point on the horizon line of an artist's canvas.

The subject of projective geometry pertains to three primary elements: points, lines and planes, and the properties of these elements that are preserved under projective transformations. Projectively equivalent objects and images are considered to be identical in projective geometry, just as congruent figures are indistinguishable in Euclidean geometry. The three primary elements are considered as separate entities; a line is not considered in the axioms of projective geometry to be a sequence of points, but an entity in itself. Also, all points and lines are considered to have equal status. Therefore, the line at infinity has as much reality as any other line. We are also justified in saying that every pair of lines has a single point in common; when the lines are parallel the common point is a point on the line at infinity. In fact, we shall see that one of the primary values of projective geometry is its ability to deal concretely with elements at infinity.

It is well known that depending on how a right circular cone is sliced by a plane, the boundary of the section is either a circle, ellipse, parabola, or hyperbola, the so-called conic sections. Since the vertex of the cone can be thought of as being a point source of light, all conic sections are projectively equivalent. Each conic can be viewed projectively as a circle by singling out a special line in the plane, as shown in Fig. 20. Fig. 20a represents a projective view of an ellipse, Fig. 20b a parabola, Fig. 20c an hyperbola, and Fig. 20d represents a circle when a special line is at infinity. When the special line is mapped to infinity, the usual pictures of the conic reveal themselves. The richness of projective geometry, in contrast to Euclidean geometry, is due to the fact that in Euclidean geometry the special line is always taken to be the line at infinity; as a result, conics always assume their familiar forms, whereas projective

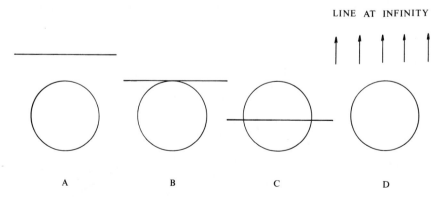

Fig. 20. When the special line is projectively mapped to infinity, the circles become (a) an ellipse, (b) a parabola, (c) an hyperbola, (d) a circle.

geometry always has a representational flexibility. Yet, all the theorems of Euclidean geometry are found to hold in the more general context of projective geometry.

Crucial to an understanding of projective geometry is the concept of duality. The axioms are so constructed that in two-dimensions, their validity is unaltered whenever point and line are interchanged in any statement or theorem, or in three-dimensions point and plane are interchanged while line is retained. Thus the statement, "any two lines contain a common point" is dual to the statement, "any two points contain a single line (i.e., through any two points a single line can be drawn)," while "three planes with no common line contain a unique point" is equivalent to "three points not all on the same line contain a unique plane (i.e., three non-colinear points define a plane)."

The remainder of Sec. 4 deals with the mathematics of projective transformations and its application to generating a family of curves known as path curves. According to the research of Lawrence Edwards as described in his books, *Field of Form*[3] and *Projective Geometry*[10] and in Ref. 11, these curves are close approximations to the spiral shapes of plants and other biological forms, as well as the watery vortex.

4.2 *Perspective Transformations of Points on a Line to Points on a Line*

One of the most elementary transformations in all of mathematics is the mapping of the points on line x to points on line x' from a point O not on lines x and x', as shown in Fig. 21. A typical point A on line x and

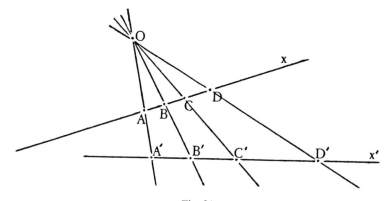

Fig. 21

A' on line x' share a common line from the pencil of lines centered at the point of projection, O. Such a projective transformation is called a perspectivity. In Fig. 22 the dual perspectivity is shown. Here two pencils of lines centered at X and X' are projected onto line o. Lines a and a' meet at a point on the line of projection, o.

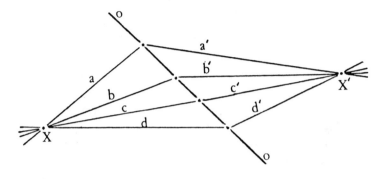

Fig. 22

It is easy to see that a perspectivity is entirely specified by arbitrarily choosing two points A and B and their transformed points A' and B'. In this case x is the line through AB, x' is the line through $A'B'$, and O is the meeting point of AA' and BB'.

By referring to Fig. 21, certain special points of the transformation are evident. The intersection of x and x' is mapped to itself and is the only fixed point of the transformation. Also, the line through O parallel to x maps the point at infinity on line x to the point P' where this line intersects x'. Likewise, the point at infinity on x' is mapped from point Q on x, where the line through O parallel to x' intersects x.

Although metric properties such as distance between points are not generally preserved by perspectivities, a somewhat obscure relation between any four points A, B, C, D, and their transforms is preserved, namely:

$$\lambda = AB/BC : AD/DC. \qquad (1)$$

This relationship is known as the cross-ratio. It is of fundamental importance to projective geometry. In fact, projective transformations can be defined to be those transformations of points and lines that preserve cross-ratio. It should be mentioned that this definition of the cross-ratio assumes successive points are in the order *ABCD*. There are 24 different orderings of these four points, and the cross-ratio in each of these definitions of cross-ratio are also preserved. As a matter of fact, we will find the ordering *DBAC* and its cross-ratio

$$\lambda = DB/BA : DC/CA \qquad (2)$$

most relevant in what follows.

4.3 Projective Transformations of Points on a Line to Points on a Line

The notion of the projective transformation of the points on line x to the points on line x' can be made more general by first relating the points on lines x and m by a perspectivity with respect to point O, and then relating the points on m and x' by a perspectivity with respect to O' as shown in Fig. 23. In a similar manner, x can be mapped to x' via a series of intermediate lines m, m', m'', etc. Any such series of perspectivities is called a projectivity.

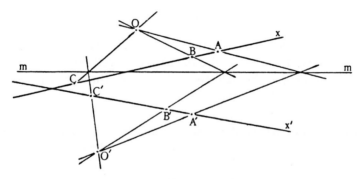

Fig. 23

It can be shown that any three points arbitrarily chosen on lines x and x' can be mapped to each other under a projectivity. It follows from the previous section that the cross-ratio between any four points of line x and their transforms are preserved by projectivities. In the event that corresponding lines from the pencil of lines through O and O' meet on a common line, the projectivity reduces to the transformation depicted in Fig. 22, a perspectivity. However, lines *OA* and *O'A'*, *OB* and *O'B'*, *OC* and *O'C'* do not generally meet on a common line. It is of fundamental importance to the study of projective geometry that these pairs of lines do meet on a conic section.

For a special class of projectivity called co-basal projectivity x and x' are the same line as shown in Fig. 24. It is evident that for these transformations two points are generally fixed, the point X where lines m and x intersect and the point Y where OO' intersects x. Since OA and $O'A'$ in general meet on a conic, the fixed points can also be pictured as the

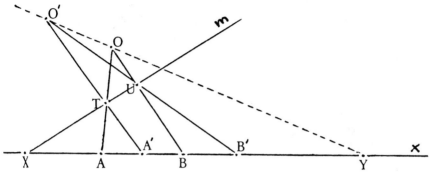

Fig. 24

intersection points of a line with a conic (circle), as shown in Fig. 25, where A' is projectively transformed to B', B' to C', etc. on line m. Conversely, a co-basal projectivity can be generated from an arbitrary conic by choosing two points O and O' on the conic and relating the intersection points on the base line of pairs from O and O' that meet on the conic. It follows from glancing at Fig. 25 that co-basal projectivities have only one

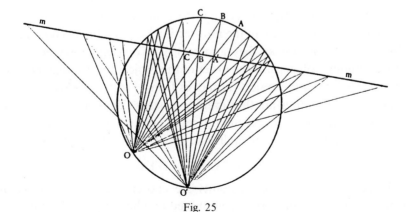

Fig. 25

fixed point when the line is tangent to the conic (circle in this figure) or no real fixed points when the line and conic have no real intersection points.

4.4 Growth Measures

Now that we have defined a co-basal projective transformation of a line, let's see what sequence of points is the result of applying such a transformation repeatedly to an arbitrary point A on line x. Such a sequence of points is called the trajectory of A under this transformation. Referring to Fig. 26, point A transforms to A', which we call B by first

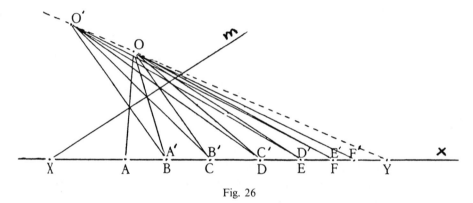

Fig. 26

projecting to line m through O and then projecting the resulting point back to x through O'. Point B, in turn, transforms to C and C to D, etc. Such a trajectory is called a growth measure. If A begins near the left fixed point X and moves toward Y, then the trajectories start out with small step sizes, which increase in the mid-section between the fixed points and then decrease in size as they approach Y. To reach Y would take an infinite number of steps. If the order of the projections through O and O' are reversed, then the trajectory moves from A to the left fixed point X, which it reaches after another infinity of steps. Thus, the entire growth measure represents a doubly infinite set of points. Using the points $XABY$ to define the cross-ratio, or multiplier as we shall refer to it, from Eq. 2:

$$\lambda = BX/YB : AX/YA. \qquad (3)$$

Therefore, once the fixed points are chosen, the cross-product and hence the entire trajectory is determined by choosing the positions of A

and B. The multiplier of the reverse transformation is determined by the sequence $YBAX$, and it is the inverse, $\frac{1}{\lambda}$, of the original multiplier.

Let's see what the effect on the growth measure is if we project the right fixed point Y to infinity. This is done by drawing an arbitrary line b and drawing a line through Y parallel to b. If we project the trajectory of point A onto b from an arbitrary point on the line through Y, then X projects to a finite point X_1 on b while Y projects to the infinite point on b and the trajectory on x projects to another trajectory on b. On this trajectory the cross-ratio or multiplier reduces to $X_1B_1/X_1A_1 = X_1C_1/X_1B_1 = \ldots$, which identifies the trajectory as a familiar geometric series. In other words, growth measures can be viewed as double geometric series seen in perspective.

Next, consider the case of a growth measure in which the line through O and O' meets at the left fixed point X. Here, the two fixed points coalesce into one and the growth measure is called a step measure. This corresponds to the case in Fig. 25 of a tangent line to the circle. In a way similar to what we did for growth measures with two fixed points, we can project a step measure onto an arbitrary line, so that the double point is projected onto the point at infinity. We then discover that a step measure is the perspective image of an evenly spaced trajectory of points on a line.

So we see that, even though projective transformations do not have obvious metric properties, they represent geometric models of multiplication in the case of growth measures and addition in the case of step measures. From the point of view of geometry, growth measures are far more likely to occur than the limiting case of a step measure. It is also interesting that it is the geometric series that manifests in the organic world. It is well known that shells of sea animals such as the Nautilus and the horns of horned animals grow according to equiangular spirals. Equiangular spirals are governed by the principle that radii from the center of the spiral at equally spaced angles about the center form a geometric series, as shown in Fig. 27. These curves were referred to by Theodore Cook as "curves of life."[12] We will have more to say about this important curve in Sec. 4.6.

There is one kind of growth measure that deserves special commentary. It is depicted in Fig. 28. Here, O and O' are on opposite sides of line m, and they are arranged so that point A transforms to B and B transforms back to A. Such a transformation is called an involution. By

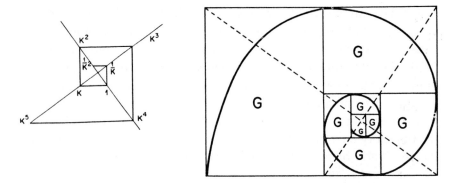

Fig. 27. Vertex points of an equiangular (logarithmic) spiral lie at a double geometric series of distances from the center.

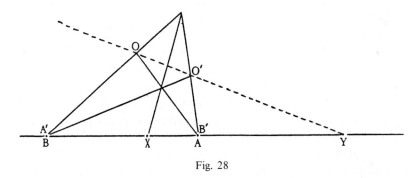

Fig. 28

reversing the order of O and O', we obtain another involution that transforms B to A. Both of these transformations are called breathing involutions, since the movement is back and forth across either of the fixed points with related points close to one of the fixed points, playing the role of shallow breaths nested within the deep breaths of the point pairs more distant from the fixed point. Notice that the lines OA, $O'A$, OB, $O'B$ form a quadrilateral and that the fixed points X, Y are the intersections of the diagonals of this quadrilateral with the base line. A and B are, in this case, said to be harmonic with respect to X and Y.

It is an interesting fact of projective geometry, and one that can be tested by construction, that if a single pair of points are found to be in involution then all points of the line are also in involution or harmonic with their transforms. Also, the cross-ratio of $BXAY$ is equal to -1.

Involutions carry with them another metric property, namely, the point harmonic to the point at infinity with respect to X and Y is the midpoint between X and Y. Involutions provide the key to understanding growth measures when the fixed points are not visible.

Given two fixed points X and Y there are an infinity of growth measures that move between the two points. For example, any point A' between A and B of the original growth measure leads to another trajectory with the same fixed points. Likewise, points O and O' can be moved to different locations so long as OO' passes through point Y, or equivalently, O and O' can be moved to different positions on the conic (circle) in Fig. 25. In fact, the collection of growth measures with fixed points X and Y form what is known in mathematics as a group. We have here a dynamic picture of a line. The line is all motion, with points forever changing their positions yet with the configuration of the trajectory remaining unchanged. Only the fixed points are motionless. Also, with respect to a given pair of fixed points there corresponds two distinct involutions: One from right to left, i.e., the direction of X, and the other from left to right, i.e. the direction of Y, so that involutions can be directly correlated with fixed points.

4.5 *Circling Measures*

To complete the picture we must also account for growth measures of co-basal projectivities, for which there are no real fixed points. Edwards calls such growth measures circling measures. They are set up by transforming an evenly spaced set of lines from a pencil of lines centered at a point. For example, the 18 lines in Fig. 29 are spaced 10° apart and each line undergoes a transformation of 60° in a clockwise direction. A circling measure is set up by the points of intersection of an arbitrary line such as the one shown in Fig. 29, with this pencil. It would require more space than we have in this short article to describe these projectivities, so we direct the reader to Edwards's book, *Projective Geometry*[10] and summarize some of the important results.

We have already seen in Fig. 25 that a conic and a line cutting it sets up a growth measure on the line with the intersection points as the fixed points. If the line and the conic do not intersect in real points, they intersect in imaginary points. We shall now see how these imaginary fixed points come about. Consider a unit circle, $x^2 + y^2 = 1$ and the line,

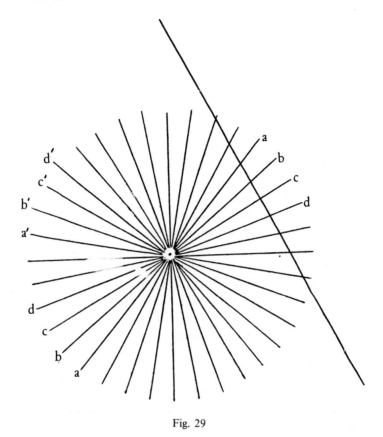

Fig. 29

$y = 2$. Solving these equations for the intersection points, we find that $(X,Y) = (\pm i, 2)$. So we can say that, in some sense, the circle and the line share these two points. Let's now consider a circle and a series of parallel lines going to infinity. Each line shares a pair of points with the circle. We shall denote the two imaginary points on the line at infinity as I and J. Of course, all circles share equally well these same points I and J.

Circling measures on a line induced by a circle or other conic can be created by the identical construction shown in Fig. 25. The fixed points of this measure will be the two imaginary intersection points, and there is, once again, a group of circling measures that have the same two fixed points. In order to make these imaginary points and lines tangible, Edwards prefers to deal with the two involutions set up on this line, instead of the imaginary points that correspond to them. In Fig. 29, transformations of a line through 90° in either a clockwise or counterclockwise direction results in an involution (can you see why?), known as a circling involution.

4.6 Path Curves

Up to now we have been considering projective transformations of points on a line to points on a line, or lines in a point to lines in a point. Projective transformations of the plane that map points to points and lines to lines are called collineations. It is fundamental to projective geometry that collineations leave, in general, three points (real or imaginary) invariant. If the fixed points do not all lie on the same line, they define a triangle. Under a collineation any line in the pencil of lines centered at one of the fixed points is mapped to another line in that pencil. The point in which a line through a fixed point intersects the line of the opposite side of the triangle sets up a growth measure on that line. The boundary lines of the triangle are invariant in the sense that any point on one of them is transformed by the collineation to another point on the same line, although not generally to the same point. This invariant triangle is the setting for a remarkable set of curves, known as path curves.

Given a set of fixed points $A, B,$ and C and invariant lines a, b, and c as shown in Fig. 30, a collineation is completely determined by choosing an arbitrary point M within the triangle and its transform M' under the collineation. This is obvious, since the projected points P, Q on line a and P', Q' on line b determine growth measures on those lines as we saw in Sec. 4.4. Therefore, the next point of the trajectory determined by the collineation is the point in Fig. 30 at the junction of the lines AR and CR' where R and R' are the next points of the growth measures set up on their respective lines. Thus, the action of the collineation upon the initial point M sets up a trajectory of points, and these points lie on a set of invariant curves, known as path curves. As you can see, the path curves cut across the diagonals of the grid of quadrilaterals determined by the growth measure, i.e., the diagonal between M and M'. The path curves are invariant since any point that lies on such a curve is transformed to another point on the same curve. In general, any other pair of points not on the same path curve generates, by the same construction, another

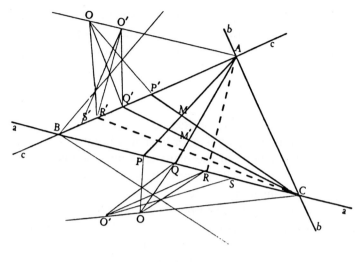

Fig. 30

family of path curves. In fact, all path curves keeping the same triangle invariant form a group that mathematicians refer to as a Lie group.

One set of path curves is shown in Fig. 31. All path curves have in common with this figure the fact that they pass through two of the fixed points but not the third and lie tangent to two of the invariant lines but not the third. Edwards has the following to say about path curves:

> We have a plane in which everything is moving. What can live, can hold itself intact, within the flux? It is the whole set of path curves, and nothing else! Quantitatively we have a similar situation in any living organism; the substance of which it is made was not in it yesterday, and will not be in it tomorrow; as far as its matter is concerned it is in a state of continual flux; the substance flows in and flows out; if the organism was simply its subtance we would not be able to recognize it from one day to another. Yet its being and largely its form are invariant from one moment to another, and from one day to another. The form can live within the flux.

A family of path curves can also be determined by specifying the multipliers, given by Eq. 3 of the growth measures on any two of the invariant lines, say α on line a and γ on line c, along with the directions of the trajectories on these lines, say counterclockwise. The growth measure on the remaining line b is the product or composition of the growth measures on lines a and c, and has the effect of inducing a growth measure in the opposite sense (clockwise) with multiplier equal to the

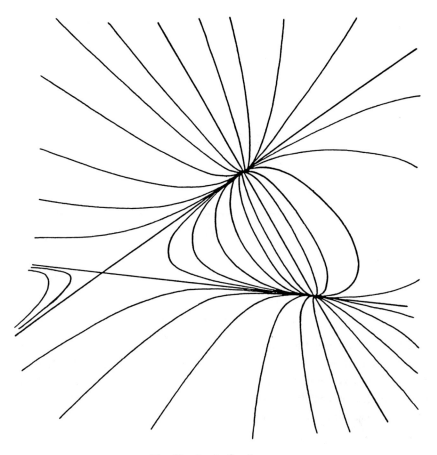

Fig. 31 A set of path curves.

product of the two, i.e., $\alpha\gamma$. Since reversing the sense of a growth measure results in inverting its multiplier, then $\beta = \frac{1}{\alpha\gamma}$ and α, β, and γ are related by:

$$\alpha\beta\gamma = 1, \qquad (4)$$

when all growth measures have the same sense (counterclockwise) as shown in Fig. 32.

The equations of path curves are as elementary as the primeval nature of the geometry would suggest. If we work in homogeneous coordinates, then the equation of the family of path curves is given by:

$$x^a y^b z^c = k, \quad \text{where } a + b + c = 0 \qquad (5)$$

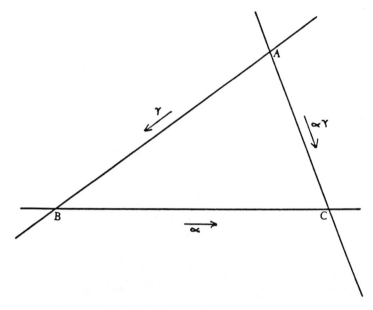

Fig. 32. The invariant triangle.

and a, b, and c are the logarithms of multipliers α, β, and γ and each value of k pertains to a different path curve of the family. The fact that this equation is what mathematicians call homogeneous means that the shape of the curves depends not on the values of α and β but on their "exponential ratio" given by:

$$\Lambda = \frac{\log \alpha}{\log \beta}.$$

Thus if $\alpha = 16$ and $\beta = 4$ the curves would have the same shape as if $\alpha = 9$ and $\beta = 3$, only the step size of the trajectory along the curves corresponding to the smaller values would be smaller.

Three special cases of path curves are of interest to us:

Case 1: Two multipliers are identical, but in opposite senses, say $\gamma = \frac{1}{\alpha}$. In this case the path curves look much like the ones in Fig. 31 except that they are conics. But since, according to Eq. 4, $\beta = 1$, the growth measure on the other line is the identity transformation, the pencil of lines centered on B must be another family of path curves of the invariant triangle. Together the two families form a grid of path curves.

Case 2: In Fig. 33, line *b* of the invariant triangle is mapped to the line at infinity, while different multipliers are taken on lines *a* and *c* in directions going counterclockwise from *C* to *B* and *B* to *A*. Notice that the path curves turn out to be egg-shaped forms, sharper at one end and blunter at the other.

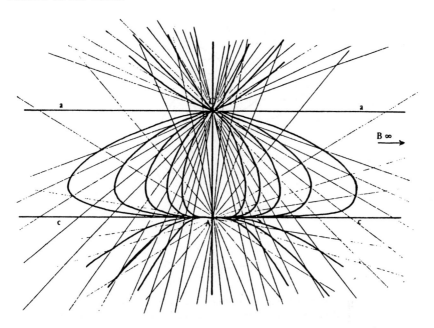

Fig. 33. Point *B* of the invariant triangle is mapped to infinity and the path curves take on forms with organic appearances that are sharp at one end and blunt at the other. Reprinted by permission from Floris Books.

Case 3: Points A and B are mapped in Fig. 34 to the points *I* and *J* on the infinite circle, while *C* remains fixed at a real point. In this case, the line of the invariant triangle connecting *C* to a point at infinity is real, while the other two lines of the triangle are imaginary. If the growth measures on these imaginary lines are identical, then the path curves are conics and lines as in case 1, only now the conics are concentric circles and the lines radiate from *C*. Also, the multiplier of the circling measures set up on the circle results in points on the infinite circle (line) at equally spaced angles, while the growth measure on the real line forms a geometric series (see Sec. 4.4). Now, if the two points on opposite

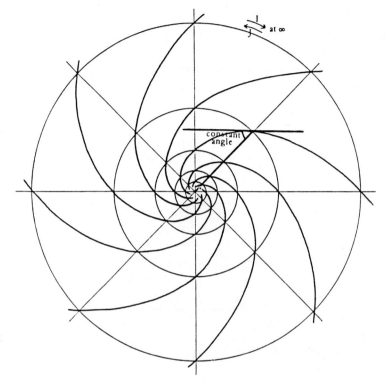

Fig. 34. Points *A* and *B* of the invariant triangle are mapped to *I* − and *J* on the infinite circle and the path curves become equiangular spirals. By permission of Floris Books.

vertices of a curved quadrilateral of the grid in Fig. 34 are used to initiate a new family of collineations, the resulting path curves are equiangular spirals. These spirals are clockwise or counterclockwise, depending on the diagonal chosen. So we see that Cook's "curves of life" are actually a family of path curves.

4.7 *Path Curves in Three Dimensions*

In three dimensions, collineations leave four points (real or imaginary) invariant. In general, these four points define a tetrahedron. With the exception of these four points, all points of the plane are in motion under iterations of the collineation. However, the four planes and six lines that define the tetrahedron are invariant in that the trajectory of a point that begins on such a line or plane remains there. Also, each of the fixed points defines a pencil of lines centered at that point. Under the collineation, a line of the pencil must transform to another line, and the trajectory of the intersection of these lines with the opposite plane maps out a path curve on that plane.

Once again, by specifying the location of an arbitrary point within the tetrahedron and its transform, the entire transformation is fixed. The trajectory of this arbitrary point traces out a path curve through space. The path curves wind through space with each point lying on an osculating plane defined by the tangent vectors of two infinitesimally close points on the curve. Starting with the points on an arbitrary curve through space, the path curves of the points on this curve map out an invariant surface in space made up of path curves. These surfaces can often be quite beautiful. If the starting curve is the tangent line to a point of one of the path curves, then the path curves envelop a conic on the osculating plane at that point. If the starting point is an arbitrary line not tangent to a path curve, then that line is carried by the projectivity into an invariant surface made up of a winding sequence of lines, called a ruled surface.

Most of the surfaces of path curves of interest to Edwards's studies of organic forms arise from tetrahedra, two of whose fixed points are the imaginary circling points I and J while the other two fixed points X and Y are finite and real. The tetrahedron can then be visualized as being made up of two real infinite planes and two imaginary planes, and two real lines, one joining X to Y, and the other one being the line at infinity which carries points I and J. The path curves on the two real planes contain a family of equiangular spirals such as the ones shown in Fig. 34, where the planes in edge view are shown in Fig. 3b. There are two possible kinds of path curves that can result from this kind of projectivity. If a projective transformation sends points on the upper plane spiraling inward while the points on the bottom plane spiral outward then spirals such as the one in Fig. 35a result. On the other hand, if both spirals wind in the same direction as in Fig. 35b, i.e., inward from the infinite periphery and inward to the other finite pole then the spirals have a vortex appearance.

Edwards's fundamental surfaces are created by choosing a pair of congruent spirals on each of these real planes with spirals that lead out

(a)

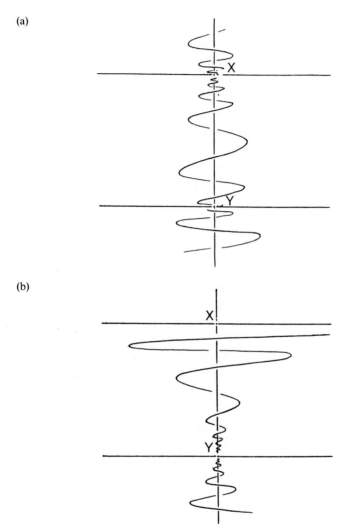

(b)

Fig. 35. (a) A typical three-dimensional path curve within the invariant tetrahedron, (b) a type of path curve that takes on a vortex-like appearance. Reprinted by permission from Floris Books.

of the bottom pole and into the top pole, as in Fig. 36. He then develops an invariant surface of path curves by starting with a circle centered at a point on *XY* and in a plane perpendicular to this axis. The trajectories of the points on this circle form an invariant surface with an egg-shaped

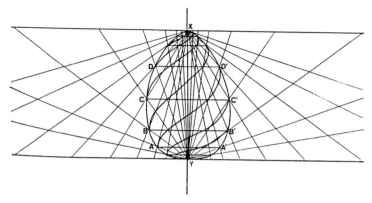

Fig. 36. Spiral path curves lying on an egg-shaped surface. Reprinted by permission from Floris Books.

form, and the path curves on this surface are spirals of the type shown in Fig. 36. Edwards then characterizes the shape of these surfaces by a parameter much like Λ but suitable for three-dimensions, not mentioned here due to space limitations. Edwards has the following to say about the forms that can be derived from these fundamental surfaces:

> Although there are many possible variations of this form, the range of possibilities are strictly limited. The degree of blunting and sharpening at the two ends is exactly balanced; one will never get a form which is very blunt at one end and slightly sharp at the other; nor will one ever find one slightly blunt and very sharp; nor can they ever be blunt at both ends. In fact the list of things which they cannot be is much longer than that which they can be.

4.8 *Field of Form*

According to Edwards:

> When our attention is drawn to the various path curve surfaces previously described, and especially the egg-like forms which occur with the semi-imaginary tetrahedron, we immediately become aware that forms very similar to these are to be found in at least four situations in the plant world (illustrated in Fig. 37): the numerous families of pine cones and related seed formations; tightly packed bunches of flower buds (e.g., rhododendron, flowering currant) in which the separate buds are nearly always arranged in spiral formations; leaf buds of deciduous trees (oak, beech, elm, etc.) in which the little leaflets are themselves set in spirals; the large domain of flower bud where in a large proportion of cases the petal edges climb spiralwise around the egg-shaped form of the bud itself.

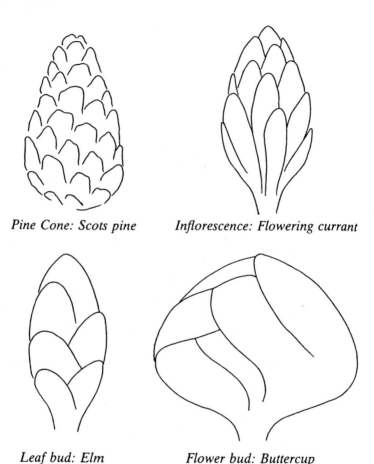

Fig. 37. Four ways in which path curves are to be seen in the plant kingdom. Reprinted by permission from Floris Books.

Edwards then sets out to meticulously measure, using statistical methods, the outer shapes for many such plants, and he has been able to make a convincing argument for their being path curves. A typical path curve within the tetrahedron takes the form of a spiral on the surface of an egg-shaped form. When two of the points of the tetrahedron are mapped to points at infinity, the path curves take on forms remarkably like the shape of plants, buds, and other organic forms. He has also applied his methods to studying the shape of eggs of different species of animals, the shells of sea animals, the shape of the hearts of animals and the living human heart as seen through an angioigram. All of these have corroborated his ideas about the relation of path curves to living forms.

Edwards reports on an odd form that resisted all his techniques of analysis, namely, the seed chamber buried within the depths of the rose bud, called the rose hip. After much thought he was able to correlate this form with yet another fundamental idea of projective geometry that had been anticipated by Rudolph Steiner and studied by George Adams.[13] He projected one of the remaining finite fixed points of the semi-imaginary tetrahedron to infinity and applied a transformation which he calls a pivot transformation in which the original transformation that generated the rose bud forms the basis of a new transformation between elements from the dual spaces of planes (positive space) to elements of the space of points (negative space). In this transformation, shown in Fig. 38, the

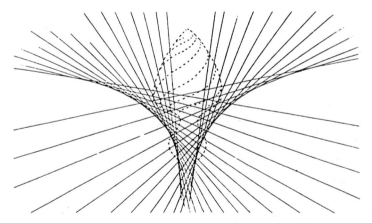

Fig. 38. The form of the rose led to the concept that the bud would mediate between a planewise vortex and the form of the rose hip. Reprinted by permission from Floris Books.

plane at infinity, the absolute point of the positive space, representing the cosmic realm is related to the pole at point Y, the absolute point of negative space, which represents the seed. The points on the infinite plane are dual to a cone of planes centered at the absolute point.

The shape of the family of path curves of this transformation resembles a watery vortex, and this family was so named by Edwards. These path curves have proven very good at describing the shape of the ovaries of plants. Much to his amazement, using an apparatus for measurement designed by a colleague, Edwards discovered that this form fits exactly to the shape of actual watery vortices. This connection to flowing water is being studied as a way to improve the design of the Flowforms described in Sec. 3.7, and as a way of providing the empirical studies of Wilkes with a theoretical foundation based in projective geometry. The lemniscate curves characteristic of Wilkes's flows (see Sec. 3.7) have also been described projectively by Adams[13] and may provide another key to connecting vortex phenomena to geometry.

5. Comparison Between the Three Systems

Three systems of thought were presented in Secs. 2, 3, and 4: Guidoni's analysis of some primitive myths and rituals, Schwenk's observations of the living forces within water, and Edwards's "field of form." Although these systems are quite different in their representations of the natural world, they also have much in common.

The most striking similarity between them is their dynamical vision of nature. For example, the act of conception is presented in the Bamba ritual of the Australian aborigines as a movement toward a sacred center. All creation in the Fali myth is represented as a series of vibrations. Schwenk represents the genesis of form in terms of movement of waves, vortices, and meanders. The particles of water continuously change but the outward form is stable. The same holds for Edwards's path curves, which can be viewed as the trajectories of a moving series of points.

The three systems agree with each other and are at variance with scientific theory in their view of the role that cosmic influences have on earthly events. The sacred pole of the Achilpa serves as a direct link to transmit some form of life energy from heaven to earth to nourish the tribe. The vortex, by opening up sensitive membranes of water to external influences, plays the same role in Schwenk's system. These influences are sucked toward the vortex center by negative pressure. Edwards's watery vortex transformation places a plane at infinity to absorb cosmic influences that pass along spirals to a fixed center at the base of the plant's ovaries.

Each system is built on the notion of sacred centers with no fixed location within an otherwise undifferentiated chaos. Wherever the chief of the Achilpa places the sacred pole, that is where the center lies. Each watery vortex functions as a closed system and carries the center of its own "universe," complete with its built-in direction to the "fixed stars." Edwards also sees every plant as being a closed system with its center located at the base of the ovaries.

The equiangular spiral plays a key role within each of these systems. In the Bamba ceremony, the dancers crawl toward the sacred center along equiangular spirals. Each of the fundamental patterns of water manifests in the natural world in spiral formations. The fundamental surfaces from which Edwards develops organic forms are derived from equiangular spirals. Even meandering streams can be looked at as helices that have been flattened onto a planar surface.

The mythic beginnings of the Fali people go back to two cosmic eggs. Schwenk feels that the forces of water are instrumental in the development of the embryo from the egg. The path curves of Edwards always generate egg-shaped forms with spiral striations.

In Fali myth, all creation comes about through alternate and opposing movements. These alternate and opposing movements are evident in the oppositely directed spirals of vortex trains, and are also incorporated in the spongy structures of joint formations of humans and animals. Involutions that characterize the group of growth measures can be viewed as alternating and opposing movements or vibrations, and these growth measures are the key to deriving Edwards's "field of form."

Finally, every element of Fali society either partakes in a positively or negatively directed motion or is a fixed center. As we saw, the "feminine" cylindrical walls of their huts are positively directed while the "masculine" conical roofs are negatively directed with respect to the fixed center at the vertex of the cone. Is it far-fetched to imagine a connection between this image and Edwards's pivot transformation? Here the infinite plane and the center of this transformation are conceived of as positive and negative dual spaces. Points of the positive space are transformed to planes of the negative space that envelop a cone about the center.

Many other associations can be found between these systems, but I leave them to the reader.

6. Conclusion

We have shown that spiral forms are ubiquitous in the natural world. Primitive people, in close contact with the natural order, understood the importance of the spiral as an expression of nature. Projective geometry and the mathematics of the spiral may help to bridge the enormous chasm between ancient systems of thought and the modern world of science. There are also benefits to be gained for science by bridging this gap. Perhaps new paradigms will be found that will better explain areas of knowledge not well explained by present scientific models.

Acknowledgement

I would like to thank Rudolf Steiner Press for their generosity in permitting me to reprint many figures from *Sensitive Chaos* by Theodor Schwenk and *Projective Geometry* by Lawrence Edwards.

References

1. E. Guidoni, *Primitive Architecture.* New York, Abrams (1978).
2. T. Schwenk, *Sensitive Chaos.* New York, Schocken Books (1976).
3. L. Edwards, *Field of Form.* London, Floris Books (1982).
4. M. Eliade, M. *The Sacred and the Profane.* New York, Harper and Row (1961).
5. A. C. Crombie, Descartes. In *Mathematics: An Introduction to its Spirit and Use,* ed. M. Kline, New York, Freeman (1979).
6. J. Purce, *The Mystic Spiral.* London, Thames and Hudson (1974).
7. T. Schwenk and W. Schwenk, *Water: The Element of Life.* Hudson, NY, Anthroposophic Press, (1989).
8. M. Riegner and J. Wilkes, Art in the service of nature: The story of flowforms. *Orion Nature Quart.* **7,** No. 1 (Winter 1988).
9. J. Green, personal communication.
10. L. Edwards, *Projective Geometry.* London, Rudolph Steiner Inst. (1985).
11. O. Whicher, *Projective Geometry.* London, Rudolph Steiner Press (1971)
12. T. Cook, *The Curves of Life.* Dover, New York (1979).
13. G. Adams, *The Lemniscatory Ruled Surface in Space and Counterspace.* London, Rudolph Steiner Press (1979).

DOES THE GOLDEN SPIRAL EXIST, AND IF NOT, WHERE IS ITS CENTER?

Arthur L. Loeb and William Varney

1. The Golden Fraction and Dynamic Symmetry

The *golden ratio* or *golden fraction* ϕ is a positive number defined recursively by the equation

$$\phi = 1/(1 + \phi), \tag{1a}$$

which is equivalent to

$$\phi^2 + \phi = 1. \tag{1b}$$

These equations, when solved, produce an expression for ϕ in terms of $\sqrt{5}$, but like ϕ, $\sqrt{5}$ is an irrational number that is less interesting and no more significant than ϕ. Eqs. (1a) and (1b) provide properties of ϕ that are useful in the present geometrical and visual discourse.

Fig. 1 shows an arbitrary rectangle $ABCD$ whose longer sides are AB and CD. A diagonal AC is drawn, a perpendicular is dropped from B onto AC, intersecting it at O, and extended till it meets CD at E. Finally, a line is drawn through E parallel to the shorter sides BC and AD, which intersects the side AB at F. The line EF divides the rectangle $ABCD$ into two new rectangles, of which one, $BCEF$, is geometrically similar to $ABCD$. The other rectangle, $AFED$, is called the gnomon of $ABCD$.

The line segment BE, already drawn, is a diagonal of the rectangle $BCEF$, and the line AC is perpendicular to BE. If we call the point where AC intersects EF G, then a line parallel to BF drawn through G will

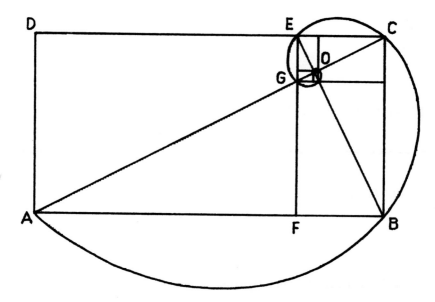

Fig. 1. Nested rectangles related by dynamic symmetry together with their proper logarithmic spiral.

divide the rectangle BCEF in turn into a rectangle similar to ABCD and to BCEF, and a gnomon similar to AFED. This process may be continued indefinitely, so that a series of nested rectangles is generated in which each member is related to the previous one by a rotation through 90°, accompanied by a linear reduction in size by a factor **f**, equal to the ratio of the lengths of the shorter to the longer side of each rectangle. This relation between these nested rectangles is called *dynamic symmetry*.[1]

A logarithmic spiral may be drawn through the points A, B, C, E, G, etc. Such a spiral has the same function in dynamic symmetry as has the circle in rotational symmetry. In point of fact, if the rectangle ABCD had been a square, the spiral would have been a circle. A logarithmic spiral is characterized by the fact that the angle between a tangent to the spiral at a given point and the line joining that point to the spiral center is constant.

The point O is invariant under dynamic symmetry: in each rectangle it is the intersection of a diagonal with a perpendicular dropped onto it from a vertex. This point O is called the *center* of the logarithmic spiral through A, B, C, G, etc., and a center of dynamic symmetry. The radius of the spiral, i.e., the distance of a point on the spiral from the spiral's center, is reduced by a factor **f** every time the angular coordinate is increased by 90°. We shall call a logarithmic spiral that passes through corresponding vertices of a series of nested dynamically symmetrical rectangles a *spiral proper to* these rectangles.

There is a particular rectangle whose gnomon is a square. If, in Fig. 1 we set the length of the longer side of the original rectangle, *AB*, equal to unity, then the shorter side, *BC*, will have a length **f**. In turn, the length of the shorter side of the next rectangle, *CE*, is $\mathbf{f}^2$. The gnomon *AFED* therefore has one side, *AF* whose length equals 1-$\mathbf{f}^2$, whereas the length of another side, *EF* equals **f**. The rectangle *AFEB* is accordingly a square if:

$$1 - \mathbf{f}^2 = \mathbf{f}, \; i.e.,$$

$$\mathbf{f}^2 + \mathbf{f} = 1. \tag{2}$$

Comparing Eqs. (1) and (2) we find that for a rectangle whose gnomon is a square the length of the shorter side is ϕ times that of the longer side. Such a rectangle is called a *golden rectangle*.

If *ABCD* is a golden rectangle, then the tangent of angle *CAB* equals ϕ. Fig. 2 shows a golden rectangle *ABCD*, the line *EF* dividing *ABCD* into the rectangle *BCEF* similar to *ABCD*, the square gnomon *AFED*, both diagonals of *ABCD*, *AC* and *BD*, which intersect each other at *X*, the spiral proper to *ABCD*, and *O*, the center of that spiral. The angle *BXC* equals twice angle *BAC*:

$$\text{Angle } BXC = 2 \text{ arc tan } \phi.$$

Using the formula $\tan 2\alpha = 2 \tan \alpha / (1 - \tan^2 \alpha)$:

$$\tan BXC = 2\phi / (1 - \phi^2).$$

From Eq. (1b) we know that $1 - \phi^2 = \phi$, so that $\tan BXC = 2$. Therefore the length of OB is exactly twice that of *OX*. This result leads to the following conclusions:

   a. *Theorem 1:* The diagonals of each golden rectangle intersect at an angle whose tangent equals 2. The center of a spiral proper to a golden rectangle lying on one of its diagonals is exactly twice as far from the nearer end of the other diagonal as from the intersection of both diagonals.

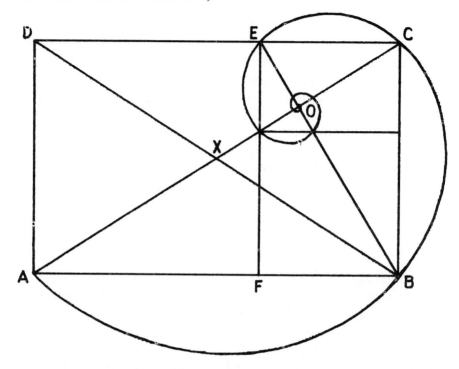

Fig. 2. Nested golden rectangles having square gnomons.

b. *Theorem 2:* The golden fraction is given by the expression

$$\phi = \tan\left(\frac{1}{2}\operatorname{arc tan} 2\right). \tag{3}$$

Eq. (3) gives an explicit expression for the golden fraction. Admittedly it uses transcendental trigonometric functions that may not be considered superior to the use of the square root of five. However, Eq. (3) allows a simple construction of a golden rectangle: the tangent of the angle between the diagonals equals 2; hence that angle is easily drawn. Thus the diagonals of the golden rectangle should be drawn first. The vertices of such a rectangle lie on a circle whose center lies at the intersection of the diagonals. If the length of the diagonals is given, then the diameter of that circle should equal the length of the diagonals. On the other hand, if the desired golden rectangle is to have a given side length, then the angle between the diagonals should be bisected, and the desired length measured off along the appropriate bisector.

2. A Golden Triangle

Fig. 3 shows an isosceles triangle *ABC*, whose sides *AC* and *BC* are both of length unity. The length of the base *AB* is $\mathbf{f} < 1$. A point *D* is chosen on *BC* such that *AB* = *AD*. Since triangles *ABC* and *BDA* are isosceles and have a common base angle, they are geometrically similar, so that the length of the line segment *BD* equals $\mathbf{f}^2$. Therefore the length of *CD* equals $(1-\mathbf{f}^2)$. Triangle *ADC* is called the *gnomon* of *ABC*; the gnomon of *ABC* is isosceles if *AD* = *CD*, i.e., if $\mathbf{f} = 1 - \mathbf{f}^2$. Again, this condition tells us that $\mathbf{f} = \phi$.

The isosceles triangle whose gnomon is also isosceles is called a golden triangle; its gnomon is called the golden gnomon. The length of the base of the golden triangle is ϕ times its sides, whereas the length of the side of the golden gnomon equals ϕ times its base length. If the value of angle *ACB* is called α, then, because the sum of the angles of triangle *ADC* equals 180° and that triangle is isosceles, angle *ADC* equals (180° - 2α), and therefore angle *ADB* equals 2α. As triangles *ABC* and *BDA* are geometrically similar, angle *BAD* equals α, and *ABD* equals 2α. Therefore:

$$2\alpha + 2\alpha + \alpha = 180°, \text{ hence } \alpha = 36°.$$

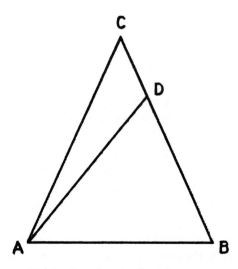

Fig. 3. Isosceles triangle and gnomon.

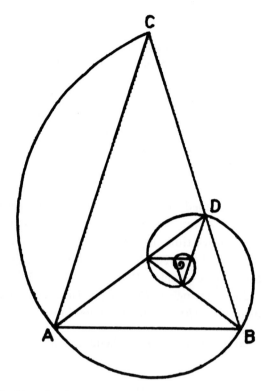

Fig. 4. Nested golden triangles and golden gnomons together with their proper logarithmic spiral.

Accordingly, the angles of a golden triangle are 36°, 72°, and 72°; those of a golden gnomon are 36°, 36°, and 108°. These two triangles are the only finite triangles all of whose angles are integral multiples of 36°.[2]

Triangle *BDA*, a golden triangle, may be in turn subdivided into a golden triangle and a golden gnomon, and the process continued indefinitely (Fig. 4). This figure exhibits dynamic symmetry; in this instance a rotation of 108° is accompanied by a reduction in size by a factor ϕ. A logarithmic spiral can be drawn through successive dynamically related points, but whereas in the case of the golden *rectangle* a *90°* rotation is accompanied by a scaling factor of ϕ, for the golden *triangle* the same scaling occurs over a rotation by *108°*. It is therefore not possible to apply the name *golden spiral* uniquely *both* to the logarithmic spiral proper to the golden rectangle and to that proper to the golden triangle.

In contrast to the case of the spiral proper to the golden rectangle, it is not at all obvious where the center of the spiral proper to the golden triangle is located. Huntley[3] states that it lies on the median line joining a base vertex to the center of the opposite side, but does not provide a proof. If this assertion is indeed correct, then corresponding medians in nested golden triangles will intersect at the center of the spiral, as this center necessarily lies on the corresponding median of each of these nested golden triangles. We shall return to this assertion presently, but will first consider the golden parallelogram.

3. A Golden Parallelogram

Fig. 5 shows a parallelogram *PQRS* having angles 108° and 72° and sides whose lengths are in the golden ratio. As we shall be considering this parallelogram in various orientations and scalings, we shall call the lengths of its sides *PS* and *PQ* a and $a\phi$ respectively.

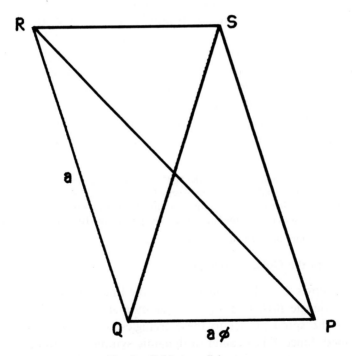

Fig. 5. Golden parallelogram.

From the law of cosines we calculate the diagonals of this so-called golden parallelogram:

$$PR^2 = a^2(1 + \phi^2) + 2a^2\phi \cos 72°$$
$$QS^2 = a^2(1 + \phi^2) - 2a^2\phi \cos 72°.$$

As 72° is the value of the base angles of a golden triangle, and the ratio of base length to side length of the golden triangle is ϕ, $\cos 72° = \frac{1}{2}\phi$, so that:

$$PR = a\sqrt{(1 + 2\phi^2)}$$
$$QS = a.$$

Accordingly, the golden parallelogram is divided by its short diagonal into two golden triangles. As two diagonals of a parallelogram bisect each other, the long diagonal is a median line in each of these golden triangles.

The long diagonal divides the golden parallelogram into two mutually congruent triangles which are not as familiar as are the golden triangles, but which are nevertheless quite remarkable, and will be helpful in locating the center of the spiral proper to the golden triangle. Because of their special relationship to the golden triangle, we shall call them *golden supplements*. Triangle PRS has an angle of 108°; the sides of this angle have lengths related by the golden ratio. The short diagonal of the golden parallelogram is a median line of the golden supplement. This median line has length $\frac{1}{2}a$, and divides the 108° angle into one angle of 72° and one of 36°. Hence:

Theorem 3: The median line joining the 108° angle of the golden supplement to the center of its opposite side has a length exactly half that of that obtuse angle's longer side, and divides that obtuse angle into two unequal portions of which the smaller is exactly one half as large as the latter. The side opposite the 108° angle has a length $\sqrt{(1 + 2\phi^2)}$ times that of that angle's longer side.

4. Return to the Golden Triangle

Fig. 6a shows the golden triangle ABC, the line AD subdividing it into a golden triangle BAD and a golden gnomon ADC, and a point Y, the center of the spiral proper to these triangles, whose location is to be determined. Since Y is a center of dynamic symmetry, and the points C and A are respective apices of triangles related by dynamic symmetry,

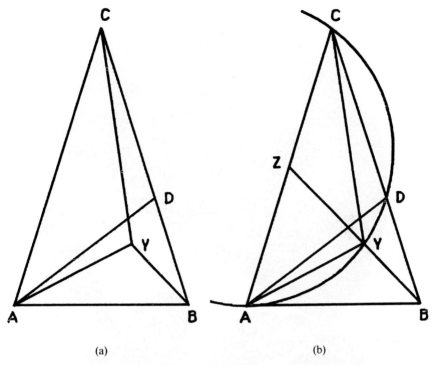

Fig. 6. Golden triangle, golden gnomon and the center of their proper spiral.

angle $AYC = 108°$, and $YA/YC = \phi$. Accordingly, triangle AYC is a golden supplement. By the same reasoning it is concluded that the triangle BYA is a golden supplement.

The point Y subtends a circular arc of 108° at the line AC. The same is true of the point D. Therefore A, Y, D and C lie on a common circle (Fig. 6b), the circle circumscribing the golden gnomon ADC. The same reasoning applies to the entire set of nested golden gnomons, and leads to:

> *Theorem 4:* In a set of nested golden triangles and golden gnomons related by dynamic symmetry, all circles circumscribing the golden gnomons pass through a common point which is the center of the logarithmic spiral proper to the nested golden triangles and gnomons.

This theorem provides a simple construction method for locating the center of our spiral. It provided the inspiration for artist Damian

Fig. 7. Construction by Damian Bagdan based on the observation that all circles circumscribing a set of nested golden gnomons pass through the center of their proper logarithmic spiral. (From the Teaching Collection in the Carpenter Center for the Visual Arts, Harvard University. Reproduced by permission from the Curator. Photography C. Todd Stuart.)

Bagdan's design (Fig. 7). We have, however, not yet fully exploited the properties of the golden supplement. To do so, extend the line BY to meet AC at point Z (Fig. 6b). Since angles AYC and BYA both equal 108°, the angle BYC equals 144°. Therefore, angle ZYC equals 36°, with the result that angle ZYA equals 72°. Since we have seen that triangle AYC is the golden supplement, the line YZ, which divides the angle AYC into portions of 72° and 36°, is a median of triangle AYC. Therefore, $AZ = CZ$, so that the line BZ is a median line of triangle ABC, and Huntley's assertion is proven correct. Since the side AC of the golden triangle ABC was set at unity, we find from Theorem 3 that the sides of golden supplement AYC have lengths $CY = 1/\sqrt{(1 + 2\phi^2)}$, $AY = \phi/\sqrt{(1 + 2\phi^2)}$ and $AC = 1$, and that the length of its median line YZ equals $1/2\sqrt{(1 + 2\phi^2)}$.

5. Return to the Golden Parallelogram

These results may be quite lucidly summarized in terms of a golden parallelogram $ABCE$ constructed so that triangle ABC is the original golden triangle, and E is the intersection of a line drawn through the point C parallel to AB with a line drawn through the point A parallel to BC (Fig. 8). Triangle ECA is, of course, golden. The long diagonal BE passed through the centers of two logarithmic spirals, one, Y, proper to golden triangle ABC, the other, X, proper to golden triangle ECA. As in the case of the golden rectangle, either of these spirals may be called proper to the golden parallelogram $ABCE$, and from the values found above for distances AY, CY and YZ we derive the following theorem:

Theorem 5: The center of a logarithmic spiral proper to a golden parallelogram lies on the longer diagonal of that parallelogram, exactly twice as close to the intersection of the two diagonals as to the farther end of the shorter diagonal.

There is a striking similarity between this theorem and Theorem 1; before too fundamental a connection is suspected, note, however, that in Theorem 1 it is the *nearer*, in Theorem 5 the *farther* end of the other diagonal which is referred to. It *is* notable, however, that, in spite of the irrationality of the golden fraction, these figures still possess remarkable integral ratios.

We noted above that the spiral center Y, subtending an angle of 108° at the points A and C, lies on the same circle which circumscribes the

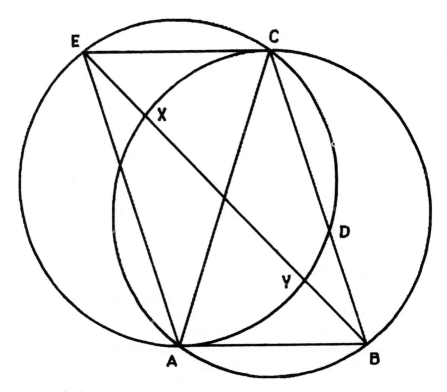

Fig. 8. Golden Parallelogram and its constituent twin golden triangles.

golden gnomon *ADC*. The point *E* subtends an angle of 72° at points *A* and *C*, and therefore also lies on the circle passing through points *A, Y, D* and *C.* Thus:

> *Theorem 6:* The twin golden triangles constituting a golden parallelogram are paired so that the spiral center proper to one lies on the circle circumscribing the other.

This last observation leads to a simple construction method for locating the center of the logarithmic spiral proper to a golden triangle. Remember that the center of the circle circumscribing a triangle is at the intersection of the perpendicular bisectors of the sides of that triangle. If we consider golden triangle *ABC,* and recall that the circle circumscribing its gnomon *ADC* will also circumscribe the twin golden triangle *ECA*, then we conclude that the center of that circumscribing circle lies at the intersection of the perpendicular bisector of *AC* with the perpendicular to *AB* drawn through *A*, the latter being also the perpendicular bisector of *CE*. Thus we have used the concept of the golden parallelogram without actually needing to draw it explicitly in the construction process.

The construction method is accordingly as follows: (cf. Fig. 9)

Step 1: Draw the perpendicular bisector of *AC*, calling its intersection with *AC Z*. (Fig. 9a)

Step 2: Draw *BZ*. (Fig. 9b)

Step 3: Draw a perpendicular to *AB* through *A*. (Fig. 9b)

Step 4: Call the intersection of this latter perpendicular with the perpendicular bisector of *AC U*. (Fig. 9b)

Step 5: With *U* as center and *UC* as radius, draw a circle. Call the intersection of this circle with *BC D*, and its intersection with *BZ Y*. Triangle *ADC* will be the first nested gnomon in *ABC*, and *Y* the center of their proper logarithmic spiral. (Fig. 9c)

6. Summary and Conclusions

We have noted that a logarithmic spiral circumscribing a set of nested figures related to each other by dynamic symmetry may not be generally called the golden spiral, for in the case of a golden rectangle this spiral decreases its radius by the golden fraction after a 90° rotation, whereas in the case of a golden triangle the same relative change in radius occurs after a 108° rotation. We propose therefore to refer to "the (logarithmic) spiral proper to the golden rectangle/triangle/parallelogram."

We showed that the diagonals of a golden rectangle make an angle arc tan 2 with each other, an expression in which the golden fraction does not appear explicitly. From this property we derived the fact that the center of a spiral to a golden rectangle lies twice as close to the intersection of that rectangle's diagonals as to one of that rectangle's vertices.

We further considered golden triangles and golden gnomons, and locations of the centers of their proper spirals. On the way we introduced two new polygons having interesting properties, a golden parallelogram and a golden supplement. We found a proof for Huntley's assertion that the center of a golden spiral proper to a golden triangle lies on a median line of that golden triangle. We showed that all circles circumscribing a set of nested golden gnomons pass through the center of the proper spiral,

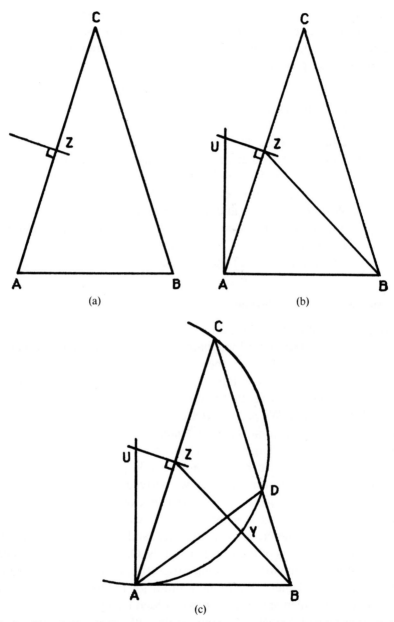

Fig. 9. Construction of the center of the logarithmic spiral proper to a nested set of golden triangles and golden gnomons.

and we combined the last two findings to propose a simple construction method for finding such a center.

Lastly, we showed that the center of the spiral proper to the golden triangle is twice as close to the midpoint of one of the equal sides of that triangle as to its apex.

References
1. E. Edwards, *Patterns and Design with Dynamic Symmetry*. Dover, New York (1967 reprint of the 1932 edition).
2. A. L. Loeb, The magic of the pentangle: Dynamic symmetry from Merlin to Penrose, *Computers Math. Applic.* **17** (1989) pp. 33–48 and in *Symmetry 2: Unifying Human Understanding*, I. Hargittai, ed. Pergamon Press, Oxford (1989).
3. H. E. Huntley, *The Divine Proportion*. Dover (1970) pp. 170–171.

PYTHAGOREAN SPIRALS

Ernest J. Eckert

1. Pythagorean Spirals

When viewing Fig. 1 the eye is deceiving the brain into seeing a spiral. The construction of the figure depends on a sequence of Pythagorean triangles (a, b, c) characterized by $c - b = 1$. At the corners of the squares the angle between the line, $\ominus$ = constant, and the tangent to the perceived spiral is 45°, reminding one of a logarithmic spiral. What the eye sees, however, is not a logarithmic spiral. Serendipity provided the pleasant surprise of the spiral when Hugo Haagensen discovered the Pythagorean tiling of the plane depicted in Fig. 1.

Each of the four Pythagorean triangles surrounding the innermost 1×1 square in Fig. 1 is a (3, 4, 5) triangle. In Fig. 2, which is an enlargement of the central part of Fig. 1, the four triangles form a 5×5 square. Reflecting each triangle in its hypotenuse, we obtain a 7×7 square. Surround this square by four (5, 12, 13) Pythagorean triangles to form a 13×13 square. Reflecting each triangle in its hypotenuse, we obtain a 17×17 square. The process may be continued indefinitely by appropriately choosing the sequence of Pythagorean triangles:

$$(a_1, b_1, c_1) = (3, 4, 5),$$
$$(a_2, b_2, c_2) = (5, 12, 13),$$
$$(a_3, b_3, c_3) = (7, 24, 25), \ldots, (a_n, b_n, c_n), \ldots.$$

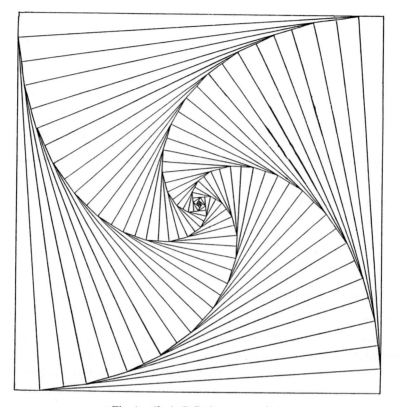

Fig. 1. (3, 4, 5) Pythagorean spiral.

From Fig. 2 it is evident that the key to success is:

$$b_n - a_n = b_{n-1} + a_{n-1}. \qquad (1)$$

Eq. (1) will be satisfied if we take:

$$(a_n, b_n, c_n) = (2n + 1, 2n^2 + 2n, 2n^2 + 2n + 1), n \geq 1.$$

In fact, $b_n - a_n = b_{n-1} + a_{n-1} = 2n^2 - 1$. Note also that $c_n - b_n = 1$, a constant.

To construct a spiral, S_1, we connect the points C_n and C_{n+1} with a continuous curve. We choose a circular arc with center at A_n, radius $|A_n C_n|$ and radian measure $2 \tan^{-1} \frac{a_n}{b_n}$. Another spiral, S_2, may be obtained by drawing circular arcs, concentric with the above arcs, with radii $|A_n B_n|$. The radial distance between such concentric arcs is 1. In this sense the two

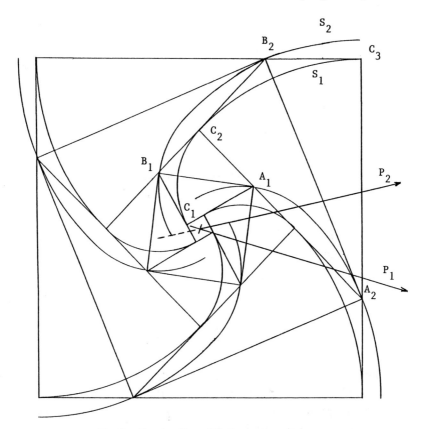

Fig. 2. Construction of Pythagorean spiral.

spirals are parallel. Clearly, four such pairs of spirals may be drawn.

We may imagine the 7×7 square with vertex C_2 being obtained by a sevenfold magnification of the 1×1 square followed by a clockwise rotation through $2\tan^{-1}\frac{3}{4}$. The 17×17 square with vertex C_3 may be obtained from the 1×1 square by a magnification factor of 17 and a rotation of $2\tan^{-1}\frac{3}{4} + 2\tan^{-1}\frac{5}{12}$. In general, the $(a_n + b_n) \times (a_n + b_n)$ square with vertex C_{n+1} is obtained from the 1×1 square by a magnification factor of $(a_n + b_n)$ and a clockwise rotation through $2\sum_{k=1}^{n}\tan^{-1}\frac{a_k}{b_k}$. This process makes it is easy to obtain the polar coordinates, (r, Θ), of a vertex of any square. For example, with the polar axis P_1 located as indicated in Fig. 2, we obtain for the vertex C_{n+1}:

$$r_{n+1} = \frac{\sqrt{2}}{2}(a_n + b_n) = \frac{\sqrt{2}}{2}(2n^2 + 4n + 1),$$

$$\Theta_{n+1} = -\pi - 2\sum_{k=1}^{n} \tan^{-1}\frac{a_k}{b_k} = -\pi - 2\sum_{k=1}^{n} \tan^{-1}\frac{2k+1}{2k^2 + 2k}, \quad n \geq 0.$$

Since $\sum_{n=1}^{\infty} \tan^{-1}\frac{a_n}{b_n}$ diverges, the spiral S_1 winds infinitely often about the origin.

Polar coordinates for the vertices B_n, on spiral S_2, may be obtained in a similar way. Using polar axis P_2 we obtain:

$$r_n = \frac{\sqrt{2}}{2}c_n = \frac{\sqrt{2}}{2}(2n^2 + 2n + 1),$$

$$\Theta_n = -\pi - 2\sum_{k=1}^{n-1} \tan^{-1}\frac{a_k}{b_k} - \tan^{-1}\frac{a_n}{b_n}$$

$$= -\pi - 2\sum_{k=1}^{n-1} \tan^{-1}\frac{2k+1}{2k^2 + 2k} - \tan^{-1}\frac{2n+1}{2n^2 + 2n}, \quad n \geq 1.$$

Other sequences of Pythagorean triangles may be used to construct spirals as above. It seems fitting to name them Pythagoreans spirals. In Fig. 3 the innermost square is 7×7. The sequence of triangles to be used in the construction of spirals is:

$(a_1, b_1, c_1) = (8, 15, 17),$
$(a_2, b_2, c_2) = (12, 35, 37),$
$(a_3, b_3, c_3) = (16, 63, 65), \ldots,$
$(a_n, b_n, c_n) = (4n + 4, 4n^2 + 8n + 3, 4n^2 + 8n + 5), n \geq 1.$

Here $b_n - a_n = b_{n-1} + a_{n-1} = 4n^2 - 4n + 1$, so that Eq. (1) is satisfied. We note that $c_n - b_n = 2$, a constant, so that the two parallel spirals are a distance of 2 apart.

Arguing as before, we can easily determine the polar coordinates of the vertices of the squares. If we think of Fig. 2 as applying to the present situation, the polar coordinates of C_{n+1} are:

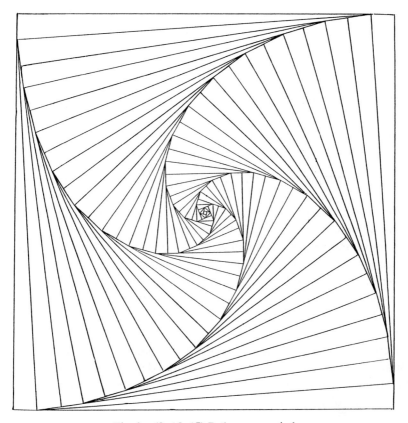

Fig. 3. (8, 15, 17) Pythagorean spiral.

$$r_{n+1} = \frac{\sqrt{2}}{2}(a_n + b_n) = \frac{\sqrt{2}}{2}(4n^2 + 12n + 7)$$

$$\Theta_{n+1} = -\pi - 2\sum_{k=1}^{n} \tan^{-1}\frac{a_k}{b_k} = -\pi - 2\sum_{k=1}^{n}\frac{4n+4}{4n^2+8n+3}, n \geq 0.$$

Again, the spirals wind infinitely often about the origin.

2. Matrix Generation

Another way of constructing sequences of Pythagorean triangles giving rises to Pythagorean spirals is given in Ref. 1. The linear transformation A given by the matrix

$$\begin{pmatrix} 1 & -2 & 2 \\ 2 & -1 & 2 \\ 2 & -2 & 3 \end{pmatrix}$$

sends the cone, $x^2 + y^2 = z^2$, into itself, as is easily verified. If (x, y, z) has positive integral components, so that (x, y, z) is a Pythagorean triangle, then A sends (x, y, z) into another Pythagorean triangle, (u, v, w). Additional properties of A are:

$$|v - u| = x + y \text{ and } w - v = z - y,$$

exactly the properties which guarantee the successful construction of Pythagorean spirals.

To obtain a sequence of Pythagorean triangles defining a Pythagorean spiral we may start with any Pythagorean triangle (a, b, c) with $a < b$. The successive applications of A will yield a sequence of Pythagorean triangles which gives rise to a Pythagorean spiral.

Let $(a_1, b_1, c_1) = (20, 21, 29)$ for which $c_1 - b_1 = 8$. Repeated applications of A yield the sequence: $(20, 21, 29)$, $(36, 77, 85)$, $(52, 165, 173)$, $(68, 285, 293), \ldots, (a_n, b_n, c_n) = (16n + 4, 16n^2 + 8n - 3, 16n^2 + 8n + 5), \ldots$.

Here, $b_n - a_n = b_{n-1} + a_{n-1} = 16n^2 - 8n - 7$, and $c_n - b_n = 8$. The parallel spirals, constructed as above, will be separated by a distance of 8 (see Fig. 4). Accepting, again, Fig. 2 to represent the situation at hand, the polar coordinates of C_{n+1} are:

$$r_{n+1} = \frac{\sqrt{2}}{2}(a_n + b_n) = \frac{\sqrt{2}}{2}(16n^2 + 24n + 1),$$

$$\Theta_{n+1} = -\pi - 2\sum_{k=1}^{n} \tan^{-1}\frac{a_k}{b_k}$$

$$= -\pi - 2\sum_{k=1}^{n} \tan^{-1}\frac{16n + 4}{16n^2 + 8n - 3}, n \geq 0.$$

The spirals wind infinitely often about the origin.

3. Continuous Pythagorean Spirals

A Pythagorean spiral is pieced together of circular arcs with increasing radii and decreasing radian measures. A Fibonacci spiral is pieced

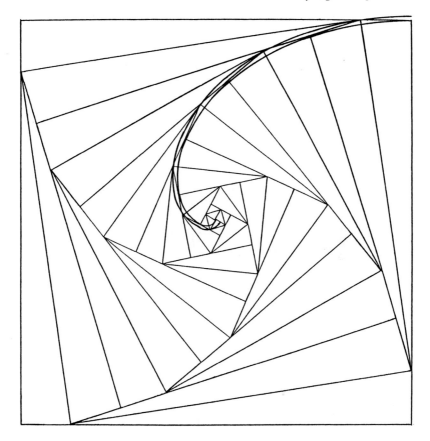

Fig. 4 (20, 21, 29) Pythagorean spiral.

together of circular arcs with increasing radii and constant radian measure, $\frac{\pi}{2}$. A Pythagorean spiral may be perceived as a refinement of a Fibonacci spiral. The refinement may be carried a step further by demanding a spiral pieced together of circular arcs of continuously increasing radii and infinitesimal radian measures. An appropriate name for such a spiral might be: continuous Pythagorean spiral.

Starting with a 1×1 square, as in the first section, let us take the size (i.e., the half-diameter) of the magnified square to be $\frac{\sqrt{2}}{2} c_t = \frac{\sqrt{2}}{2}(2t^2 + 2t + 1)$, and a clockwise rotation rate of

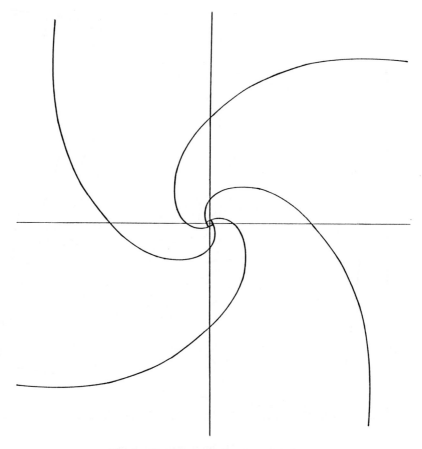

Fig. 5. Continuous Pythagorean spiral.

$$2 \tan^{-1} \frac{a_t}{b_t} = 2 \tan^{-1} \frac{2t + 1}{2t^2 + 2t}.$$

In polar coordinates we may then write the parametric representation of the curve:

$$r(t) = \frac{\sqrt{2}}{2}(2t^2 + 2t + 1),$$

$$\Theta(t) = -\pi - 2\int_0^t \tan^{-1}\frac{2v+1}{2v^2+2v}\,dv,\ t\geq 0.$$

The initial point of the spiral is $(r(0), \Theta(o)) = (\frac{\sqrt{2}}{2}, -\pi)$, and the initial rotation rate is $-\pi$ (see Fig. 5).

The functions $r(t)$ and $-\Theta(t)$ are continuous and increasing on $[0, \infty)$, hence they have continuous inverses which are also increasing functions. It follows that r is an increasing function of $-\Theta$, so that the curve is indeed a spiral. Note that $\lim_{t\to\infty}\Theta(t) = -\infty$, so that the spiral winds infinitely often about the origin. Clearly, one may draw four spiral arms, 90° apart, to complete the analogy with Pythagorean spirals. For each Pythagorean spiral there is a continuous Pythagorean spiral.

The particular continuous Pythagorean spiral in Fig. 5 has an interesting property. If we think of the spiral as being traced out by a moving particle starting at $(r(0), \Theta(0)) = (\frac{\sqrt{2}}{2}, -\pi)$, then the radial acceleration is constant, $2\sqrt{2}$, and the angular acceleration is inversely proportional to the distance r from the origin to the particle. In fact, $\frac{d^2\Theta}{dt^2} = \frac{-2\sqrt{2}}{r(t)}$. Since $\frac{d^2r}{dt^2} = 2\sqrt{2}$, we may express this as $\Theta''(t) = \frac{-1}{r(t)}r''(t)$. Which particles move in this way? What other properties does the spiral have?

Since there is a continuous Pythagorean spiral for each of the infinitely many Pythagorean spirals one may wonder which properties the infinitely many continuous spirals have in common. Is it true, for example, that each such spiral has the above mentioned properties of radial and angular accelerations? There is a unified description of the set of logarithmic spirals, $r = ae^{b\Theta}$. Is there a unified theory of continuous Pythagorean spirals? Do continuous Pythagorean spirals occur in nature?

References
1. E. J. Eckert and H. Haagensen, A Pythagorean tiling of the plane, *Math. Mag.* **62** (1989) 175.

DYNAMICAL SPIRALS

Arun V. Holden

1. Introduction

Spiral structures are common throughout nature, on the length scale of galaxies to the more modest length scales characteristic of living systems. Although these structures appear permanent they are in fact transitory: a fossil *Nautilus* may be some hundreds of millions of years old, while a spiral galaxy may be some billion years old. Both these structures have been formed and will eventually decay.

The formation of the *Nautilus* shell during the life of the animal is by the successive addition of compartments, while spiral galaxies, spirals in chemical and some biological systems (the aggregation of social amoebae, spiral waves in heart tissue) arise by a more general process, that of a coupling of a nonlinear, reaction process with diffusive transport in a spatially extensive medium.[1]

Whatever the process that produces and maintains the spiral structure, it will finally disappear, either because the spiral itself collapses (as in a model of spiral waves in cardiac tissue,[2] or because the medium within which the spiral is embedded erodes. Structures are not permanent, they change and dissolve with time, and so have some kind of dynamics. Since this may be on a very slow time scale it can be described as evolutionary dynamics.

In this paper I am considering dynamics on a very much faster time scale, where the phenomenon that is apparent is change, not structure.

Further, spatial processes will not be considered, only processes that change with time. These dynamical systems may be represented by differential equations, but I will only consider graphical representations of their behaviour.

Although mathematicians study dynamical systems for their own reasons, a large part of the interest in nonlinear dynamics arises from its applications: there is a strong belief that an understanding of the behaviour of dynamical systems can provide a means of describing, and even understanding, natural phenomena. The aim of a description is to accurately reproduce observations, the aim of understanding is to find "structures" that account for the observations. These structures are not real, spatially extensive, physical entities, but are abstractions that inhabit state space, where the state space is an abstract space whose "space-like" coordinates represent the different variables of the system.

If one is observing a variable from a real dynamical system, say an electrical signal from a single nerve cell,[3] then it may be steady, oscillating, or changing irregularly. This behaviour will be generated by several interacting variables. These maintained behaviours, and changes between them, are often associated with spiral structures in state space.

2. Dissipative Systems and Attractors

If you were to scatter iron filings over a sheet of paper, and bring up one end of a long bar magnet under the paper, then on tapping the paper the iron filings would all be attracted to a region over the pole of the magnet. Think of each grain of iron as a starting point, or initial condition — each grain will follow a path or trajectory, finally arriving over the pole of the magnet. The area over the pole acts as an attractor, and has trajectories leading into it, but not out of it. During the process, iron filings initially spread over a large area are contracted into a smaller area.

This provides an analogy for a dissipative process, within which trajectories starting at different initial points in state space are finally drawn into an attractor, accompanied by a contraction of state space "volume." Most real, physical systems are dissipative, in that a final condition can be reached from a large number of initial conditions. Such a convergence imparts a high degree of regularity on the real world.

The types of trajectory and attractor that can be found for a dynamical system depend on how many variables the system has.

If there is only one variable, the state space is a line and so all trajectories are confined to this line. The only possible trajectories are to stay in one place (an equilibrium solution), or to move toward one of these equilibria. The equilibrium points can be stable, in which case they are attractors, or unstable, in which case any small deviation away from the equilibrium grows. The attractors are points. If, in addition to the single variable, there is also a parameter, then the stability of an equilibrium can depend on the value of the parameter. An equilibrium can lose its stability at a particular value of the parameter: this is a bifurcation point, where a qualitative change in behaviour occurs.

3. Spiral Transients

In a two-variable system, there is also the possibility of oscillations, where the attractor is a simple closed curve. Thus the attractors can be points or closed curves. Since there are only two variables, the trajectories can be plotted on paper, with each axis representing a variable. In such a display the trajectory cannot cross itself, as, with a given set of parameters and a given set of initial conditions, the solutions of a deterministic dynamical system are unique. If the trajectories crossed then at the point of crossing the same system, with the same values for the two variables, it would be following two different trajectories at once. Every point in the two variable plane has a trajectory passing through it, and trajectories can end on either a point or closed curve attractor: under these circumstances it is not surprising to find spiral trajectories (see Fig. 1). These spiral trajectories would appear as damped oscillations: the spiral structure is not apparent in an observation of one of the variables changing with time, but underlies the trajectory and can be reconstructed from it.

At some parameter value a stable equilibrium can lose its stability, and small periodic oscillations emerge. This bifurcation from an equilibrium to periodicity is commonly known as a Hopf bifurcation,[5] at which small amplitude, stable or unstable sinusoidal oscillations emerge. If the small amplitude oscillations are unstable, then the unstable small amplitude oscillations provide a pathway to a new, stable solution corresponding to a new attractor. For the dynamical systems that represent nerve[3] and heart muscle membrane,[6] which have considerably more than two variables, the unstable periodic oscillations provide a pathway to larger

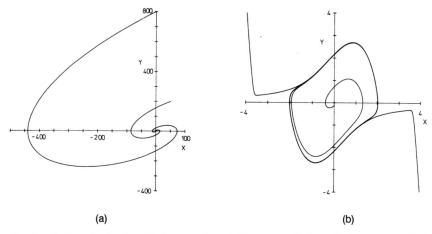

Fig. 1. Trajectories leading (a) into a sink and (b) out to a limit cycle for a van der Pol oscillator. From Ref. 4.

amplitude, stable oscillations that correspond to a rhythmic discharge of nerve impulses, or the periodic electrical activity of the heart.

At a parameter value far from a bifurcation point, the approach to an attractor is relatively rapid; however, close to a bifurcation point the transient approach to an attractor can be greatly prolonged. This is equivalent to the critical slowing down seen close to a phase transition in physics. An example of such a prolonged, oscillatory transient is shown in Fig. 2, for equations that describe excitability in nerve fibre membrane. Although the equations in fact have more than two variables, the spiral nature of the trajectory is apparent.

In these examples the spiral trajectories are transients, either approaching an equilibrium, or spiralling away from an equilibrium. A simple spiral cannot be an attractor, it can only describe motion into or away from an attractor.

4. Spiral Attractors

If there are three or more variables, there is the possibility of quasi-periodic and chaotic attractors. Quasi-periodicity arises when there are two frequencies of oscillation, and these frequencies cannot be expressed as a ratio of whole numbers. If there were only one frequency,

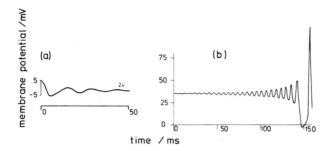

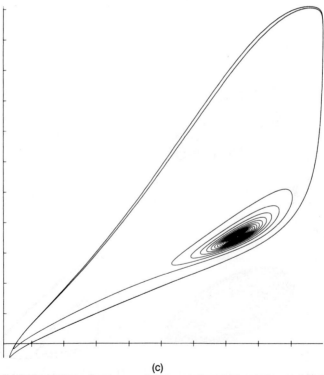

Fig. 2. Numerical solutions of nerve membrane equations (a) just below and (b) just above a bifurcation from a stable equilibrium into autorhthymicity. From Ref. 7.
(c) State space projection of the trajectory of (b), showing its spiral structure. These are numerical solutions of the Hodgkin-Huxley membrane equations[8] with the maximal potassium conductance reduced to 3.9 mS cm^{-2}.

one would have a simple period periodicity; with two frequencies, the faster frequency winds around the slower frequency. If there is a simple ratio, say 3:2, then the faster frequency will wind round the attractor three times while the slower frequency goes around two times — the attractor will be a closed curve, and when projected onto a two-dimensional piece of paper will appear as a looped, closed curve, or pattern of florets. If the frequencies are not in an integer ratio, but are irrationally related, then the pattern never repeats itself: instead of a closed curve the trajectories wind around the surface of a torus.

Examples of quasi-periodicity are shown in Fig. 3, for equations representing a sinusoidally forced glycolytic oscillator (the enzyme catalysed oxidation of glucose, at different values of the forcing parameters). Although quasiperiodity can generate interesting patterns, it is

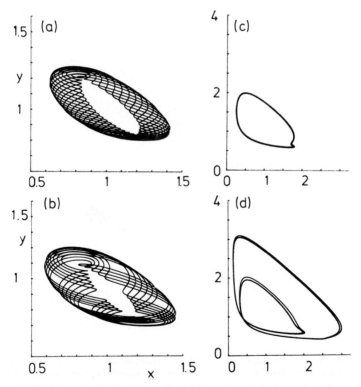

Fig. 3. Quasi-periodicity (a,b) and period doubling (c,d) in numerical solutions of equations representing a glycolytic oscillator. From Ref. 4.

of little significance in the real world as it occurs under very special circumstances. However, the simple periodicity and period doubling illustrated in Figs. 3c, d, although not as visually interesting, in fact are of more significance. Periodic solutions often lose their stability in a period-doubling bifurcation, where the frequency is divided by two and the state space projection onto a plane doubles its number of loops. A cascade of period doubling bifurcations often develops, with the bifurcations occurring at smaller and smaller increments in parameter until the period becomes infinite and so the oscillation never repeats itself and is therefore no longer periodic — it is irregular and its long-term future behaviour is unpredictable. This irregularity produced by deterministic nonlinear systems is chaos[9] and is important in mathematics,[10] physics,[11] chemistry,[12] and biology.[13]

The irregularity is produced by sensitivity to initial conditions: trajectories that start very close together separate rapidly. There is a local expansion in state space, as if it were being stretched. However, chaotic motion is not explosive, it is ordered and confined to some region of state space, a chaotic attractor. For there to be an attractor there must be an overall contraction of volume in state space: although there is stretching, the trajectories must also be folded back. Such a process of stretching and folding forms a powerful physical analogy for the processes generating chaotic motion.[14]

Period doubling into chaos for the Rössler equations, a simple system of three ordinary nonlinear differential equations that were designed[15] as a model for chaos, is illustrated in Fig. 4 as a parameter is increased. The trajectories spiral out and are folded back. All chaotic flows are produced by this combination of expansion and contraction and so will have the characteristics of a spiral; however, projections of attractors for the Rössler equations provide the clearest example of this spiral nature. Further increases in the parameter lead into an image very similar to that of the shell of *Nautilus*.

5. Conclusion

When one thinks of spirals in nature, it is usual to imagine the spiral forms of real objects. Examples are plentiful and often possess an intricate beauty. However, underlying many of the changing process in

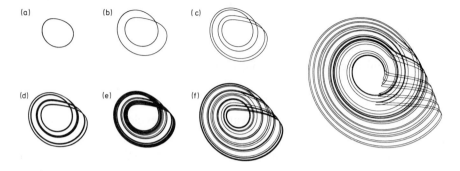

Fig. 4. Period doubling into spiral chaos. From Ref. 4.

nature, both transients and maintained chaotic irregularity, are trajectories and attractors that have a spiral form in state space.

References

1. A. V. Holden, M. Markus, and H. G. Othmer (eds.), *Nonlinear Wave Processes in Excitable Media*. Plenum, New York (1990).
2. A. V. Panfilov and A. V. Holden, Spatiotemporal irregularity in a two-dimensional model of cardiac tissue, *Bifurcation and Chaos* **1** (1991) 219–225.
3. A. V. Holden and W. Winlow. The nerve cell as dynamical system, *IEEE Trans SMC* **13** (1983) 711–719.
4. A. V. Holden and M. A. Muhamad, *A Graphical Zoo of Strange and Peculiar Attractors*. In Ref. 13 (1986).
5. B. D. Hassard, N. D. Kazarinoff, and Y-H. Wan, *Theory and Applications of Hopf Bifurcation*. Cambridge University Press, Cambridge (1981).
6. A. V. Holden and M. Lab, Chaotic behaviour in excitable systems. *New York Acad. Sciences Annals* **591** (1990) 303–315.
7. A. V. Holden and W. Winlow, Comparative neurobiology of excitation. In *The Neurobiology of Pain*, A. V. Holden and W. Winlow, eds. Manchester University Press, Manchester, UK (1984).
8. A. L. Hodgkin and A. F. Huxley, A quantitative description of membrane current and its application to conduction and excitation in nerve. *J. Physiol* **117** (1952) 500–544.
9. A. V. Holden (ed.) *Chaos*. Manchester University Press, Manchester, UK; Princeton University Press, Princeton, New Jersey (1986).
10. M. Tabor, *Chaos and Integrability in Nonlinear Dynamics*. Wiley-Interscience, New York (1989).
11. H. G. Schuster, *Dynamical Chaos: An Introduction*. VCH Physik-Verlag, Weinheim (1988).
12. G. Nicolis and I. Prigogine, *Exploring Complexity*. Freeman, New York (1989).

13. H. Degn, A. V. Holden, and L. F. Olsen (eds), *Chaos in Biological Systems*. Plenum, New York (1988).
14. J. M. Ottino, *The Kinematics of Mixing*. Cambridge University Press, Cambridge (1989).
15. O. E. Rössler, An equation for continuous chaos. *Phys. Lett.* **57A** (1976) 397.

RANDOM SPIRALS

W. A. Seitz and D. J. Klein

1. Introduction

It has long been recognized that spiral structures arise in a great diversity of situations in nature. Witness for instance T.A. Cook's book[1] *The Curves of Life*, D'Arcy Thompson's classic[2] *Growth and Form*, or the present volume. But typically the spirals considered have been deterministic. Admittedly this seems perfectly appropriate for the beautiful precise spirals of the Nautilus, the phyllotactic spirals on the heads of sunflowers, or the molecular helices of DNA.

Surely too there must be circumstances where random spirals occur. For instance, Gause[3] has observed *Bacillus mycoides* in (both left- and right-handed) spiral forms of a random sort, as in Fig. 1. Other species have exhibited[4] random spiral forms too.

Another conceivable circumstance involving random spirals involves polymer chains. Suppose the monomer units are chiral with one side preferentially adsorbing to a (plane) surface. Then a preferential bending from one monomer to another should result in some sort of random spiral. Perhaps most simply, bending in one direction (e.g., clockwise) might be prohibited, while a straight-ahead step or a bend in the opposite direction is of equal energetic favorability. The same sort of circumstance could conceivably arise with filamentous algae with chiral cells, one side of which preferentially adheres to an underlying substrate.

Here we review this model, in context with other work in the area of

Fig. 1. Sketch of a counterclockwise twisting of the growing filaments of colonies of Gause's[3] bacillus. (A few branches in Gause's drawing that overlaid one another are omitted.)

polymer statistics. In Sec. 2 we review the concepts of self-avoiding walks and of "universality classes," which identify certain characteristics of simple models to be independent of modifications in the model that might be added to give an otherwise more faithfully realistic model. Sec. 3 then addresses the spiral case and reviews the exact results that have been obtained. Sec. 4 presents a heuristic discussion that indicates some features of random spirals we anticipate should persist in more general circumstances than has been treated exactly.

2. Self-Avoiding Walks

The standard simplest model for linear-chain polymers views them as random walks. If the effects of "volume exclusion" are to be included, then the walks are self-avoiding, in the sense that the path is not to intersect itself. But the set of all walks (not necessarily self-avoiding) is much easier to deal with mathematically, so that the question of differences in properties of the two sets arises. Furthermore, the question of how yet further decorations affect the results arises. Such decorations could, for example, provide different probabilities or (Boltzmann) weights for different walks depending upon angles between successive steps, or variable length steps, or numbers of close contacts between different positions along the walk. Yet others concern the effect of restricting the walk to a lattice, or of adding branches. Notably there appear to be (large) *universality classes* of models for which certain critical exponents remain fixed. Thence the characterization of these universality classes becomes of fundamental interest, as perhaps first emphasized in a general text by deGennes.[5]

Two general properties for a polymer are the number of accessible conformations and their mean extension in space. Usually the count of (possibly weighted) conformations depends upon N, the number of monomers (each monomer corresponding to a step), as

$$\#_N \approx C N^\alpha (\log N)^\beta \kappa^{(N^\mu)} \qquad (2.1)$$

at least for large N. Typically $\mu = 1$ and the logarithm of this count is a size-extensive entropy. Then the "connective constant" κ is a measure of the number of choices at each step and is a (nonuniversal) model-dependent feature. The proportionality factor C is also model dependent. But the exponents μ and α are believed to be characteristic of general universality classes. The exponent β seems usually to be believed to be 0. A mean spatial extent usually depends asymptotically upon N as

$$<r>_N \approx A (\log N)^\rho N^\nu. \qquad (2.2)$$

This spatial extent might be the span distance between head and tail of the walk, or it might be the root of the mean of the squares of the distances of the monomers in a polymer conformation away from the center of mass. The exponent ν is believed to be independent of which of several such choices one might take. Here too, the logarithmic exponent ρ seems generally to be believed to be 0.

Evidence indicates that the exponents μ and ν (and perhaps α) are independent of many decorations, but then they seem abruptly to change with others. In particular they appear to be independent of variations in monomer size, of variations in (many) angle preferences between successive monomers, and of (weak) self-interactions of the polymer with itself. On the other hand, they do appear to depend on whether or not the walk may intersect itself, and, if self-avoiding, they then also depend on the dimension d in which the walk occurs. That is, while $\alpha = 0$, $\mu = 1$, and $\nu = 1/2$ independent of d for the nonself-avoiding case, (plausible) values for the self-avoiding case are indicated in Table 1. Of course, the $d = 1$

Table 1. Exponents for all Self-Avoiding Walks

d	μ	ν	α
1	1	1	0
2	1	3/4	11/32
3	1	$\simeq 0.60$	$\simeq 0.17$
4	1	1/2	?

values are exact. A nonrigorous derivation of the values for $d = 2$ has been given by Nienhaus.[6] The $d = 3$ values simply are plausible estimates, supported by much numerical work.

Generally there seems to be great difficulty in treating self-avoiding walk problems. Very few rigorous results are known. Indeed proofs seem to be lacking even for such obvious statements (for walks on a regular lattice) as:[7]

conjecture 1 — The mean end-to-end separation of N-step self-avoiding walks is no less than that for N-step random walks on the same lattice.
conjecture 2 — The mean end-to-end separation of N-step self-avoiding walks increases with N.

Thence a great number of different approximation schemes have been pursued, as discussed elsewhere.[5,8] Exact solutions for special subclasses of walks are then of interest too.

3. Spiralling Walks

We define a (counterclockwise outward) *spiralling* (self-avoiding) walk as one that satisfies three conditions:

(a) It is self-avoiding.
(b) Each step is either straight ahead or through an angle ϕ to the left.

(c) Every N-step walk can be extended to one of arbitrary length satisfying conditions (a) and (b).

For walks on a regular 2-dimensional lattice, ϕ is fixed at one of the angles: $\pi/3$, $\pi/2$, or $2\pi/3$. The third condition precludes counterclockwise inward spiralling walks and also *scrolling* walks, which the first two conditions would otherwise allow. An example of such a scrolling walk is shown in Fig. 2. They can[9,10] be obtained from suitable pairs of spiralling

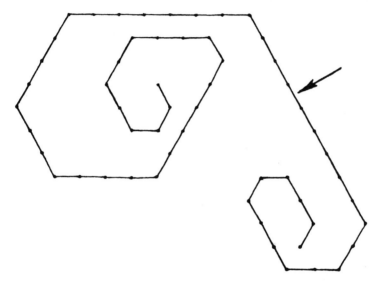

Fig. 2. A scrolling self-avoiding walk (with $\phi = \pi/3$), which may be separated (e.g., at the position of the arrow) into two oppositely spiralling walks.

walks—one left and one right spiralling. In any case these scrolling walks exhibit the same counting exponent μ and the same size-extension exponent ν as in the spiralling ones. It is to be emphasized that condition (c), unlike (a) and (b), is "nonlocal," so that the referenced link between scrolling and spiralling walks is important in allowing us throughout the following to reasonably deal with just the simpler spiral case.

Spiral walks were evidently first considered on a square-planar lattice (with turning angle $\phi = \pi/2$). The first work by Melzak[11] in 1958 seems to have been a mathematical exercise. Somewhat later (in 1983) Privman[12] independently reconsidered them as a special type of linear-chain polymer model. He made exact enumerations and attempted to find from this the asymptotic form, as in Eqs. (2.1) and (2.2). Soon following this, several authors[13,14,15] pointed out that $\mu = \nu = 1/2$—the value for μ being rather "anomolous." Blöte and Hilhorst[13] obtained all the constants in the forms of Eqs. (2.1) and (2.2), deriving $\alpha = -1/2$, $\beta = 0$, and $\rho = 1$, as well as values for C, κ, and A. Joyce[16] obtained the complete asymptotic expansion, involving correction terms beyond that indicated in Eqs. (2.1) and (2.2).

All the rigorous results established for spiralling walks depend upon relations to partitions of N. These relations to partitions of N (into sums of positive integers) can be viewed as arising in different ways. One elegant correspondence[17] arises through a consideration of the various maximal straight-line segments, here called *legs*. We label the ith horizontal or vertical leg by its length, x_i or y_i, as indicated in Fig. 3.

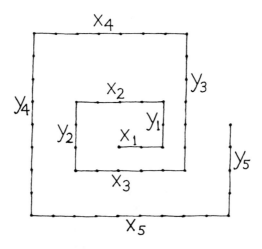

Fig. 3. Labelling for a representative ($\phi = \pi/2$) spiralling walk of $N = 50$ steps. The successive legs are of lengths $x_1 = 2$, $y_1 = 2$, $x_2 = 4$, $y_2 = 3$, $x_3 = 5$, $y_3 = 6$, $x_4 = 7$, $y_4 = 8$, $x_5 = 9$, and $y_5 = 4$.

Then we imagine a corresponding structure with unit-width strips glued together in the herringbone fashion of Fig. 4. With the exception possibly of the last leg, such a structure is called a *Young diagram* and is in correspondence with a partition, namely that with components that correspond to the lengths of the rows of this diagram. Thus, for example, with the exception of the last leg, the example of Fig. 4 corresponds to

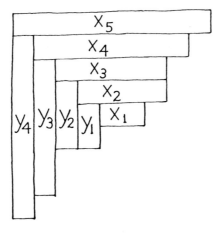

Fig. 4. Construction of a Young diagram corresponding to all but the last leg of the spiral of Fig. 3.

the partition $1 + 2 + 2 + 4 + 6 + 7 + 7 + 8 + 9 = 46$. As a consequence, the various length N spiralling walks are seen to be in one-to-one correspondence with partitions of M for $M < N$. Thence

$$\#_N = \sum_{k=1}^{N} p(N-k) \tag{3.1}$$

where $p(M)$ is the number of partitions of M, k is the length of the last (exceptional) leg, and we define $p(o) = 1$. Indeed this result was first obtained by Melzak,[11] though by a much longer, more involved argument. Standard asymptotic formulas for $p(M)$, as in Ref. 18, then readily lead to rigorous results for the spiralling walks. One measure of the size of a spiral is of the mean size of the components of the corresponding partition, and this may then be done too to yield $\mu = 1/2$ and $\rho = 1$.

Spiral walks for a couple other turning angles ϕ have been rigorously treated too. For $\phi = 2\pi/3$ a correspondence to partitions also leads[19] to a count the same as in Eq. (3.1), except that $p(N-k)$ is replaced by $p'(N-k)$, the number of partitions of $N-k$ with distinct parts. For $\phi = \pi/3$, the rigorous treatment[10] leads to an evidently somewhat less standard number-theoretic problem, for which the asymptotic form of $\#_N$ but not $\langle r \rangle_N$ has been obtained. For either $\phi = 2\pi/3$ or $\pi/3$, $\mu = 1/2$, and for ϕ

= $2\pi/3$, ν = 1/2. For other choices of ϕ (other than $\pi/3$, $\pi/2$, or $2\pi/3$), the spiral walks turn out not to be on a lattice, unless the step lengths are allowed to vary with direction (e.g., for ϕ = $\pi/4$, lengths 1 or $\sqrt{2}$ in axial and diagonal directions, respectively). For these other cases rigorous results seem not yet to be obtained.

4. Heuristic Argumentation

An earlier heuristic argument[14] for ϕ = $\pi/2$ can be extended to say that the exponents μ = ν = 1/2 are independent of the turning angle ϕ ($<\pi$). To do this, consider a typical random spiral developed through N steps, and consider its further extension through an additional angle θ. The circumstance is indicated in Fig. 5. If the radius of the spiral is denoted

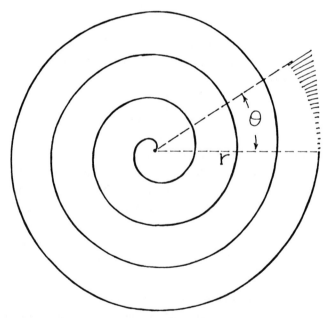

Fig. 5. A typical (Archimedean) spiral and the identification of quantities to be used in the extension, through a further angle θ.

by r, then the number of additional steps needed to make such an extension is roughly $r\theta$. And during this extension no more than about θ/φ turns (through θ) should be made, for otherwise the walk would be aimed to turn in on itself and violate condition (c) in our definition of spiralling walks. Thence the number of different ways to make this extension should scale as $\sim (r\,\theta)^{\theta/\varphi}$. But also this number should be obtainable from the form Eq. (3.1), thus

$$\frac{\#_{N+r\theta}}{\#_N} \sim \frac{\kappa^{(N+r\theta)^\mu}}{\kappa^{N^\mu}} \approx \kappa^{\mu r \theta (N^{\mu-1})}, \tag{4.1}$$

where we have reasonably assumed that r/N becomes very small as N becomes very large. Thence equating these two counts of the number of ways of making the extension, and taking logarithms, we obtain

$$\frac{\theta}{\phi} \log r\theta \sim \mu r \theta N^{\mu-1} \log \kappa, \tag{4.2}$$

or after a little manipulation,

$$r \sim N^{1-\mu} \log r. \tag{4.3}$$

Evidently we must have

$$\mu + \nu = 1, \tag{4.4}$$

and indeed there is even a much less reliable suggestion that $\rho = 1$. Next the maximum number of choices as to the possibility of turning is $\sim N^\mu$ and so is a maximum when ν is a minimum. But the most compact self-avoiding walks (spiral or not) in two dimensions are of size $\sim \sqrt{N}$, corresponding to $\nu = 1/2$. Thence we have

$$\mu = \nu = 1/2, \tag{4.5}$$

independent not only of ϕ, but of the possibility of different length steps in different directions, of a distribution of turning angles, or of a distribution of step lengths. These exponents evidently identify a fairly general universality class of random walks. In Table 2 we give the known

Table 2. Exponents for Spiralling Self-Avoiding Walks

ϕ	μ	ν	α	ρ	β
$2\pi/3$	1/2	1/2	−1/4	1	0
$\pi/2$	1/2	1/2	−1/2	1	0
$\pi/3$	1/2	(1/2)	−3/2	?	1
other	(1/2)	(1/2)	?	?	?

exponents for spiralling walks, with the values in parentheses being those surmised via the present heuristics.

5. Summation and Outlook

An overall picture of random spirals now emerges. The typical spiralling self-avoiding walk of length N evidently has a diameter $\sim \sqrt{N}$ with $\sim \sqrt{N}$ turns, so that the spacing between successive turns scales as $\sim N^0$. That is, the spiral of Archimedes is typical of this class. Moreover the number of different members of the class should scale $\sim \kappa^{\sqrt{N}}$, where the value of κ is determined by further details (e.g., ϕ) of the particular class. Evidently these spiralling walks, though random, are rather ordered, with a conformational entropy per site vanishing in the limit of long chains. Rigorization and more quantitative embellishment of this picture could be of use.

The idea of random branched structures should also be important. Recall the bacterial structure of Fig. 1. Work on such problems[20] is even less complete than for the unbranched chain. So far, evidence seems to be primarily "numerical." Understanding better the branched case would presumably be of value too.

Acknowledgement

The authors acknowledge the support of the Welch Foundation of Houston, Texas.

References

1. T. A. Cook, *The Curves of Life.* Constable & Co., London (1914) or Dover Pub., New York (1979).
2. D'Arcy Thompson, *On Growth and Form,* 2nd Edn., Cambridge University Press, Cambridge (1943).
3. G. F. Gause, On the relation between the inversion of spirally twisted organisms and the molecular inversion of their protoplasmic constituents, *Biodynamica* **3** (1940) 125–143.
4. R. V. Jean, *Mathematical Approach to Pattern and Form in Plant Growth.* Wiley, New York (1984).
5. P. G. deGennes, *Scaling Concepts in Polymer Physics.* Cornell University Press, Ithaca, New York (1979).

6. B. Nienhaus, Exact critical point and critical exponents of $O(n)$ models in two dimensions, *Phys. Rev. Lett.* **49** (1982) 1062–1065.
7. F. T. Wall and D. J. Klein, Self-avoiding random walks on lattice strips, *Proc. Natl. Acad. Sci. USA* **76** (1979) 1529–1531.
8. See for example,
C. Domb, Self-avoiding walks on lattices, pp. 229–260 in *Stochastic Processes in Chemical Physics*, K. E. Shuler, ed. Wiley, New York (1969).
D. J. Klein and W. A. Seitz, Graphs, polymer models, excluded volume, and chemical reality, pp. 430–445 in *Chemical Applications of Topology and Graph Theory*, R. B. King, ed. Elsevier, Amsterdam (1983).
9. A. J. Guttman and N. C. Wormald, On the number of spiral self-avoiding walks, *J. Phys. A.* **17** (1984) L271–L274.
10. G. Szekeres and A. J. Guttman, Spiral self-avoiding walks on the triangular lattice, *J. Phys.* **A 20** (1987) 481–493.
11. Z. A. Melzak, Partition functions and spiralling in plane random walk, *Can. Math. Bull* **6** (1963) 231–237.
12. V. Privman, Spiral self-avoiding walks, *J. Phys. A* **16** (1983) L571–L573.
13. H. W. J. Blöte and H. J. Hilhorst, Spiralling self-avoiding walks: An exact solution, *J. Phys.* **17A** (1984) L111–L115.
14. D. J. Klein, G. E. Hite, T. G. Schmalz, and W. A. Seitz, Spiralling self-avoiding walks, *J. Phys.* **A 17** (1984) L209–L214.
15. S. G. Whittington, The asymptotic form for the number of spiral self-avoiding walks, *J. Phys.* **17A** (1984) L117–L119.
16. G. S. Joyce, An exact formula for the number of spiral self-avoiding walks, *J. Phys.* **17** (1984) L463–L467.
17. A. J. Guttmann and M. Hirschhorn, Comment on the number of spiral self-avoiding walks, *J. Phys. A* **17** (1984) 3163–3614.
K. Y. Lin, Spiral self-avoiding walks (pp. 57–70) in *Progress in Statistical Mechanics*, C.-K. Hu, ed. World Scientific Pub., Singapore (1988).
18. See, for example,
G. H. Hardy and E. M. Wright, Chap. 19 in *Introduction to the Theory of Numbers*. Oxford University Press, London (1938).
G. E. Andrews, *The Theory of Partitions*. Addison-Wesley, Reading, Massachusetts (1976).
19. K. Y. Lin, Spiral self-avoiding walks on a triangular lattice, *J. Phys. A* **18** (1985) L145–L148.
G. S. Joyce and R. Brak, An exact solution for a spiral self-avoiding walk model on the triangular lattice, *J. Phys* **A 18** (1985) L293–L298.
K. C. Liu and K. Y. Lin, Spiral self-avoiding walks on a triangular lattice: End-to-end distance, *J. Phys* **A 18** (1985) L647–L650.
K. Y. Lin and K. C. Liu, On the number of spiral self-avoiding walks on a triangular lattice, *J. Phys. A* **19** (1986) 585–589.
20. T. C. Li and Z. C. Zhou, Spiral bond animals—ratio approach, *J. Phys. A* **18** (1985) 67–72.
I. Bose and P. Ray, Spiral-lattice-site animals: An exact enumeration study, *Phys. Rev.* **B35** (1987) 2071–2074.
I. Bose, P. Ray, and D. Dhar, Rooted spiral trees in dimensions 2, 3 and 4, *J. Phys* **A 21** (1988) L219–L225.

S. B. Santra and I. Bose, Statistics of spiral lattice animals with loops, *J. Phys.* **A 22** (1989) 5043–5049.

SPIRAL GALAXIES

Bruce G. Elmegreen

If spiral symmetry were the theme of an art contest, then galaxies would win the prize for size. Of the more than one million galaxies that have been catalogued, one-third have spirals and two-thirds of these are more-or-less symmetric over at least half of their visible disks. The most regular of the galactic spirals wind between one and two times around the galaxy's center and span a total length in excess of 1,000,000,000,000,000,000 kilometers (one billion billion). Fig. 1 shows an example, Messier 74, in the constellation Pisces.

To an astronomer, galactic spirals have a mechanical as well as an aesthetic beauty. The symmetric spirals in most galaxies are generally believed to be compression waves moving through the stars and gas as if they were in a continuous fluid. The waves are driven by the force of gravity. According to Julian and Toomre,[1] the origin of the waves might be understood intuitively as follows. Imagine a rotating galaxy without any spirals at some starting time. The stars tend to move in near-circular orbits at about the same speed throughout most of the disk, independent of radius, although they have slower speeds in the far inner parts. This speed law causes the angular speed around the center to decrease approximately inversely with increasing radius, just as the runners on an inside track overtake the more distant runners even though they have about the same speed. Thus there is angular shear in the galactic fluid.

Imagine putting a massive object at some radius, rotating around with

Fig. 1. A digitized photograph of the galaxy Messier 74, taken with a blue filter on the 4-meter telescope at the Cerro Tololo Interamerican Observatory by Debra Elmegreen. The calibration bars in this and in the following photographs are, on the left and right, 100 pixels and the characteristic size of the galaxy, respectively; the characteristic size is given by the isophotal radius at 25 magnitudes per square arc second.

the stars. The stars at smaller radii overtake the object from behind because they have higher angular rates, and the stars at larger radii are overtaken by the object. When each star passes, it feels the gravitational pull of the object and its motion gets a little jolt. This causes the star to oscillate a little around its old orbit after it passes, or is passed by, the object. Now, the stars at different radii oscillate at slightly different rates, and they also drift by the object in their orbits at slightly different angular speeds, so their oscillations mix them together a bit, first causing a slightly greater star density near the object, and then causing a slightly lower star density beyond. Eventually the oscillations mix up and the star density returns to normal, but near the perturbing object, a spiral of higher-than-average stellar density appears.

An example is shown is Fig. 2. This diagram contains pieces of 40 orbits, all starting on the radial line through the perturber (the big dot). The frame of reference for the diagram is rotating around with the

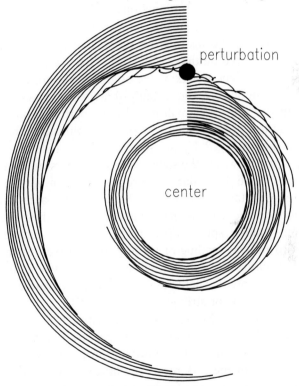

Fig. 2. An example of a spiral formed by the crowding of numerous stellar orbits that have been perturbed by a massive object at the position of the dot. Each curve is an orbit with a slight radial and azimuthal oscillation. The spiral appears as a slightly shaded region within the whole system of lines. The frame of reference for this image is taken to be rotating clockwise around the galaxy center with the angular rate of the perturber; thus the inner orbits appear to be circling clockwise and the outer orbits counterclockwise.

perturber; then the inner stars move clockwise and the outer stars move counter-clockwise in this frame. The orbits farthest from the object move at the highest relative angular rates, so they cover the largest angles in the same amount of time. The orbits are mostly circular, but at the start they are given a slight oscillation in both angular and radial positions, according to the rules of galactic dynamics. These oscillations bring the orbits closer together and farther apart in a periodic fashion, leading to the perception of a faint spiral pattern trailing both outward and inward from the perturber. Secondary spirals appear where the phases of the stellar oscillations make other conjunctions.

At first, the mass of the perturbing object is what causes the spiral wake, but after the wake begins to grow, there is an even larger object — the wake itself — centered on the original one. This causes an even larger response and has the effect of making the original wake grow. After about one revolution of the galaxy, the initially short wake grows into a long spiral arm. Galaxies can have small initial mass perturbations amplified into big spirals in this fashion. There are other amplification theories too.

How far can the wake go? Theory suggests that a stellar wave cannot penetrate into a region where the stellar velocity dispersion is high or the disk mass is low. There the stellar pressure from random motions greatly exceeds the self-gravitational forces on the wave. Galaxies with large three-dimensional bulges of stars have such barriers in their inner parts. Inward moving trailing spiral waves turn around at such a barrier, forming outward moving leading or trailing spiral waves. Such outward moving waves eventually reach corotation again — where the original perturbation was located. There, as "new" mass perturbations, they further amplify into new or stronger trailing spirals.

According to Lin[2] and Mark,[3] the most symmetric spiral galaxies, including those with only two main spiral arms trailing away from the center on opposite sides, have the additional property that the reflected wave from each side of the disk meets the other wave at the corotation resonance. Then each wave reinforces the other. The result is a rapid growth of a symmetric two-arm spiral from initially small and random perturbations in the middle part of the disk. The corotation radius will be determined automatically by the constraint that each reflected wave amplifies the other at corotation.

We have found that computer-enhanced images of symmetric spiral galaxies show the theoretically expected interference pattern caused by the superposition of inward and outward moving waves.[4] Fig. 3 shows a rectified and enhanced version of the galaxy Messier 81. The symmetrically placed bright parts of the main arms are where the (imperceptible) outward-moving leading arms meet the more obvious, inward-moving trailing arms. At these intersection points, the arms have a greater stellar density because the two waves add together. Between the intersection points but still on the main arms, the stellar density is lower, almost the

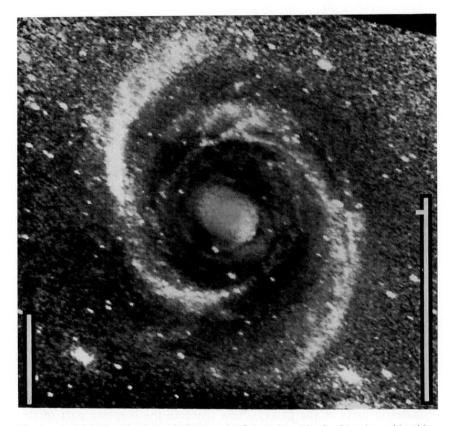

Fig. 3. A digitized and enhanced photograph of the galaxy Messier 81, taken with a blue filter on the 1.2-meter Schmidt telescope at Palomar Observatory by Debra Elmegreen. The enhancement was made by rectifying the original galaxy image to a face-on orientation, subtracting the average radial light profile from the image, and then normalizing the result to have a constant rms deviation with radius.

same as if there were no spiral at all. A sketch of this pattern is shown in Fig. 4. The bright trailing spirals are denoted by solid lines, the weak leading spirals by dotted lines, and the intersection points on the bright arms, where the stellar density is larger, are drawn thicker. A more quantitative discussion of the spiral components in Messier 81 can be found in Lowe and Lin.[5]

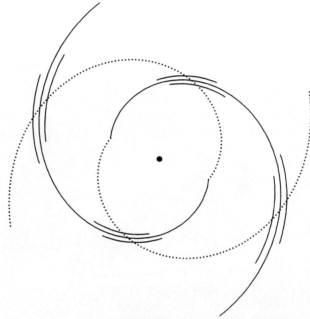

Fig. 4. A sketch of the two components of the spiral in Messier 81, with solid lines and triple-hatched regions showing the main spiral and its intensity peaks, and dotted lines showing the leading spirals that presumably reflected off of the inner bulge.

Computers can be used to illustrate the purely symmetric parts of galactic spirals in order to study the component spiral wave modes. This is done as follows. Suppose that the (digitized) image of a galaxy is denoted by the letter G, and a rotated version of the same galaxy, rotated by 180°, is denoted by G'. Then $G-G'$ sets to zero all of the perfectly symmetric parts of the galaxy. Now truncate the negative part of this difference $G-G'$, so the result is only positive. The truncated version of $G-G'$, call it $(G - G')_T$, is all of the original image that is not symmetric. Now subtract $(G - G')_T$ from the original image G to get the symmetric part of the galaxy.

Two symmetrized galaxies are shown in Figs. 5 and 6. Fig. 5 is the same galaxy as in Fig. 1, enhanced by subtraction of the average radial light distribution and then made symmetric as described above. Fig. 6 illustrates the regular, enhanced, and symmetrized versions of another galaxy, Messier 33.

Fig. 5. (a) Enhanced (see above) and (b) symmetrized (see next page) versions of the galaxy Messier 74, also shown in Fig. 1. (a) is reprinted by permission from *Galactic Models*, J. R. Buchler, S. Gottesman, and J. H. Hunter, eds., courtesy of the New York Academy of Sciences.

102 Bruce G. Elmegreen

Fig. 5b.

We can learn a lot about spiral waves from the symmetrized images. For example, we can see the full extent of a wavemode better than in the conventional sky view, and the radial limit of the wave reveals something about the angular pattern speed. We can also highlight the symmetric spurs that branch off from the main arms in some galaxies (e.g., Messier 74). These spurs are thought to be located at the positions where the oscillation period of the random stellar motions equals one half of the relative angular period of the wave.

Fig. 6. Three views of the galaxy Messier 33: (a) the conventional sky view, (b) the enhanced version, and (c) the symmetrized version. The original photograph was taken by Debra Elmegreen through a blue filter on the 1.2 meter Schmidt telescope at Palomar Observatory, and then digitized with a microdensitometer.

The locations of such spur resonances also tell us the rates at which the spiral patterns move around the galaxies.

Spiral waves may prove to be an important tool for determining the distribution of the mass and random energy of the stars in a galaxy. In

Fig. 6b.

this sense, galactic spirals are analogous to the spectrum of an atom: they satisfy a quantum-like condition that two oppositely directed wavetrains reinforce each other, and in doing so, they depict the fundamental modes of oscillation in the system. In the future, we might be able to learn more about the dark halos of galaxies from spiral-like spectra in their disks than we can learn from direct observations of faint halo light.

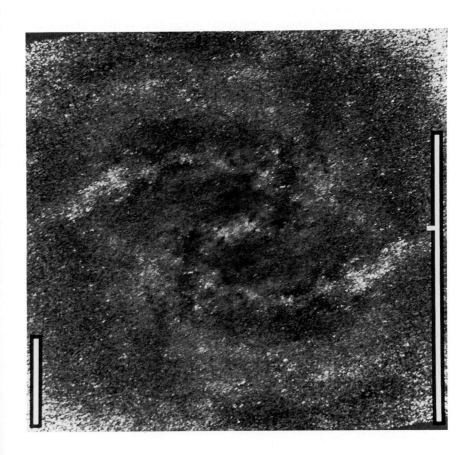

Fig. 6c.

References

1. W. H. Julian and A. A. Toomre, Non-axisymmetric responses of differentially rotating disks of stars, *Astrophysical Journal*, **146** (1966) 810–830.
2. C. C. Lin, Interpretation of large-scale spiral structure, in *Proceedings of the International Astronomical Union Symposium No. 38, The Spiral Structure of Our Galaxy*, W. Becker and G. Contopoulos, eds. (pp. 377–390). Dordrecht, Reidel (1970).
3. J. W.-K. Mark, On density waves in galaxies III: Wave amplification by stimulated emission, *Astrophysical Journal*, **205** (1976) 363–378.

4. B. G. Elmegreen, D. M. Elmegreen, and P. E. Seiden, Spiral arm amplitude variations and pattern speeds in the grand design galaxies M51, M81, and M100, *Astrophysical Journal,* **343** (1989) 602–607.
5. S. A. Lowe and C. C. Lin, Modal approach to the morphology of spiral galaxies III: Comparison with Observations, *Astrophysical Journal* (in preparation).

SPIRAL-BASED SELF-SIMILAR SETS

Keith Wicks

1. Introduction

In Ref. 1 Hutchinson laid the foundations of a certain concept of self-similarity, the basic notion being that of an object made up of a number of smaller images of itself. It follows that these smaller images will be made of still smaller images of the original object, and so on ad infinitum, typically resulting in detail at all levels of magnification, a trait commonly associated with objects referred to as *fractals*. Since the spiral-based sets of this paper fit into Hutchinson's framework, I will first of all describe some of its main features, along with the random iteration algorithm for obtaining pictures of the objects in question.

2. Self-Similar Sets

Throughout, let X denote the Euclidean plane $\mathbb{R}^2$ and let $d(x, y)$ denote the distance between points x and y of X.

A central concept in the following is that of a contraction. A *contraction of* X is an operation f on X that reduces all distances by at least some fixed factor; more precisely, there must exist $r<1$ such that for all points x,y of X, $d(fx, fy) \leq rd(x, y)$. For example, the operation of shrinking the plane by a factor $r<1$ about some point p is a contraction (under which all distances are reduced by *exactly* factor r); but there are other types too.

Suppose we have a finite set $F = \{f_1, \ldots, f_N\}$ of contractions of X. We'll say a subset A of X is *self-similar with respect to F* if it is composed of its images under the operations in F, that is, if $A = f_1 A \cup \ldots \cup f_N A$. By a beautiful application of Banach's contraction mapping theorem, Hutchinson proved that there is exactly one nonempty compact subset of X that is self-similar with respect to F (note: "compact" here means closed and bounded; for the formal definition of these terms consult any basic book on metric spaces). This set then, call it K, is completely determined by F. In this sense it is simple to describe. Yet *visually* it may be extremely complex, typically having detail at all levels of magnification, because not only is K made of the smaller images $f_i K$ of K but each of the latter are in turn made of still smaller images of K and so on ad infinitum. Let me make this more explicit, because it will underlie why the "random iteration algorithm" for obtaining a picture of K works.

For $n \geq 1$, an nth-*level image of K* is a set of the form $g_n \ldots g_1 K$, where each operation g_i belongs to F; in other words it is something obtained by starting with K and successively applying n operations (not necessarily distinct) each belonging to F. To obtain $g_n \ldots g_1 K$, g_1 is applied first, then g_2, and so on. Each nth-level image $g_n \ldots g_1 K$ of K is a subset of K since K is closed under all the operations in F. And it is composed of $(n+1)^{\text{th}}$-level images of K as shown by the following:

$$g_n \ldots g_1 K = g_n \ldots g_1 (f_1 K \cup \ldots \cup f_N K)$$
$$= (g_n \ldots g_1 f_1 K) \cup \ldots \cup (g_n \ldots g_1 f_N K)$$
$$= \text{the union of all } (n+1)^{\text{th}}\text{-level images}$$
$$\text{of } K \text{ of the form } g_n \ldots g_1 f_i K.$$

Using this, it is easy to see by induction that for all n, K is composed of its nth-level images. For, we know it's true for $n = 1$ as $K = f_1 K \cup \ldots \cup f_N K$, whilst if it's true for n then it's true for $n+1$, since each $g_n \ldots g_1 K$ (where $g_n, \ldots, g_1$ belong to F) is composed of the $(n+1)^{\text{th}}$-level images $g_n \ldots g_1 f_i K$ (where f_i belongs to F), and the possible sequences of the form $g_n \ldots g_1 f_i$ above comprise *all* the possible sequences of the form $h_{n+1} \ldots h_1$, where each h_j belongs to F. These few basic facts about n^{th}-level images were proved in Ref. 1.

We'll be using one more fact which roughly speaking is this: since contractions make sets smaller, the nth-level images of K become smaller and smaller as n increases. We can phrase this more precisely as follows. Let $r < 1$ be such that each f_i contracts by at least factor r, that is, for all

x,y in X, $d(f_ix, f_iy) \leq r\, d(x, y)$. The *diameter* of a nonempty compact set A is defined to be the largest possible distance between points in A:

$$\text{diam } A = \max \{d(x, y) : x \text{ and } y \text{ belong to } A\}.$$

It is easily seen that $\text{diam } f_i A \leq r\, \text{diam } A$; that is, each contraction f_i reduces diameters of sets by at least factor r. Hence if we successively apply n of the contractions from F, the diameter of A is reduced by at least factor r^n. In particular then, the diameter of any n^{th}-level image $g_n \ldots g_1 K$ of K is at most r^n times the diameter of K:

$$\text{diam } g_n \ldots g_1 K \leq r^n\, \text{diam } K.$$

And as n increases, r^n decreases toward zero, hence so do the diameters of the nth-level images of K, as claimed.

We now move on to a method of generating a picture of K.

3. The Random Iteration Algorithm

This algorithm was introduced by Barnsley and Demko in Ref. 2. Imagine a sheet of paper before you representing a region of the plane in which K lies. Suppose that some point x_0 lying in K is indicated on the sheet by a black dot. Choose one of the contractions $f_1, \ldots, f_N$ from F at random according to some fixed set of probabilities $p_1, \ldots, p_N$ (that is to say, choose f_i with probability p_i), say it is f_{i_0}, apply it to x_0, and plot the resulting image $x_1 = f_{i_0} x_0$ by another black dot. Again choose one of the contractions at random, says it is f_{i_1}, apply it to x_1 and plot the image $x_2 = f_{i_1} x_1$. Continue in the same fashion, repeatedly choosing a contraction at random according to the probabilities $p_1, \ldots, p_N$, applying it to the last plotted point x_m, and plotting the image point x_{m+1}. In this way a sequence $x_0, x_1, x_2, \ldots$ of points is generated and a growing pattern of dots develops on the paper showing where the sequence has been. And the remarkable thing is that this growing pattern becomes (with probability one) a better and better picture of K as more and more dots appear; the pattern *converges* to K in a certain mathematical sense.

To mathematicians, the sense is that letting A_m denote the set $\{x_0, \ldots, x_m\}$, the sequence $A_0, A_1, A_2, \ldots$ converges to K with respect to the Hausdorff metric h on the set of nonempty compact subsets of X. The practical import of this is the following. Suppose your visual resolution is such that you cannot resolve two points when they're distance δ or less apart. We'll imagine then that if two sets A and B are such that each point of A is within distance δ of some point of B and each point of B is within distance δ of some point of A, you wouldn't be able to tell A and B apart; formally we'll say that A and B are δ-*indistinguishable*. Then what the above convergence implies is that for sufficiently large m (i.e., if you wait long enough), A_m will be δ-indistinguishable from K. So at that point you'll have as close a picture of K as your powers of resolution allow.

How can we prove the above claim that with probability one, for sufficiently large m, A_m is δ-indistinguishable from K? Since all the points $x_0, x_1, x_2, \ldots$ lie in K (as x_0 does and K is closed under the contractions f_i) it really only remains to show that with probability one there will come a stage m such that every point of K is within distance δ of some point of A_m (because then this set A_m and all later sets A_{m+1}, A_{m+2} etc. will be δ-indistinguishable from K). And we can prove this using the few results on n^{th}-level images of K in the last section. First of all, take n such that $r^n \text{diam } K \leq \delta$. From the last section we know then that every n^{th}-level image of K has diameter $\leq \delta$. And regarding an nth-level image $g_n \ldots g_1 K$ of K, note that if as the random iteration algorithm progresses you should happen to choose in succession $g_1, \ldots, g_n$ then a point in $g_n \ldots g_1 K$ will be plotted, because if the last point plotted was say x then after applying $g_1, \ldots, g_n$ you end up plotting the point $g_n \ldots g_1 x$, which is in $g_n \ldots g_1 K$ as x is in K. Now since the sequence $g_1, \ldots, g_n$ is finite and all the probabilities p_i were taken to be non-zero, then with probability one it *will* occur sooner or later as an actual succession of choices of contractions in the random iteration algorithm, and correspondingly a point of $g_n \ldots g_1 K$ will be plotted. Since this goes for *each* of the nth-level images of K (of which there are only *finitely* many), it follows that with probability one there will come a stage m when a point from each nth-level image of K has been plotted, i.e., A_m contains a point from each nth-level image of K. And at such a stage, every point of K is within distance δ of some point of A_m because it belongs to some n^{th}-level image of K (recalling that K is the union of its nth-level images) and this has diameter $\leq \delta$. This concludes the proof.

Readers may be wondering how the probabilities p_i are chosen. The role they play is the practical one of determining how evenly the picture of K develops as the dots appear. What one usually does is be guided by the rule that the more f_i contracts compared with the other contractions, the lower should be p_i compared with the other probabilities. In this way the dots will tend to distribute themselves more evenly over K as opposed to growing densely in one place and sparsely in another, and as a consequence a good picture of K is likely to emerge more quickly than otherwise.

Drawing all those dots manually is of course not recommended. What we need to do is replace the piece of paper by a computer screen and have the *computer* run the algorithm. In Sec. 5 the outline of a suitable program doing this is given which is specifically for drawing the spiral-based self-similar sets about to be described.

4. Spiral-Based Self-Similar Sets

Consider an operation f_1 on the plane that both rotates points about the origin by some angle and also takes them closer in. Repeatedly applied, such an operation will make any point spiral in toward the origin. Combined with a secondary "feeder" operation f_2 which can be thought of as feeding points to the primary operation f_1 to work on, we can produce self-similar sets based on the spiral orbits induced by f_1.

In more detail, suppose f_1 rotates by an angle θ anticlockwise about the origin then contracts about the origin by a factor r_1, whilst f_2 contracts about the origin by a factor r_2 then translates right by a distance $1 - r_2$ (equivalently f_2 contracts about the point $(1, 0)$ by factor r_2). Since they're both contractions we know from the framework described in Sec. 2 that there's a unique nonempty compact set K that is self-similar with respect to $F = \{f_1, f_2\}$. Such a set is what I'll refer to as a *spiral-based self-similar set*. The parameters θ, r_1, and r_2 determine K, and an attractive spiral appearance typically results when r_1 is taken close to 1 and r_2 is relatively small. Two pictures suggestive of the type of set described here appear on pages 37 and 99 of Ref. 3. A selection of examples follows, along with mention of some features you may like to investigate. All the pictures were generated using a random iteration algorithm program similar to the one outlined in the next section.

Fig. 1 shows four examples in which r_1 is 0.98 and r_2 is 0.15, θ being respectively 18°, 47°, 76°, and 113° in (a), (b), (c), and (d). The squares are centred on the origin and have side-length 2.04, which gives a suitable region of the plane to be looking at because every spiral-based self-similar set lies therein. Moreover, since the unit disc (i.e. the disc of

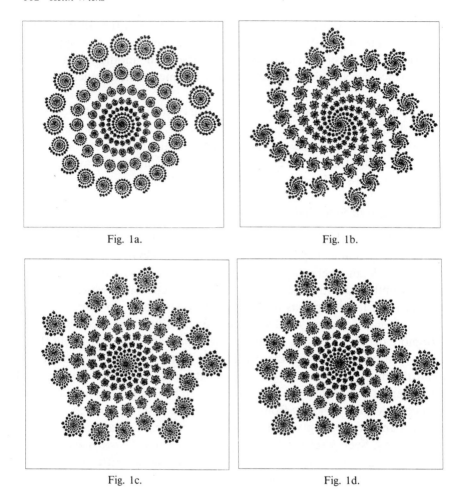

Fig. 1a.

Fig. 1b.

Fig. 1c.

Fig. 1d.

Fig. 1. Spirals for $r_1 = 0.98$, $r_2 = 0.15$, and (a) $\theta = 18°$, (b) $\theta = 47°$, (c) $\theta = 76°$, (d) $\theta = 113°$.

radius 1 centred on the origin) is closed under f_1 and f_2, it follows from (8) on page 727 of Ref. 1 that K is a *subset* of the unit disc; and in fact K touches the edge of the disc only at the rightmost point $(1, 0)$, which is the unique fixed point of f_2. In each of (a), (b), (c), and (d), you should be able to identify the two first-level images $f_1 K$ and $f_2 K$ of K, which together form K. Here, $f_1 K$ and $f_2 K$ are disjoint, with the consequence that, in mathematical parlance, K is totally disconnected.

Fig. 2 shows a zoom-in sequence in which r_1 and r_2 are again 0.98 and 0.15, whilst θ is 82°. Small copies of K (each of them some n^{th}-level image of K) appear without limit as we zoom further and further in, and by (d) we've arrived at a copy that is about a twentieth the size of K.

In Fig. 3, r_1 is 0.99 and r_2 is 0.15 with θ being respectively 11.75°, 58.5°, 83°, and 111° in (a), (b), (c), and (d). (a) is reminiscent of an ammonite, and with angles around 174°, 124°, or 93° for example, one

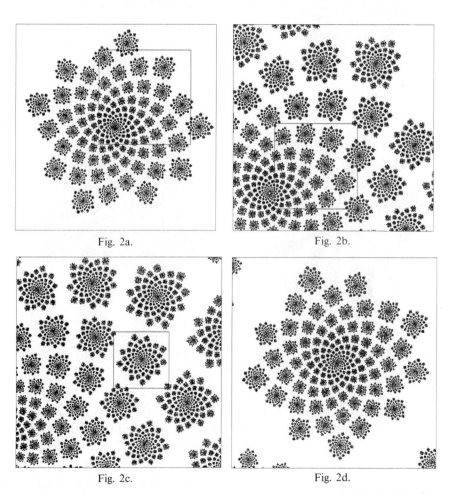

Fig. 2a. Fig. 2b.

Fig. 2c. Fig. 2d.

Fig. 2. Zoom-in sequence on the spiral with $r_1 = 0.98$, $r_2 = 0.15$, and $\theta = 82°$. Small copies of the original spiral appear without limit.

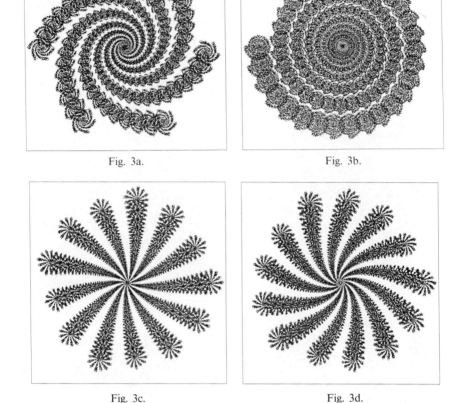

Fig. 3a. Fig. 3b.

Fig. 3c. Fig. 3d.

Fig. 3 Spirals for $r_1 = 0.99$, $r_2 = 0.15$, and (a) $\theta = 11.75°$, (b) $\theta = 58.5°$, (c) $\theta = 83°$, (d) $\theta = 111°$.

can obtain more unusual "ammonites" with 2, 3 or 4 arms spiralling in. The feathery appearances of (c) and (d) can be obtained with other angles too, giving different numbers of arms; for example, try 77°. In each of the cases in Fig. 3, the two first-level images of K overlap, with the consequence (by theorem 4.6 of Ref. 4) that K is connected. Another consequence is that when we zoom in for closer looks the structure can be rather chaotic. This is illustrated in Fig. 4, where θ is 61° and r_1 and r_2 are again 0.99 and 0.15.

Now in general, a small variation of the parameters produces only a small variation in K; so if the parameters vary smoothly so will the corresponding set K. A particularly natural thing to do is to keep r_1 and r_2 fixed and see how K varies with θ. An example is shown in Fig. 5, in which r_1 is 0.99, r_2 is 0.1, and θ is increased from 360°/12 = 30° to

Fig. 4a.   Fig. 4b.

Fig. 4c.   Fig. 4d.

Fig. 4. Zoom-in sequence on the spiral with $r_1 = 0.99$, $r_2 = 0.15$, and $\theta = 61°$, revealing a rather chaotic structure of interfering small copies.

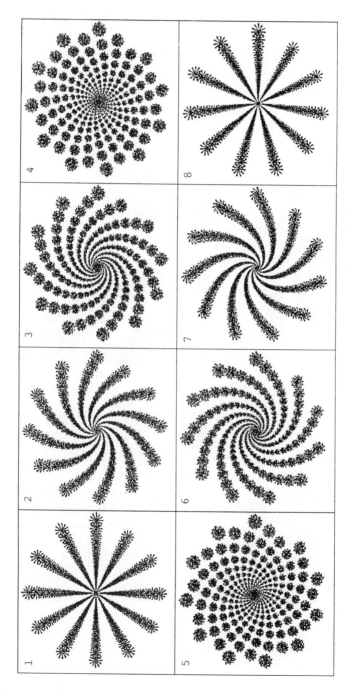

Fig. 5. Sequence showing the smooth variation of a spiral with θ as θ increases from $360°/12 = 30°$ to $360°/11 \approx 32.72°$ in equal steps, during which the initial twelve arms transform into eleven. Here, $r_1 = 0.99$ and $r_2 = 0.1$.

$360°/11 \approx 32.72°$ in equal steps. The way in which the number of arms changes from 12 to 11 is particularly interesting. As θ continues to increase, one encounters 10 arms, 9 arms, 8 arms, and so on, reaching 2 arms by the time θ is around 180°. Along the way though, other numbers of arms are encountered that emerge during the main transition periods between the above states. This evolution of K with θ is perhaps the most central aspect to study, and to readers interested in pursuing this I'd suggest first taking $r_2 = 0$ which should simplify matters. With $r_2 = 0$, K is just made up of the orbit of $(1, 0)$ under the repeated action of f_1, together with the origin. With a sufficiently fast computer you can make *movies* of how K evolves with θ by having the computer draw one picture after another very quickly, increasing θ by a fraction of a degree each time. An alternative is to *print* the successive pictures and form them into a flick-book. Seeing K change dynamically in this way can impart a much better feel for what's going on, quite apart from being fascinating to watch.

5. A Suitable Computer Program

Outlined in this section is a program that generates pictures of spiral-based self-similar sets by the random iteration algorithm. We'll assume the pixels of the computer screen are square, and that their horizontal and vertical coordinates increase as you go righward and upward respectively. If your computer differs you'll need to make a few minor changes to the program.

The first thing the program does is to find out what area of the screen you wish the picture to appear on (which it then clears, i.e., colours white) and what area of the plane it should represent. The first area, the "screen window", is taken to be square and is described by its lower left pixel (*imin, jmin*) and the number *numpix* of pixels on a side. The second area, the "real window", is likewise taken to be square, and is described by its lower left corner (*xmin, ymin*) and its side-length *sidelength*. Imagining the screen window lying exactly over the real window, each pixel covers a real square of side *pixlength* = *sidelength/numpix*, and the conversion from real coordinates (x, y) to pixel coordinates (i, j) is easily worked out as

$$i = imin + \text{integer part of } ((x - xmin)/pixlength)$$

$$j = jmin + \text{integer part of } ((y - ymin)/pixlength)$$

where (i, j) is the pixel in which (x, y) lies. This conversion formula is used in the subroutine "plot(x, y)" which plots (x, y) on the screen by a black dot.

Next, the contractions f_1 and f_2 are described by inputting *theta*, *r*1, and *r*2. *Theta* is assumed to be in degrees and the program converts it to radians for internal use. The computer is later going to have to represent the actions of f_1 and f_2 and to do this it uses the fact that, representing points (x, y) of the plane by column matrices $\begin{bmatrix} x \\ y \end{bmatrix}$, the actions of f_1 and f_2 can be described in the matrix form

$$f_i \begin{bmatrix} x \\ y \end{bmatrix} = \begin{bmatrix} a[i] & b[i] \\ c[i] & d[i] \end{bmatrix} \begin{bmatrix} x \\ y \end{bmatrix} + \begin{bmatrix} e[i] \\ f[i] \end{bmatrix} = \begin{bmatrix} a[i]x + b[i]y + e[i] \\ c[i]x + d[i]y + f[i] \end{bmatrix},$$

where the matrix entries $a[i]$, $b[i]$ and so forth are as assigned in the program.

Next the program must be told the probabilities *p*1 and *p*2 with which it is to choose f_1 or f_2 at each step of the algorithm. As a user-convenience, you can instead input the *relative* probabilities *relativep*1 and *relativep*2, and the program automatically calculates the *actual* probabilities *p*1 and *p*2. The relation is that *p*1/*p*2 = *relativep*1/*relativep*2. For example, if you want *p*1 to be 30 times bigger than *p*2 just input 30 and 1; the program then calculates that *p*1 = 30/31 and *p*2 = 1/31, saving you the bother.

The last piece of information the program asks for is the number *numits* of iterations it should perform, i.e. the number of times it chooses a contraction and applies it.

The program then begins the random iteration algorithm, starting with the point (1, 0). Note that it only plots those points that are "visible," i.e. that lie strictly within the real window. To choose one of the contractions at random, the program first obtains in some manner a random floating-point number *u* between 0 and 1. Although your machine may have a function that will supply one directly, it will probably be many times faster to instead obtain *u* by taking a machine-supplied random *integer* and dividing it by the largest possible such integer.

The program outline follows, preceded by a list of the variables used:

Integer variables: imin, jmin, numpix, numits, n, i, j.
Floating-point variables: xmin, ymin, sidelength, xmax, ymax, pixlength, theta, r1, r2, relativep1, relativep2, p1, p2, x, y, newx, newy, u.

Floating-point arrays: a[2], b[2], c[2], d[2], e[2], f[2].
Floating-point constants: PI = 3.14159265359
Program
{
"OBTAIN AND CLEAR SCREEN WINDOW"
input imin, jmin, numpix
clear screen window (imin, jmin, numpix)
"OBTAIN REAL WINDOW"
input xmin, ymin, sidelength
xmax = xmin + sidelength
ymax = ymin + sidelength
pixlength = sidelength/numpix
"OBTAIN CONTRACTIONS"
input theta, r1, r2
theta = theta*PI/180
a[1] = r1* cos (theta)   a[2] = r2
b[1] = - r1* sin (theta)   b[2] = 0
c[1] = r1* sin (theta)   c[2] = 0
d[1] = r1* cos (theta)   d[2] = r2
e[1] = 0   e[2] = 1 - r2
f[1] = 0   f[2] = 0
"OBTAIN RELATIVE PROBABILITIES AND CONVERT TO ACTUAL ONES"
input relativep1, relativep2
p1 = relativep1/(relativep1 + relativep2)
p2 = relativep2/(relativep1 + relativep2)
"OBTAIN NUMBER OF ITERATIONS"
input numits
"START AT (1, 0), PLOTTING IT IF IT'S VISIBLE"
x = 1
y = 0
if (xmin<x<xmax and ymin<y<ymax) plot (x, y)
"NOW PERFORM THE ITERATIONS, PLOTTING ANY VISIBLE POINTS"
for n = 1 to numits
    {
    "OBTAIN A RANDOM ELEMENT i of {1, 2} ACCORDING TO p1 AND p2"
    u = unirand ( )   "u IS A RANDOM NUMBER BETWEEN 0 AND 1"
    if u≤p1 then i = 1, else i = 2
    "APPLY CONTRACTION f_i TO GET THE NEXT POINT"
    newx = a[i]*x + b[i]*y + e[i]
    newy = c[i]*x + d[i]*y + f [i]
    x = newx
    y = newy
    if (xmin<x<xmax and ymin<y<ymax) plot (x, y)
    }
}
plot (x, y) "THIS SUBROUTINE PLOTS (x, y) ON THE SCREEN"
    {
    i = imin + integer part of ((x - xmin)/pixlength)

```
j = jmin + integer part of ((y - ymin)/pixlength)
colour pixel (i, j) black
}
```

Here are some hints on the use of the program.

Naturally, the larger the value of *numpix*, the more detail can be displayed but the longer the picture will take to form. The pictures of the last section all used *numpix* = 600 except those in Fig. 5, which used *numpix* = 287, but reasonable detail can still be obtained with *numpix* = 100 for example. Regarding the real window, for a picture of the whole set one could take the window I did which has $xmin = ymin = -1.02$ and *sidelength* = 2.04. But smaller windows can also be taken in order to magnify a part of the set, such as was done in Figs. 2 and 4. However, the run time rises quite rapidly with the magnification, which in practice places a severe limit on how far one can magnify.

The number of iterations, *numits*, can be decided on by trial and error. If allowed to run indefinitely, initially a flood of dots will appear and as time passes the rate at which new dots appear reduces until eventually hardly anything changes. Sometime before this extreme stage, we'll typically have a pleasant "textured" image of K with the self-similarity quite apparent, and I'd suggest aiming to stop the program about there. In Figs. 1 and 3, *numits* was around 150,000. This was practicable because the SUN 3/60 workstation used to generate the pictures worked at almost 5,000 iterations per second. Fig. 2(d) took over an hour to produce using around 17 *million* iterations, which illustrates the remark concerning the practical limits of magnification. A last point on *numits* is that it should generally increase proportional to the square of *numpix* in order to maintain the same overall dot density, everything else remaining the same.

To give a guide regarding the probabilities, relative probabilities of 30 to 1 (for f_1 and f_2, respectively) were used in the pictures where r_1 and r_2 were 0.98 and 0.15, and around 50 to 1 in those where r_1 and r_2 were 0.99 and 0.15. The higher r_1 is, the more strongly one needs to favour f_1 in order that the central area of the spiral be well filled in. Experiment and you'll see. Incidentally, there's nothing to stop us taking $p1 = 1$ and $p2 = 0$, in which case the program will trace out the orbit of (1, 0) as it spirals in toward the origin under the repeated action of f_1, giving a very quickly produced picture of the set K for which $r_2 = 0$ (with θ and r_1 as inputted). Reference to such K was made in the last section.

A slight modification to the program will make the two first-level images $f_1 K$ and $f_2 K$ of K apparent by plotting with *two* colours instead of one. What you should do is, whenever contraction f_i is applied, plot the resulting point in colour i. Also, plot the initial point $(1, 0)$ in colour 2. Then $f_i K$ will end up coloured in colour i, with a mix of the two colours where (if at all) $f_1 K$ and $f_2 K$ overlap.

In conclusion, one last suggestion is that you might like to try running the program with $r_1 = 1$. The mathematical framework of Sec. 2 no longer applies since f_1 does not contract, now being purely a rotation by angle θ; and one doesn't get spirals. What one *does* get (with $r_2 = 0.15$, for example, and suitable probabilities) is an extensive range of remarkably realistic images of tyres, some of which give the illusion of rotating as they're being drawn (for example, try $\theta = 19°$, $38°$, $47°$, $49°$), and also some attractive ring formations (rings made of rings made of rings ad infinitum) when θ is a relatively small divisor of $360°$ and r_2 is suitably chosen. With $\theta = 120°$ and $r_2 = 0.5$ one can obtain the classic "Sierpinski gasket", whilst with $\theta = 72°$ and $r_2 = 0.5/(\cos(72°) + 1) \approx 0.382$, one can obtain a similar type of set based on the pentagon rather than the triangle, and with $\theta = 60°$ and $r_2 = 1/3$, one can obtain the hexagonally based Fig. 5.4.3 in Ref. 3. *The values $\theta = 90°$ and $r_2 = 0.5$ will give a solid square.* And with $r_2 = 0$, most values of θ will generate a circle.

References

1. J. E. Hutchinson, Fractals and self similarity, Indiana Univ. *J. Math.* **30** (1981) 713–747.
2. M. F. Barnsley and S. Demko, Iterated function systems and the global construction of fractals, *Proc. Royal. Soc. Lond,* **A399** (1985) 243–275.
3. M. F. Barnsley, *Fractals Everywhere.* Academic Press, London (1988).
4. M. Hata, On the structure of self-similar sets, *Japan J. Appl. Math.* **2** (1985) 381–414.

SYMMETRY AND SPIRALS: AN ARTIST'S PERSONAL STATEMENT

Rochelle Newman

I am not a scientist, nor am I a mathematician. I am an artist and a professor of art. My attraction for spirals came long after my concern for patterns that grew out of my intrigue with tilings and an even earlier interest in symmetry.

As a child, I was first fascinated by the image obtained when ink was poured onto a piece of paper that was then folded and reopened to produce the magic of mirror symmetry. Of course I did not know the vocabulary of the act.

In my art school training, the emphasis was much more on representation than abstraction, more on uniqueness than on repetition, more on asymmetry than symmetry. Pattern was not a strong focus of the curriculum. I took one course in wallpaper design that I thoroughly enjoyed, but certainly the language of the instructor was not the precise one of the mathematician nor the scientist. It was talk about "drops" and "half-drops" but never point, line, or plane groups. It has only been in the last dozen years or so that I have come to mathematics, especially geometry, with trepidation in my heart, to find the tool that would allow me to understand, and then explore, those topics that have always held my attention.

Few mathematicians interact much with artists and vice versa. Art and design school curricula do not generally offer courses in mathematics. Although the computer has recently brought the two together through the generation of glorious fractal images, the disciplines are seen as antithetical rather than complementary. But our common interest in patterns bonds us together. Where we differ is in the forms and aspects of expression.

I have never truly wanted to make pictures of things that are about illusion, but rather my interest has been in the reality of the plane surface and what could happen on it. Its physicality and its boundaries have been insistent constraints in my mind. What are the conditions of natural space and how can the artist explore symbolic space using the limits of the two-dimensional surface? My desire was, and is, to use the "rules" of the natural world so that my art would provide a parallel microcosm. It was the rules for patternmaking that I was after. To dance across the surface plane, interlock, and be interdependent is what pattern is about. Artists of every persuasion speak metaphorically. Visual artists use visual metaphors. Pattern is the primary visual metaphor for the expression of my belief in the interconnectedness of All in the Universe.

Symmetry and tiling are intertwined for me. Symmetry tells me there is order despite the deception of appearance. Tiling satisfies my economical bent for wanting elements contiguous to one another as well as for not having leftover "pieces" of space. It is the straight line segment that is one of the essential elements in my form vocabulary. It is the tool of the intellect, while color, another essential element, is the instrument of the heart. So, joining these two, one of the themes of my explorations has been variations on the straight-line spiral as a motif for symmetry manipulations as seen in Figs. 1, 2, 3, and 4.

Symmetry and Spirals: An Artist's Personal Statement 125

Fig. 1. Dancing Spirals (acrylic paint on illustration board).

Fig. 2. Black Baravelle (acrylic paint on illustration board).

Fig. 3. Hexagonal Spirals (acrylic paint on illustration board).

Symmetry and Spirals: An Artist's Personal Statement 127

Fig. 4. Primary Spirals (acrylic paint on illustration board).

SPIRAL STRUCTURES IN JULIA SETS AND RELATED SETS

Michael Michelitsch and Otto E. Rössler

1. Introduction

A multitude of beautiful spiral structures resulting from Julia sets is well known (e.g., Refs. 1–4). Julia sets are obtained from the complex-analytic map $z_{n+1} = z^2_n - c$ by a feedback process[1] carried out repeatedly. They are defined by fixing the parameter c ($= p$) and by varying z_o in the plane of complex numbers.

Interesting spiral structures of Julia sets appear, for example, in an equation such as

$$z_{n+1} = p \cdot (z^3_n - 1), \qquad (1)$$

with $z = x + i \cdot y$ and $p = p_x + i \cdot p_y$.

We also looked at the following modified versions of the complex logistic map, $z_{n+1} = z^2_n - p$, namely

$$\begin{aligned} x_{n+1} &= |x_n^2 - y_n^2 - p_x| \\ y_{n+1} &= 2 \cdot x_n \cdot y_n - p_y \end{aligned} \qquad (2)$$

and

$$\begin{aligned} x_{n+1} &= |x_n^2 - y_n^2| - p_x \\ y_{n+1} &= 2 \cdot x_n \cdot y_n - p_y. \end{aligned} \qquad (3)$$

Note that both are no longer analytic solely because of the absolute-value expressions introduced in the first line.

2. Results

Using the technique of plotting the escape times for different p_x, p_y - values,[1,2] we present in Figs. 1 and 2 a Julia set for Eq. (1), with

Fig. 1. A Julia set generated by Eq. (1). Parameters: $p_x = -0.844$ and $p_y = 0.03$; coordinates: $-2 < x < 2$ and $-1.5 < y < +1.5$; maximum iteration number: 400; threshold value: $x^2 + y^2 = 300$.

$p_x = -0.844$ and $p_y = 0.03$. The teddy-bear-like structure exhibits, in the "throat" and "ear" regions, self-similar spiral shapes that continue to reappear in a self-similar fashion within the spirals themselves.

Figs. 3 through 7 were obtained using Eqs. (2) and (3), respectively. Specifically, Fig. 3 (Eq. (2)) shows self-similar spirals along a border line of that Julia-analogous set. Fig. 4 gives a magnification of one of the "eyes" of those spirals, inside of which the same kind of spirals is found again. Unexpectedly, at farther distances from the center of the spiral, the double-eye structures show higher resolution, whereas the more central

Fig. 2. Magnification of the "throat" region of Fig. 1. Parameters as in Fig. 1, coordinates: $0.45 < x < 1$ and $-0.15 < y < 0.15$.

structures do not yet show this behavior.

The appearance of the observed structures depends on the iteration depth chosen numerically. The present sets therefore presumably are all analogous to a "dust-like" Julia set.

Figs. 5 and 6 are magnifications of Julia-like sets that were obtained at parameter values close to those of Fig. 3. They show double-spiral structures in each case, but the inner parts differ significantly from each other. A similar structure obtained from Eq. (3) is shown in Fig. 7.

3. Concluding Remarks

A new type of self-similar spiral has been presented for a complex-analytic map, Eq. (1), together with some equally striking spiral structures occurring in nonanalytic modifications of the complex logistic map.

Numerical and graphical investigations of nonanalytic equations can provide interesting insights, as evidenced by the Julia-analogous sets

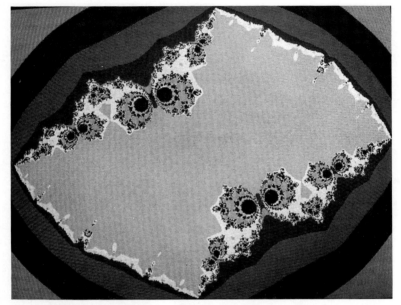

coordinates: $-1.5 < x < 1.5$ and $-1 < y < 1$; maximum iteration number: 150; threshold value: 150.

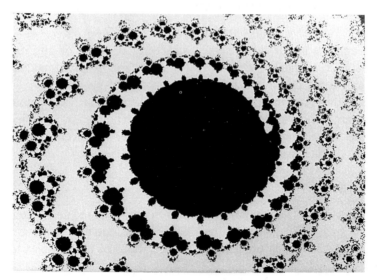

Fig. 4. Blow-up of part from Fig. 3. Coordinates: $-0.49 < x < -0.38$ and $0.27 < y < 0.36$; maximum iteration number: 270; threshold value: 270.

Spiral Structures in Julia Sets and Related Sets 133

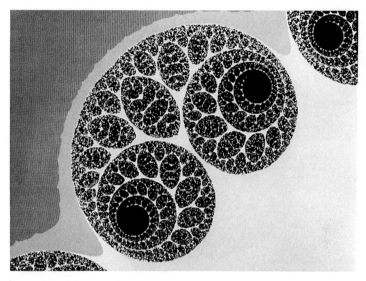

Fig. 5. A Julia set obtained from Eq. (2). Parameters: $p_x = 0$ and $p_y = 0.3$; coordinates: $-0.617 < x < -0.083$ and $0.15 < y < 0.55$; maximum iteration number: 220; threshold value: 220.

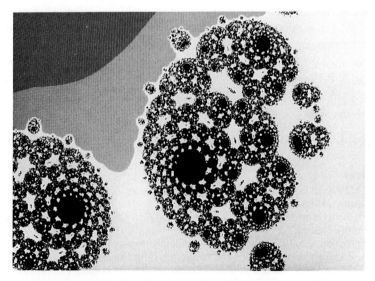

Fig. 6. A third Julia-like set from Eq. (2). Parameters: $p_x = 0.5$ and $p_y = 0.07$; coordinates: $-0.7 < x < 0.2$ and $0.07 < y < 0.715$; maximum iteration number: 120; threshold value: 120.

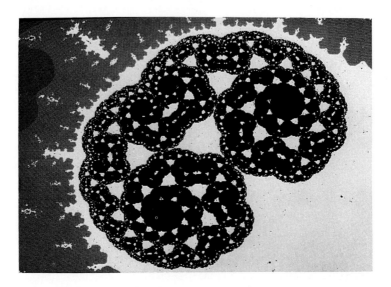

Fig. 7. Julia-analogous set generated by Eq. (3). Parameters: $p_x = 0.55$ and $p_y = 1.02$; coordinates: $-0.3 < x < 0$ and $0 < y < 0.26$; maximum iteration number: 140; threshold value: 140.

described in Ref. 5. It is presently an open problem under what general conditions self-similarity can occur in nonanalytic two-dimensional maps. The present pictures were, nevertheless, selected mostly for their aesthetic appeal. We expect simpler (C^∞) equations to show similar behavior.

Acknowledgement

We thank A. Mürle for kindly sharing with us his computer programs.

References

1. H. O. Peitgen and H. P. Richter, *The Beauty of Fractals,* Springer Verlag, Germany (1986).
2. H. O. Peitgen and D. Saupe, *The Sciences of Fractal Images,* Springer Verlag, Germany (1988).
3. M. Barnsley, *Fractals Everywhere,* Academic Press, London (1988).
4. C. A. Pickover, Mathematics and beauty: A sampling of spirals and "strange" spirals in science, nature and art, *Leonardo,* **21** (2) (1988) 173–181.
5. C. Kahlert and O. E. Rössler, Analogues to a Julia boundary away from analycity, *Z. Naturforsch.* **42a** (1987) 324–328.

THE EVOLUTION OF A THREE-ARMED SPIRAL IN THE JULIA SET, AND HIGHER ORDER SPIRALS

A. G. Davis Philip

1. Introduction

This paper will show examples of some of the many spiral forms to be found in the Julia Set, starting with a simple, three-armed spiral found in Radical 3 of the Mandelbrot Set and ending with a complex 17-turn, quad-spiral found in a region known as Elephant Valley. The evolution of spiral forms in the Julia Set will be described, showing the great complexity encountered at high magnifications.

Mathematicians have studied spirals and generated many formulae that allow real number functions to be plotted. Spirals have been studied because they appear in many forms in nature. For example, the Chambered Nautilus can be described by a logarithmic spiral of the form $r = ae^{\theta}$. An example is shown in Fig. 1a. A double-armed spiral can be generated by the equation $r^2 = a^2\theta$ (see Fig. 1b). The Lituus has the form of $r^2\theta = a^2$ (an example is shown in Fig. 1c). Spirals of these forms and others (found in the Mandelbrot and Julia Sets) will be illustrated in some of the pictures accompanying this paper.

2. The Mandelbrot Set

In even a cursory search of the patterns found in the Mandelbrot and Julia Sets, spirals of many types will be found. Before discussing specific patterns, however, a few general details will be discussed about the Mandelbrot and Julia Sets. The Mandelbrot Set is generated in the

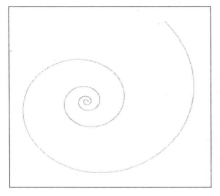
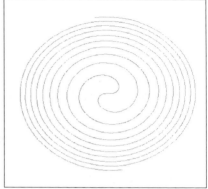

Fig. 1a. Log spiral. Fig. 1b. Fermat spiral.

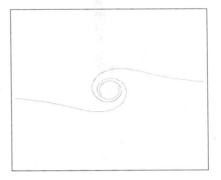

Fig. 1c. Lituus.

complex plane by iterating the equation $Z \rightarrow Z^2 + C$, where Z and C are complex numbers (of the form $x + yi$). For each pixel in the picture being generated, the value of C is the value of the position of that point in the complex plane. The initial value of Z is taken at the origin, $(0, 0i)$. Successive values of Z are calculated by substituting the new value for Z in the expression $Z^2 + C$. An astounding set of images can be found as a result of iterating these equations. Benoit Mandelbrot first saw such images, in black and white, in 1980 on the graphics terminal of a computer. He was responsible for opening a new area of mathematical study, that of fractal geometry, of which the Mandelbrot Set is a prominent and beautiful example.

Fig. 2 shows the Mandelbrot Set, extending from a real (Z) value of −2.0 (the end of the spike on the left) to a real (Z) value of approximately +0.45 on the right side. In most computer programs that generate the Mandelbrot Set, it is possible to move a rectangle to any point in the set, adjust its size, and then calculate a new diagram of the area contained inside. On IBM PCs and on Mac IIs, magnifications of about 10^{16} can be reached before roundoff errors in the last place cause major errors in the creation of the pictures. The main features shown in Fig. 2 (starting at the left) are the Spike, Head′, Head and Body. On the

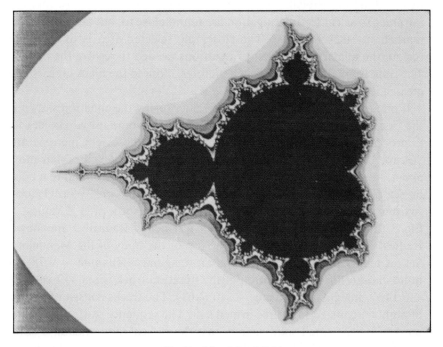

Fig. 2. Mandelbrot Set.

body and the head (and the smaller heads attached to the main head) will be found an infinite number of "atoms," small, round appendages attached to the set, pointing radially outward. Mandelbrot prefers a chemical analogy for naming parts of the set. Using this scheme, he called the round appendages attached to the body "atoms," and the whole structure around an atom is called a "radical." On each major radical there are smaller radicals, mimicking the pattern found on the head and the body. Between the body and the head can be found "Sea Horse Valley," so-named because of the many structural features that resemble sea horses and their tails. On the right side (at the back) of the body is "East Valley" or "Elephant Valley." The structures found above the radicals in Elephant Valley resemble elephants, hence the name. An article describing the naming of features on the Mandelbrot Set can be found in Philip and Philip.[1] (*Amygdala* is a newsletter devoted to articles concerning the Mandelbrot and Julia Sets and is edited by Rollo Silver, Box 219, San Cristobal, New Mexico 87564, U.S.A.) The most interesting areas to select for investigation are found close to, but not in, the area marked in black in Fig. 2. The closer the selected area is to the black region, the higher the number of iterations needed to resolve the picture. The time needed to calculate such pictures, close to the black area, is also greater.

If one monitors the trajectory of Z values (using the equation shown in the first paragraph of Sec. 2), one will find that some initial choices for C produce new Z values whose distance from the origin increase and "escape" from the set, while other choices result in Z values that move about, in sometimes very complex patterns, but that are bound to a point or set of points near their starting point. A program, "WormTracks" (written by K. W. Philip[2]), keeps track of these successive Z values. In Fig. 3 the tracks computed for a pixel just below Radical 3 are shown. Radical 3 is the largest radical at the top of the body of the Mandelbrot Set (at real (Z) value = -0.12 and an imaginary (Z) value = 0.78; the point chosen for the calculation of the iteration track is real (Z) value = -0.1437, imaginary (Z) value = -0.6261). The tracks consist of a series of points organized in a three-armed star. The sequence starts at the point marked by the small solid square, just the radical, and ends in a single point at the center of the spiral. The successive values of Z found by the iteration sequence start at the ends of the spiral arms and then wind in (counterclockwise) toward the central point.

If one of the new values for Z has distance from the origin greater than 2, then the point is said to escape from the set. Once beyond the value of 2, successive iterations of Z rapidly get larger and approach infinity. If the equation is iterated many times, and Z does not reach a distance of 2, then the point is said to be a member of the set. By convention, the

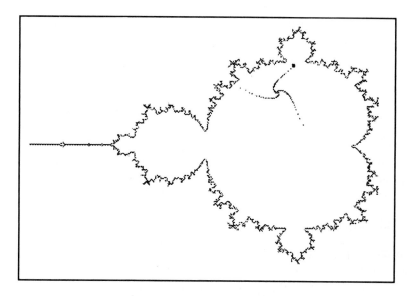

Fig. 3. Iteration tracks below Radical 3.

points that are members of the set are colored black. Other points can be colored different colors depending on how rapidly the point escapes the set. If the point escapes after only a few iterations one might color it, for example, a light blue. If the point escapes only after a much greater number of iterations, the point might be colored a darker blue, and so on. In such a picture a portion would be seen colored black (members of the set) surrounded by a series of blue bands, colored various shades of blue. All the points with the same shade of the blue would escape from the set at similar values of the iteration number (the technical term for the iteration number is the *dwell*, the number of iterations made at any point to obtain the current value of Z). The decision that a point is, or is not, a member of the set is made either because the Z value for the point reaches a value greater than 2 (and the point is not a member) or the dwell reaches the value set as an upper limit when the picture is generated (and the point is a member).

It is interesting that the Spike, the Head, and the Body of the Mandelbrot Set were presented in a paper given by Brooks and Matelski[3] in 1978 (published in 1980), but the plot was made in rough form (asterisks on computer paper) and the fractal, self-replicating features of the set could not be seen. Descriptions and pictures of Mandelbrot's work can be found in Gleick's *Chaos*,[4] Peitgen and Saupe's *The Science of Fractal Images*,[5] Peitgen and Richter's *The Beauty of Fractals*,[6] and Mandelbrot's *The Fractal Geometry of Nature*.[7]

One can find many interesting shapes by investigating the structures found in the periphery of the radicals. A common example is the spiral, the subject of this paper. Julia Sets, which were computed using a series of smaller radicals on the surface of radical 3 as starting points, reveal a series of spiral structures, starting with very simple, unwound spirals down to complex, tightly wound spirals as one moves to the higher numbered radicals.

3. The Julia Set

As mentioned earlier, the Mandelbrot Set is generated by iterating the equation $Z \rightarrow Z^2 + C$. A value of C is picked and Z is iterated until a decision is made about that point—is it a member of the set or not? Then a new value of C is chosen and the new Z is iterated, until all the pixels in the picture have been investigated. To calculate a Julia Set at the same point, the same equation is used, but this time C is held constant at a value equal to the point chosen in the Mandelbrot Set (the Julia point). Instead of starting Z values at the origin, the Z value of each point in the Julia Set being plotted is used as a starting value for that point. The Z values are iterated in the same way as in the Maldelbrot Set and the decisions about membership and the coloring of the bands surrounding the set are made the same way as in the Mandelbrot Set.

If one calculates a Julia Set near the central point of the body of Mandelbrot Set at (C = 0) the resulting diagram shows a solid, circular spot. If the starting point is moved to the head, and then to the other, smaller radicals along the body, the familiar "dragon-like" Julia sets are found. One of these is featured on the cover of *The Fractal Geometry of Nature*,[7] generated from Radical 5. A series of more complex spirals can be generated by selecting Julia points from radicals that are on the surface of the major radicals. In Fig. 4 an enlargement of Radical 3 (R3) is shown. In one scheme of naming the radicals (Philip and Philip[1]), the head of the Mandelbrot Set is named R2 and then the major radicals along the body, going into Elephant Valley are named R3, R4, R5, and so forth. The body of the set becomes R1 in this scheme. The radicals

Fig. 4. Right side of Radical 3.

on the major radicals are named in a similar fashion. The head of R3 is called R3/R2, the next large radical along the side is R3/R3, then R3/R4, and so forth. The slash indicates that the naming scheme now is identifying radicals on the surface of a radical. Fig. 5 shows the Julia Set for a point in the middle of R3/R3. In the center of the figure there is a large, diamond-like object to which are attached two smaller similar objects. If this structure is followed outward from the center, one comes to a three-armed feature, meeting at a common point. Since this Julia is computed from Radical 3, the number of arms is 3. If one computes a Julia from Radical 6, then the spiral structure will have 6 arms, if Radical 12, there will be 12 arms, and so on. This number has an important meaning; it is the cycle number of that point in the Mandelbrot Set.

A good way to describe the cycle number is to look at an iteration track. Such a diagram was shown in Fig 3. There is a three-armed spiral, which converges to a single point. This indicates that the cycle number of the point chosen, just below R3, on the body of the Mandelbrot Set, has a cycle number of 1. Since the spiral pattern has three arms, this

142 A. G. Davis Philip

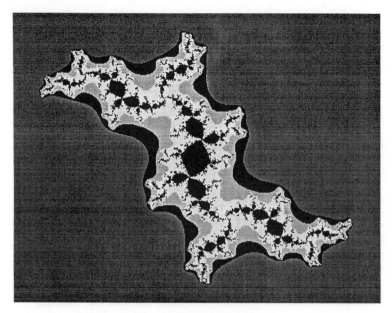

Fig. 5. Julia Set for R3/R3.

indicates that the point above the chosen point (i.e., R3) has a cycle number of 3. And this is why the radicals were named R2, R3, R4, and so forth. In each case the number of the radical is the cycle number associated with that radical. And it also indicates the number of arms in the spiral that will be found in the Julia Set of that radical.

4. Spirals from Radicals on R3

As Julia Sets are calculated from R3/R4, R3/R5, R3/R6, and higher numbered radicals, the three arms at the junction become thinner and more twisted. Figs. 6 and 7 show the spiral pattern for R3/R5 and R3/R8, respectively. In Fig. 8 an enlargement of the center of the spiral formation in R3/R8 is shown. There is a three-armed spiral and each arm has about one twist around the center. Fig. 9 shows the location of R3/R20 in the valley under R3, and Fig. 10 shows the Julia Set and spiral obtained from R3/R20. Now the spiral is much more tightly wound; each arm makes about three turns around the central point. In the Mandelbrot and Julia Sets, however, things never stay simple and this pattern soon undergoes a great change. In Fig. 11 the Julia from R3/R43 is shown and now the

The Evolution of a Three-Armed Spiral 143

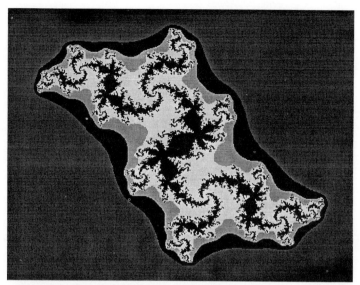

Fig. 6. Julia plot for R3/R5. The three arms begin to twist in a spiral shape. There are 5 clumps in the innermost structure and 3 arms in the spiral. This is because the Julia plot is made from Radical 5 on the surface of Radical 3.

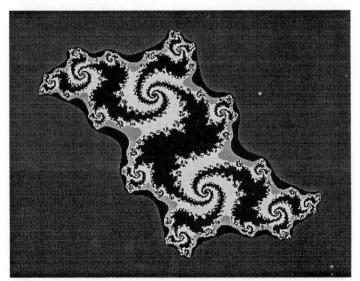

Fig. 7. Julia plot of R3/R8. The spiral is more tightly wound. An enlargement of the main spiral is shown in Fig. 8. Here there are 8 clumps in the center and 3 arms in the spiral.

Fig. 8. The central regions of the R3/R8 spiral. This spiral is not tightly wound.

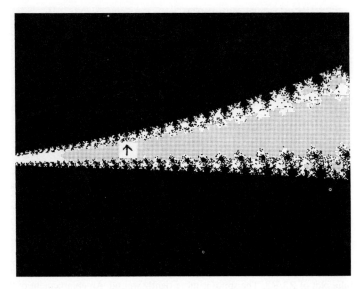

Fig. 9. The valley under R3. The radical on the far right is R9. R20 is marked with an arrow.

The Evolution of a Three-Armed Spiral   145

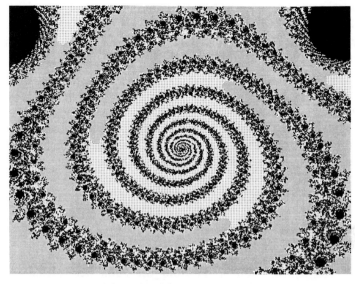

Fig. 10. The spiral formed from Radical R3/R20. The outermost arm shows a start of the pinching, which will become stronger in Figs. 12 – 14.

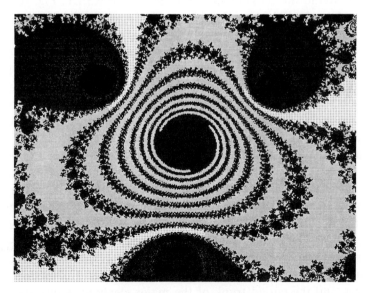

Fig. 11. The spiral from Radical R3/R43. The arms are more tightly wound. The bending of the outer spiral arms reaches the second turn.

outer spiral arms are deformed by the three "star" patterns, which have 43 spokes. A Julia Set done from a radical at Real (Z) = -0.12142 (another of the smaller radicals attached to R3) shows a quite different spiral structure. The coils are very tightly wound, but the coils near the outside bend and form a three-leaf pattern (Fig. 12). At Real (Z) = -0.12214 (Fig. 13) and Real (Z) = -0.124533 (Fig.14) the "pinching" effect becomes even stronger.

As one goes to more negative values of Real (Z) into the valley under R3, the iterations have to be done to higher and higher values. In Fig. 14 the dwell is 32,000. Thus these pictures take more time to compute. In an EGA mode (Enhanced Graphics Mode) picture, there are 640 × 480 pixels and each one can be iterated up to 32,000 times in the areas that are black. Fig. 14 took about 5 cpu hours on a 16-MHz 386 machine.

5. Julia Sets from R6, R18, R25, and R75

Fig. 15 shows the main spiral from the Julia Set calculated from R6/R6, i.e., radical number 6 on the side of radical 6 on the back of the body of the Mandelbrot Set. Now the spiral pattern has six arms and the arms are wound more tightly than the pattern for R3/R6. Fig. 16 shows the spiral from R6/R11, and Fig. 17 the spiral from R6/R18. In the upper left corner of Fig. 17, one can see a feature common to many of these pictures, a series of "spokes" coming together at the center. If one counts, the number of spokes will give the cycle number of the Julia point. However, when the number of spokes is very high it requires a high resolution picture to resolve each spoke. It also takes a high resolution to be able to see the spiral arms all the way to the center. In Fig. 16 the spiral arms cannot be seen all the way to the center, whereas in Fig. 17 they can. The dwell in Fig. 16 was 5,555; in Fig. 17 it was 15,000.

Fig. 18 shows the Julia Set from Radical 6 on the side of Radical 18, which is located in East Valley. The spiral pattern now is asymmetric, with the spiral arms below the center smaller than the spiral arms above. The Julia Set for R18/R18 is shown in Fig. 19. This Julia has a dual pattern. The spiral features are shown (enlarged) in Fig. 20 and the spirals are now very tightly wound. However, along the outside edge of the Julia pattern, "Elephants" can be seen. They face to the left (on the top of the diagram), and the curled trunk of the major elephant is just above the major spiral. The Julia Set of any place on the Mandelbrot Set

Fig. 12. Julia plot for Real (Z) = -0.12142. The inner spirals are circular but the outer spiral arms are distorted. The dwell was set at 15,000. The black circle at the center would be resolved only if a higher dwell is used.

Fig. 13. Julia plot at Real (Z) = -0.12214. The pinching effect here is very strong, affecting all the spiral arms.

Fig. 14. Julia plot at Real (Z) = -0.124533. The pinching effect is stronger than in Fig. 13.

The Evolution of a Three-Armed Spiral 149

Fig. 15. Julia Set from R6/R6. There are 6 arms in this spiral.

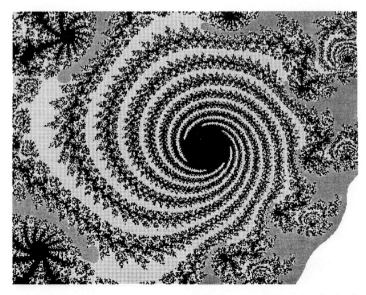

Fig. 16. Julia Set from R6/R11. There are 11 spiral arms and 11 spokes in the "stars" at the ends of the arms.

Fig. 17. The spiral from R6/R18. Now the spirals are wound so tightly that the arms cannot be distinguished as separate arms all the way to the center.

Fig. 18. The spiral from R18/R6. There are 18 arms in the spiral. The spirals below the center are smaller than those above.

The Evolution of a Three-Armed Spiral 151

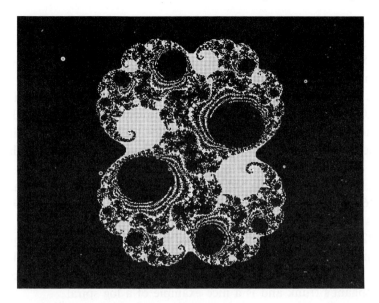

Fig. 19. Julia Set for R18/R18. Note that the outside edge of the set shows the typical "elephant" forms with their curled trunks.

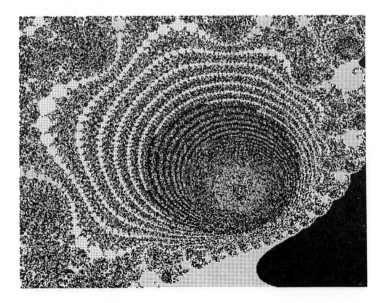

Fig. 20. The spiral in the Julia of R18/R18. The spiral arms are very tightly wound.

presents features that will be found in the Mandelbrot diagram of the same area. It is quite instructive to investigate both the Mandelbrot and Julia Sets for each region of the Mandelbrot Set. Each gives important information concerning that part of the set.

Fig. 20, showing the main spiral for R18/R18, was computed at a dwell of 30,000, and the spirals can be seen to the center of the pattern, though one needs higher resolution to be able to distinguish the very tightly-wound spirals at the center. This illustrates the need for higher and higher dwell values as one goes to higher numbered radicals.

Fig. 21 shows the Julia Set for R25/R3. It resembles the picture for R6/R6, except that now the spirals have 25 arms instead of 6. Fig. 22 shows the Julia spiral for R75/R3. The dwell in this picture was set at 22,222 and the spiral arms show all the way to the center, but in a picture this size the resolution is not high enough to separate out the central-most arms. Fig. 23 shows an area just above that of Fig. 22. This is the end of an elephant's trunk and is a nice example of a log spiral.

Fig. 21. The spiral in the Julia of R25/R3. There are 25 spiral arms.

The Evolution of a Three-Armed Spiral 153

Fig. 22. The spiral for R75/R3. There are 75 spiral arms, but in a picture this size it is not possible to resolve them all.

Fig. 23. An "elephant's trunk" above the spiral shown in Fig. 22.

154   A. G. Davis Philip

6. Double/Double Spirals

While investigating an area in Sea Horse Valley, I found a very interesting "double/double spiral" (DDS) in the Mandelbrot Set. A spiral of this type can be found by selecting a radical in Sea Horse Valley and enlarging the curl at the end of a sea horse tail. Then the end of the curl is selected and a portion of the spiral arm is enlarged until some of the spirals that exist there in great profusion can be seen. Select a spiral and enlarge it. Eventually a series of "double spirals" (DS) will be seen. An example of this is shown in Fig. 24, which shows the end of a spiral coil

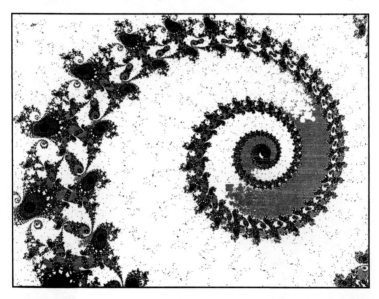

Fig. 24.   The end of a spiral in the Mandelbrot Set showing the location of double spirals (DS) and how the DS wind up as one goes closer to the end of the spiral.

at $R = 0.4052405$, $I = 0.1467907$. (This picture is from the Sea Horse Valley in the midget.) Near the inner side of the spiral arm a series of double spirals can be seen. As one moves along the main spiral arm, the orientation of the double spiral rotates clockwise. The further along the main spiral arm that a double spiral is picked, the greater the number of twists in the spiral arm surrounding the DS. Choose one of the DS and enlarge its center. After a series of enlargements a DDS can be found. The first DDS that I saw is shown in Fig. 25. The figure shows five spiral

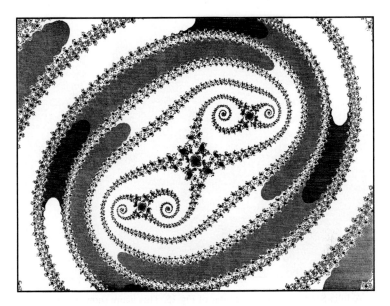

Fig. 25. A Double/Double Spiral in the Mandelbrot Set.

turns that come together in the center. Inside the center a still smaller spiral can be seen. On either side of the center there are two double spirals that have smaller spirals in their centers.

After all the enlargements were made to reach the DDS, the PC precision was close to being exceeded (the magnification for the DDS at this location is about 10^{15}). An image can be enlarged about 10^{16} times with a program written in double precision, and then no further enlargements can be made that can show the diagram clearly. Recalling that Julia Sets are easier to calculate (they take less time to calculate than the Mandelbrot Set of the same area), I decided to do a series of Julia Sets calculated on the center of Fig. 25. The next few figures show the sequence of spirals encountered as one probes deeper and deeper into the set. The Julia Set made from the center of Fig. 25 shows a symmetric pattern of multiple "elephants" in the center of which is a small single spiral. Fig. 26 shows an enlargement of this spiral. It is a two-armed, single spiral between two "eye" structures. In the outer spiral arms one can see small editions of spirals between the upper and lower eye structures. Fig. 27 shows the center of Fig. 26. This time a double spiral can be seen. There are two spiral arms, but instead of spiraling into a common center, the spiral splits into a center with a spiral on

Fig. 26. A Julia Set made from the center of Fig.25. This figure shows a single, two-armed spiral between two "eyes."

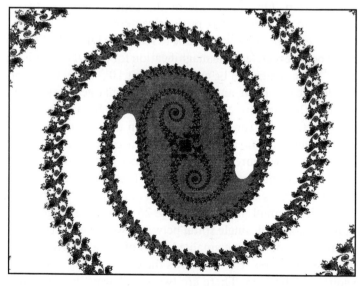

Fig. 27. A Julia Set made from the center of Fig. 26. This figure shows a double spiral and at the center is a very closely wound spiral.

either side. Fig. 28 shows the center of Fig. 27 and this is the double/double spiral. Except for the shading in between the spiral arms, this picture is identical to the DDS in Fig. 25 and is a good illustration of the statement made earlier that the Julia Set is a good map of the fundamental patterns found in the Mandelbrot Set. If the center of the small spiral in the center of Fig. 28 is enlarged another DDS will be found (Fig. 29).

As one goes deeper (i.e., to higher magnifications), structures rapidly become more complex. Fig. 30 shows an enlargement of the center of another DDS, in a spiral arm surrounding the DDS in Fig. 29. At the center of this marvelously complicated picture is a double/quad spiral. About this is a quad spiral, the arms of which each contain a DDS. The next enlargement in this sequence shows a double/octuple spiral, so it seems that this sequence would continue, 4, 8, 16, 32, and so on. But the last figure (Fig. 31) shows a new complication. This picture is an enlargement of one of the double/octuple spirals (this one found in Elephant Valley). If the number of spiral arms is counted the number comes out to eight, but the center is quite different. The very center of the figure is empty; instead there are two symmetric octuple spirals to either side of the center. If one looks closely at the right and left centers, one can see that each is made up of a double/octuple spiral. The four arms that come out from each center contain double/double spirals.

Fig. 32 shows a Julia plot of a multiturned quad spiral deep in Sea Horse Valley (first plotted in the Mandelbrot Set by KWP). This diagram is the fifth enlargement of the spiral at the center of the Julia plot and shows a 17-turn spiral winding about two separated quad spirals. At the center of each of the quad spirals are two eye-like features that are smaller reproductions of the full picture. A full description of the various types of spirals to be found in the Mandelbrot Set can be found in Philip and Philip.[8] Sec. 6 of this paper will serve as an introduction to this subject.

7. Summary

This paper presents the development of an incipient spiral feature found in the Julia Set formed from a point on Radical 3 in the Mandelbrot Set. The development continues as subradicals of higher numbers, on the surface of R3, are investigated. The changes in the spirals formed from radicals further along the body and into Elephant Valley are indicated in Sec. 5. More complex spirals (double/double spirals, double/quad spirals, and so on) are discussed in Sec. 6. As an astronomer, my primary interest,

158   A. G. Davis Philip

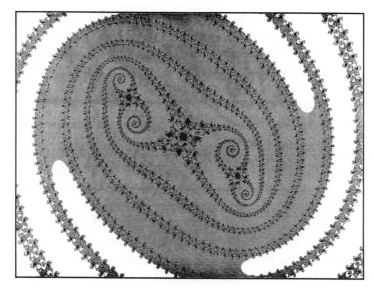

Fig. 28.   A Julia Set made from the center of Fig. 27. Each arm of the central spiral contains a double spiral, hence this is a double/double spiral.

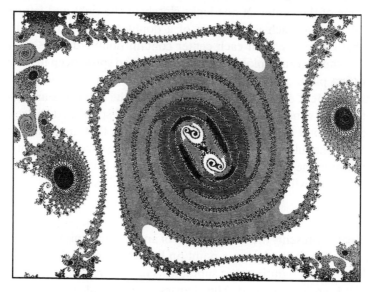

Fig. 29.   A Julia Set made from the center of Fig. 28. Another double/double spiral appears, but this one is turned some degrees clockwise relative to Fig. 28 and the twin spirals in the arms have turned a few degrees counterclockwise.

The Evolution of a Three-Armed Spiral 159

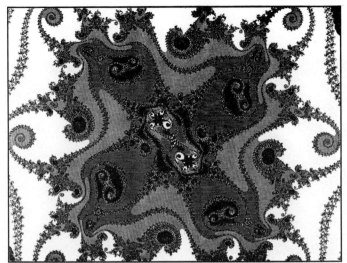

Fig. 30. A Julia Set made from another DDS, close to the DDS in Fig. 29. In the center is a double/quad spiral and about this center is a quad spiral with double/double spiral in each of the four arms.

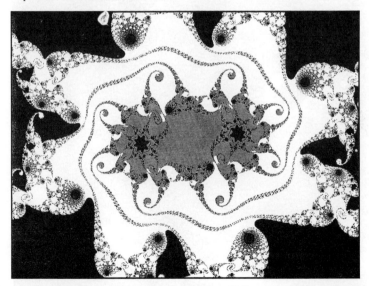

Fig. 31. A Julia Set made from a DDS in Elephant Valley. This is an enlargement of a center much like that in Fig. 30. Now there is an octuple spiral at the center, but this is a complex object. It is split into two quad spirals and each arm has a double spiral in it. At the twin centers are octuple spirals and in the centers of these spirals are twin octuple spirals of the same form as in the full picture.

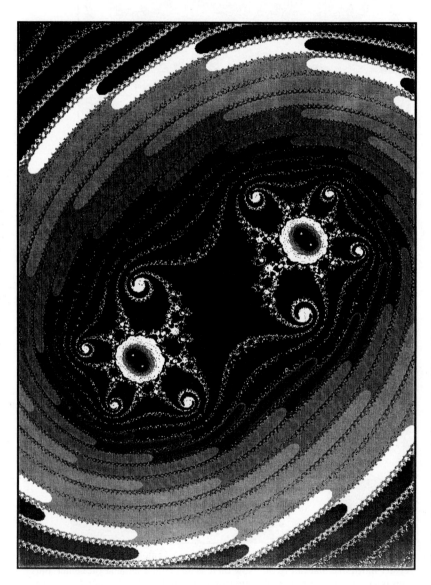

Fig. 32. A Julia Set made from a Quad Spiral in Sea Horse Valley. There are 17 turns in the spiral surrounding twin quad spirals in the center. This pattern is repeated in each of the quad centers.

it was interesting to find spiral structures in great profusion in the Mandelbrot and Julia Sets. Astronomical spirals (galaxies) are discussed by Bruce Elmegreen in this volume in a paper entitled, *Spiral Galaxies*.

As personal computers of greater power become available, it will be possible to probe deeper into East Valley and deeper into the valley under R3. It would be interesting to find out the next pattern for radicals further in than that shown in Fig. 14. And what is the spiral pattern for R1000/R3?

When studying a portion of the Mandelbrot Set, it is quite instructive to do the Julia plot of the same region — it is a good way to determine the cycle number of the point under investigation and to learn what some of the basic features of the region are. Also, as pointed out in this paper, the dwells encountered in Julia Sets are less than the corresponding areas in the Mandelbrot Set and thus the diagrams take less time to compute. And there is the important benefit that the Julia Set allows one to probe more deeply before running into the magnification limit of about 10^{16} of double precision on a microcomputer.

In the Appendix that follows, the hardware and software used in this investigation are listed and, for those who would be interested in creating some of these pictures on their own screens, there is a table of real and imaginary values for the center points of each of the Julia Sets shown in Figs. 5–32 (and the two Mandelbrot Set diagrams shown in Figs. 24 and 25). The positions are given to 16 places on the computer records, but, for most of these pictures only the first few figures are needed to reproduce the patterns, so in Table 2 the positions are given to 8 decimal places only. In the case of Fig. 25, real and imaginary coordinates are repeated below the table, listed to 16 places. for Figs. 26–30 entering the center Julia points (even to the 16 places given at the bottom of the table) does not allow one to regenerate the pictures. The way to do it is to create Fig. 25 and then recreate the Julia Sets by marking the center points as the Julia point. To obtain Figs. 26–30 just set the real and imaginary quantities to the values indicated in the last four columns and each picture can be calculated.

Acknowledgements

Discussions have been held with K. W. Philip (University of Alaska at Fairbanks) and Michael Frame (Union College) on many of the topics included in this paper, and I wish to acknowledge their help and suggestions. The fractal programs used in this investigation were written by Michael

Freeman (Vancouver, Canada). He has been very helpful in adding additional routines to the programs to make them better research tools.

Appendix

Table 1. Hardware and Software Used

Computer	A		Model 80 IBM PS/2
			8514a Video Card
			8514 Monitor
	B		XT with Intel 386 Board
			ATI VGA Wonder Card with 512 K
			Nec Multisynch 3D Monitor
Printer			HP LaserJet II
Software	A		EGAMBROT by Micheal Freeman (Vancouver, Canada)
			Pizzaz + by Applications Techniques (pictures from screen to printer, 640 × 480 pixels)
	B		V63MBROT by Michael Freeman
			LJDUMP by Michael Freeman (pictures from screen to printer, 800 × 600 pixels)

Figs. 5–23 were made with setup A. Figs. 24–32 were made with setup B.

Table 2. Positions of the Julia Sets

Figure	Imaginery	Dwell	Name	Real	Min.R	Max.R	Min.I	Max.I
5	−0.03249108	0.79245342	100	R3/R3	−1.50000000	+1.50000000	−1.28000000	+1.28000000
6	−0.03193370	0.71254855	100	R3/R5	−1.50000000	+1.50000000	−1.28000000	+1.28000000
7	−0.05868757	0.67640110	100	R3/R8	−1.50000000	+1.50000000	−1.28000000	+1.28000000
8	−0.05868757	0.67640110	500	R3/R8a	−0.14612068	−0.00646551	+0.25747126	+0.62528735
10	−0.09571678	0.65403036	1000	R3/R20a	−0.45258620	−0.02586206	+0.27218390	+0.64000000
11	−0.11112291	0.65052071	10000	R3/R43a	−0.58836206	+0.09698275	+0.13977011	+0.71356321
12	−0.12142690	0.64958287	15000	R0.12142	−0.44612068	−0.09698275	+0.26482758	+0.55908045
13	−0.12213759	0.64956060	10000	R0.12214	−0.64008620	+0.13577586	+0.13241379	+0.79448275
14	−0.12453393	0.64952013	32000	R0.12453	−0.73706896	+0.05172413	+0.13241379	+0.80919540
15	+0.37953073	0.20567030	1000	R6/R6a	+0.03232758	+0.49784482	+0.24275862	+0.64000000
16	+0.36745529	0.14230577	5555	R6/R11a	−0.12931034	+0.65948275	+0.05885057	+0.72091954
17	+0.36716188	0.14414826	5555	R6/R18a	−0.06465517	+0.64655172	+0.10298850	+0.70620689
18	+0.27803224	0.00988119	10000	R18/R6a	+0.07112068	+0.82112068	−0.02206896	+0.61057471
19	+0.27819115	0.01019907	15000	R18/R18	−1.50000000	+1.50000000	−1.28000000	+1.28000000
20	+0.27819115	0.01019907	30000	R18/R18a	−0.06465517	+0.77586206	−0.08091957	+0.64000000
21	+0.26510834	0.00357401	10000	R25/R3a	+0.31681034	+0.57543103	+0.05885057	+0.27954022
22	+0.25113433	0.00007114	32000	R75/R3a	−0.62068965	−0.36206896	−0.19862068	+0.02206896
23	+0.25113433	0.00007114	22222	R75/R3b	−0.60775826	−0.40086206	−0.01471264	+0.16183908
24	(Mandelbrot Set)		20000	WMDSL3	+0.40524036	+0.40524079	+0.14679043	+0.14679076
25	(Mandelbrot Set)		20000	Serendip	−0.74858089	−0.74858089	+0.06306469	+0.06306469*
26	−0.74858089	0.06360646	20000	JDDSOR2	−0.10763454	+0.09261576	−0.72621035	+0.07762938
27	−0.74858089	0.06360646	20000	JDDSOR3	−0.00512843	+0.00489660	−0.00364129	+0.00388376
28	−0.74858089	0.06360646	20000	JDDSOR5	−0.00008329	+0.00007707	−0.00005395	+0.00006155
29	−0.74858323	0.06307693	4000	JDDSOO3G	−0.00000441	+0.00000361	−0.00000285	+0.00000293
30	−0.74858089	0.06360646	4000	JDDSOO3G	−0.00000023	+0.00000023	−0.00000018	+0.00000023
31	0.26055650	−0.00176231	4000	JEVDS9i	−0.00000019	+0.00000018	−0.00000014	+0.00000017
32	−0.76238228	0.09556308	30000	JDSH17e	−0.00000260	+0.00000266	−0.00000177	+0.00000188

* −0.7485808956487646 − 0.7485808956487637, +0.063064691776399 + 0.0630646917776405

Note. The figures in Table 2 are truncated at 8 places and not rounded up.

References
1. A. G. D. Philip and K. W. Philip, The taming of the shrew, *Amygdala* **#19** (1990) 1–7.
2. K. W. Philip, Wormtracks, *Amygdala* **#14** (1988) 1–4.
3. R. Brooks and J. P. Matelski, The dynamics of 2-generator subgroups of PSL (2, C), in *Reimann Surfaces and Related Topics* (pp. 65–71). Princeton University Press, Princeton (1980).
4. J. Gleick, *Chaos*. Viking, New York (1987).
5. H. -O. Peitgen and D. Saupe, *The Science of Fractal Images*. Springer Verlag, New York (1988).
6. H. -O Peitgen and P. H. Richter, *The Beauty of Fractals*. Springer Verlag, Berlin (1986).
7. B. Mandelbrot, *The Fractal Geometry of Nature*. W. H. Freeman, New York (1983).
8. K. W. Philip and A. G. D. Philip, Double spirals and higher order spirals in the Mandelbrot and Julia Sets, *Amygdala*. In preparation (1991).

AUTONOMOUS ORGANIZATION OF A CHAOTIC MEDIUM INTO SPIRALS: SIMULATIONS AND EXPERIMENTS

Mario Markus

1. Introduction

In this contribution I will discuss so-called excitable or active media. For details on these media the reader may consult Refs. 1–3. In the remainder of this paragraph, I will summarize the main features and give some examples. In each point of space, an excitable medium can undergo a fast transition into an "excited" state if triggered by a threshold-transgressing perturbation coming from the neighbourhood of the point. After this transition the medium at the point becomes "refractory," slowly recovering its excitability until it becomes "receptive" and can be excited again. In this way, waves may be formed that have the following properties: they suffer no attenuation, high frequency waves annihilate low frequency waves, waves with equal frequency annihilate each other at the points of collision, and they are not reflected at the boundaries of the medium. These type of waves appear in the entire spectrum of macroscopic science. Examples are found in nerve axons,[4] the retina,[5] the surface of fertilized eggs,[6] the cerebral cortex,[7] heart tissue,[8,9] aggregating slime mould populations,[10–14] the spreading of infectious diseases,[15–17] oxidation processes on metal surfaces,[18] the Belousov-Zhabotinskii (BZ) reaction,[19–24] and spiral galaxies.[25–27]

2. Simulations

2.1 *Solving the Problem of Anisotropy*

Waves in excitable media have been simulated extensively with partial differential equations (see, e.g., Refs. 28, 29). As an efficient alternative, simulations have been performed using cellular automata. However, cellular automata suffer from the drawback that the shape of the cells in a periodic grid is reflected in the shape of the waves, so that polygonal (or nearly polygonal) waves are obtained.[8,18,25,30–35] In other words, the wave propagation obtained with a periodic grid is anisotropic. As an example, Fig. 1 illustrates how a two-dimensional cellular automaton with a square grid leads to square waves. It is assumed in this automaton that all eight neighbours of an excited cell (black) become excited in the next time step, except for the neighbouring cells that are refractory (dashed). An excited cell becomes refractory after one time step and receptive (white) after two time steps. Fig. 1a shows the propagation of

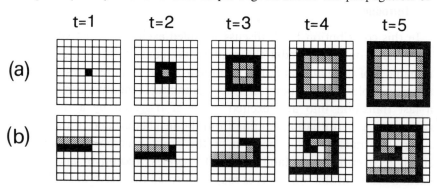

Fig. 1. Fourfold wave-symmetry obtained by a cellular automaton with square cells: (a) target pattern, (b) spiral wave. Reprinted from *Mathematical Population Dynamics*, O. Arino et al., eds., by courtesy of Marcel Dekker Inc.

a so-called "target pattern" resulting from a single excited cell, while Fig. 1b shows the development of a spiral wave in the same automaton. As a further example, Fig. 2 shows how a hexagonal grid with analogous rules (six neighbours per cell) leads to hexagonal waves.

Is the problem of anisotropy solved by a quasiperiodic Penrose[36] grid? The answer is no, as illustrated in Fig. 3. Here a target pattern is simulated by assuming that all neighbours sharing at least one corner

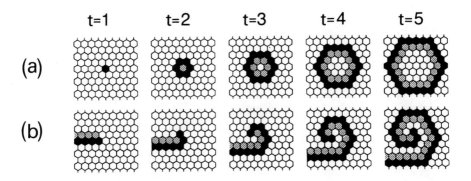

Fig. 2. Sixfold wave-symmetry obtained by a cellular automaton with hexagonal cells: (a) target pattern, (b) spiral wave. Reprinted from *Mathematical Population Dynamics*, O. Arino et al., eds., by courtesy of Marcel Dekker Inc.

with an excited cell become excited in the next time step. One sees that the wave is irregular for small times and tends to be decagonal for large times. Thus, the aperiodicity of this grid is not sufficient to solve the problem, but leads to a tenfold symmetry, as is also obtained in diffraction patterns measured with some quasiperiodic alloys.[37]

Mikhailov and Dorjsurangiyn[32,33] introduced the procedure of averaging over the cells within a neighbourhood. This procedure, which was also used in Refs. 34 and 35, leads to a propagation closer to being isotropic since the wave corners become rounded. However, the shapes of the waves still contain information about the geometry of the grid.

Recently, an algorithm was developed by the author to solve the problem of anisotropy.[38,39] This algorithm is based on a quasirandomization of the automaton geometry. In fact, the two-dimensional (resp. three-dimensional) medium is divided into square (resp. cubic) cells with side length d, and one point is placed randomly in each cell. Each of these points may assume $n + 2$ states S: $S = 0$ (receptive state), $S = 1, 2, \ldots, n$ (refractory states), and $S = n + 1$ (excited state). The states $S(t + 1)$ are determined iteratively (time $t = 1, 2, \ldots$) at each point from the states $S(t)$ as follows. First, a neighbourhood is defined by drawing a circle C (or sphere, in the three-dimensional case) with radius R around the point under consideration (in analogy with Huygen's principle). The number v of excited cells within C is counted. I will refer here only to the most

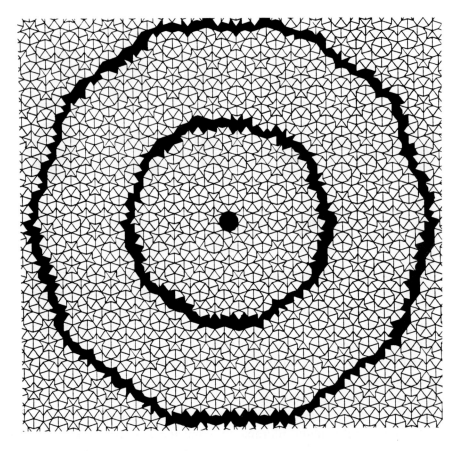

Fig. 3. Tenfold wave-symmetry obtained by a cellular automaton with a Penrose lattice consisting of "kites" and "darts."[36] The excited wavefront is shown at $t = 0$, 10, and 20.

general model of the previous work ("model II"). In this model, the "excitability threshold" m is assumed to vary linearly with $S(t)$: $m = m_o + p\,S(t)$, where $S(t) = 0,1,2,\ldots S_{max}$; m_o, p, and S_{max} are model parameters. This dependence of m on $S(t)$ describes the capability of different refractory states to become excited, taking into account wave dispersion.[22,35] S_{max} is introduced to take into account that shortly after excitation ($S = n, n-1, \ldots S_{max} + 1$), the system cannot become excited. An intermediate state $\sigma(t)$ is determined from $S(t)$ by the following rules:

(a) If $S(t) = 0,1,2,\ldots,S_{max}$ and $v \geq m_o + pS(t)$, then $\sigma(t) = n+1$.
(b) If $S(t) = 0$ and $v < m_o$, then $\sigma(t) = 0$.
(c) If $S(t) = 1,2,\ldots, S_{max}$ and $v < m_o + pS(t)$, then $\sigma(t) = S(t) - 1$.
(d) If $S(t) = S_{max} + 1, \ldots, n, n+1$, then $\sigma(t) = S(t) - 1$.

The state $S(t+1)$ is determined from $\sigma(t)$ by the rules:
(e) If $\sigma(t) = 1$ or $n+1$, then $S(t+1) = \sigma(t)$
(f) If $\sigma(t) = 2, 3, \ldots, n$, then $S(t+1) = [<\sigma(t)>]$,

where the brackets $<>$ indicate the average overall points within C and the Gaußian brackets [] indicate taking the next integer. Altogether, this model contains five control parameters: $r = R/d$, n, m_o, p, and S_{max}. Simulations can be simplified by neglecting dispersion, i.e., by setting $p = S_{max} = 0$. In that case, rule (c) is left out of consideration.

2.2 Graphical Representation

In the present work, waves simulated in two dimensions are represented as follows. One pixel (picture element) is assigned to each square cell and is given a grey value according to the state S of the point within that square. The darkness of the grey level increases with decreasing S ($S = n+1, n, \ldots, 1, 0$). Thus, the lightest areas correspond to the excited state, and the darkest areas to the receptive state.

In three dimensions, only the state $S = n+1$ (excited state) is shown, since it was found that pictures become difficult to understand if more states are displayed. Also, the medium is cut into a slice parallel to the plane of the figure, and only this slice is shown. Grey shadings in three dimensions are not used in the same way as in two dimensions, but indicate how deep the excited cells lay within the slice in the direction perpendicular to the cutting planes.

2.3 Target Patterns and Spirals

The isotropic propagation obtained with the present model is best illustrated in Fig. 4, which shows target patterns at two automaton resolutions: a = low resolution, b = high resolution. This figure clearly shows the rotational symmetry of the waves, in contrast to the polygonal (or nearly polygonal) shape shown in Figs. 1, 2, and 3, or reported elsewhere.[8,18,25,30–35]

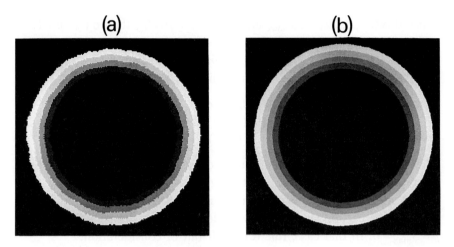

Fig. 4. Rotational symmetry obtained with the isotropic quasi-random cellular automaton presented in this work: $n = 3$, $m_o = 1$, $p = S_{max} = 0$, $t = 15$. (a) 200 × 200 cells, $r = 6$, (b) 400 × 400 cells, $r = 12$.

Different types of simulated spiral waves are shown in Fig. 5. The left parts of the Figs. 5a, b, c, and d show the initial conditions for single-, double-, three-, and four-armed spirals, respectively. The right parts of these figures show snapshots of the corresponding fully developed waves. These waves are very similar to those observed experimentally in the BZ-reaction.[19] The small white circle at the center of the single-armed spiral (Fig. 5a) is the so-called spiral core, which has been measured in the BZ-reaction[23] and in the aggregating slime mould.[14] The tip of the spiral (defined as the excited point where the normal velocity of the wavefront is zero) rotates around this core. Within the core, the system never becomes excited ($S = n + 1$), always remaining within a very narrow range of S.

The two-armed spiral (Fig. 5b) shows a remarkable feature: the distance between the tips of the two arms oscillates in time. The period of these oscillations is 9 iterations and the amplitude is equal to 8.6d. Oscillations of two-armed spirals have also been reported from experimental observations,[19] although no quantitative details have been reported as yet. The tips of three-armed spirals (Fig. 5c) oscillate connecting in pairs that alternate in time, a phenomenon that has also been observed experimentally.[19]

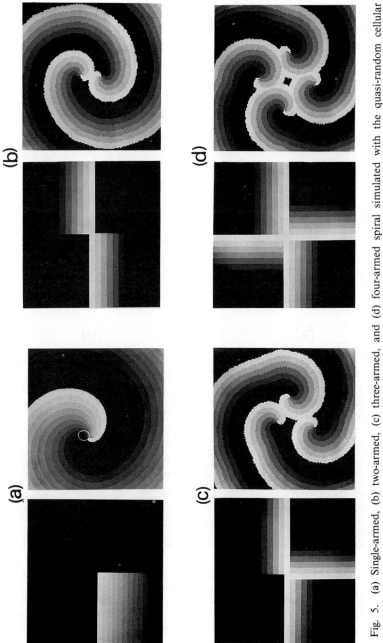

Fig. 5. (a) Single-armed, (b) two-armed, (c) three-armed, and (d) four-armed spiral simulated with the quasi-random cellular automaton. The left parts of (a) to (d) show the initial conditions, and the right parts the fully developed waves. $m = 1$, $p = S_{max} = 0$. (a) 354×354 cells, $n = 12$, $r = 12$, $t = 26$. (b,c,d,) 180×180 cells, $n = 6$, $r = 6$, (b) $t = 45$, (c) $t = 36$, and (d) $t = 50$.

Spirals may also develop in three dimensions, leading to the so-called scroll waves. They were first observed experimentally by Winfree.[40] A recent review is given in Ref. 41. The simplest scroll wave is the toroidal vortex (untwisted, unknotted scroll wave),[21,42] which is obtained by starting with an excited closed ribbon, i.e., a cylinder segment, of thickness R buffered inside by closed refractory ribbons with the same thickness and with S decreasing ($S = n, n-1, \ldots, 1$) toward the cylinder axis. On each plane passing through the cylinder axis, these scroll waves behave as the two-dimensional spiral shown in Fig. 5a.[39] This type of three-dimensional waves has been postulated from observations in thick heart muscle[43] and measured in the BZ-reaction.[20]

More complex three-dimensional spirals are the twisted scroll waves. This type of waves has been constructed geometrically (without dynamical simulations)[21] and by anisotropic automata (cubic cells).[30,31] However, they have not been observed experimentally as yet. Simulations of twisted scroll waves, as calculated with the present automaton, are given in Figs. 6 and 7. The initial conditions for these waves

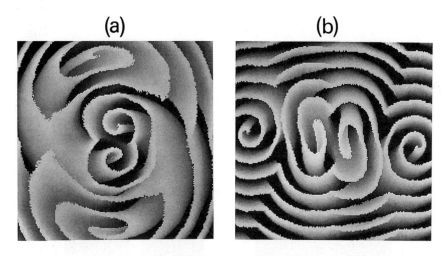

Fig. 6. Three-dimensional spiral wave obtained by starting from a ribbon with a 360° twist: $n = 3, r = 3, m = 1, p = S_{max} = 0, t = 24$. The figures show slices of a medium with 200 × 200 × 200 cells. These slices cover the 100th to the 200th cell as counted from a side of the medium perpendicular to the cylinder axis of the untwisted ribbon (a) and from a side parallel to this axis (b).

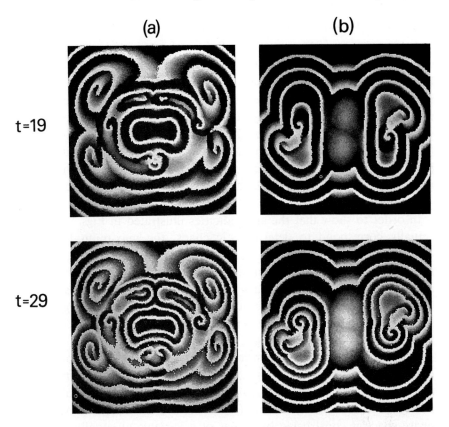

Fig. 7. Three-dimensional spiral wave obtained by starting from a ribbon with a 720° twist: $n = 3$, $r = 3$, $m = 1$, $p = S_{max} = 0$, 200 × 200 × 200 cells. The figures show slices covering the 100th to the 111th cell as counted from a side of the medium perpendicular to the cylinder axis of the untwisted ribbon (a) and from a side of the medium parallel to this axis (b).

are sets of ribbons such as that described in the preceding paragraph, with the difference that the ribbons here were given a 360° twist for Fig. 6 and a 720° twist for Fig. 7 before their ends were closed into a ring. The dependence of the wavelength on space and time in Fig. 7 is due to the collisions of wavefronts coming from different directions. In fact, such collisions cause wavefronts of different curvatures and thus different normal velocities, implying varying wavelengths (see Ref. 50).

2.4 Chaos and Self-Organization into Spirals

A spatially periodic variation of the refractory time n can cause destruction of wave periodicity, as illustrated in Fig. 8. The spiral in Fig. 8a ($t = 1$) is obtained by setting n constant in space ($n = 6$). At the next iteration, n is allowed to vary periodically in a piecewise linear manner, as shown on the left side of Fig. 9. The spiral is still recognizable in Fig. 8b ($t = 18$). In Figs. 8c ($t = 73$) and 8d ($t = 340$) aperiodicities have fully developed. This phenomenon is a spatio-temporal analogon to purely time-dependent systems, in which a periodic driving of a nonlinear periodic system can lead to aperiodic behaviour (see, e.g., Refs. 44 and 45).

In order to show that the aperiodicities illustrated in Figs. 8c and 8d actually correspond to deterministic chaos, the maximum Lyapunov exponent λ_{max} was calculated using the method by Wolf et al.[46] Phase spaces were constructed by considering 9 points p_i ($i = 1, 2, \ldots, 9$) placed at a fixed distance from each other and from the edges of the two-dimensional medium, as indicated by the crosses in Fig. 9. The phase variables v_i were defined as the average of S within circles of radius $9d$ and centered at the p_i. Fig. 10 shows λ_{max} as determined in phase spaces with embedding dimensions $D_E = 1, 2, \ldots, 9$, each space being defined by the phase variables v_j, $j = 1, 2, \ldots, D_E$. The phase spaces are constructed here analogously to the method proposed by Takens for purely time-dependent systems (see Ref. 47). In Takens's method, a phase space is constructed by values of a variable at equally displaced times $t, t + \tau, \ldots, t + k\tau$, while here a variable is evaluated at points equally displaced in space. The bars in Fig. 10 indicate the ranges within which λ_{max} oscillates after convergence in time. An example of this convergence is shown in Fig. 11. Within the error given by the bar lengths (Fig. 10) λ_{max} was independent of the parameter EVOLV (see Ref. 46) for $13 \leq \text{EVOLV} \leq 19$ iterations. The fact that λ_{max} resulting from these simulations is approximately independent of D_E for $4 \leq D_E \leq 9$ and significantly greater than zero indicates that these spatio-temporal patterns are indeed chaotic.

In the example shown in Fig. 8, n is reset to the constant value $n = 6$ at $t \geq 341$, i.e., right after Fig. 8d. This leads to self-organization of the system into spirals (Fig. 8f). In order to quantify organized structures, as compared to spatiotemporal aperiodicities, the minimum autodifference

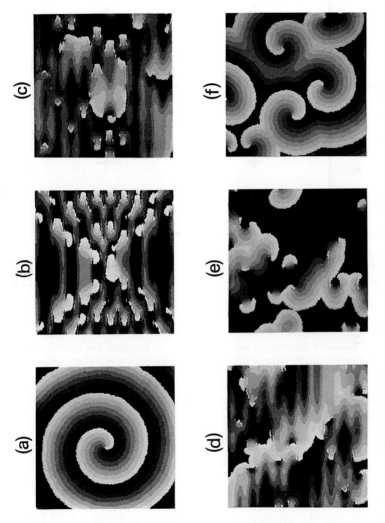

Fig. 8. Simulated transition from a spiral wave (a) to chaos (c,d) and autonomous reorganization into spirals (f). 200×200 cells, $r = 6$, $m = 1$, $p = S_{max} = 0$. (a) $t = 1$, $n = 6$ everywhere. During the time interval $2 \leq t \leq 340$, including (b) at $t = 18$, (c) at $t = 73$, and (d) at $t = 340$, n depends on the ordinate, as shown in Fig. 9, with $n_1 = 4$, $n_2 = 20$, and $\delta = 20d$. For $t \geq 341$, including (e) at $t = 344$ and (f) at $t = 389$, n is reset to $n = 6$ everywhere.

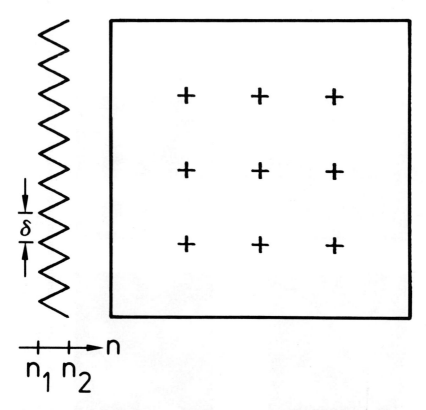

Fig. 9. Left: space dependence of the refractory time n for the induction of chaos. Right: border of the medium and points $p_1, \ldots, p_9$ (crosses). The phase variables are defined as the averages of the state S in circles with radius 9d and centered at the p_i.

function D_{min}[24] is used in the present work. The autodifference function is given here by

$$D(\Delta) = \frac{1}{R(\Delta)} \sum_{i=1}^{N} \sum_{j=1}^{N} |S(\vec{x}_i) - S(\vec{x}_j)|,$$

$$|\vec{x}_i - \vec{x}_j| = \Delta$$

where the $\vec{x}_i$ are the positions of the automaton points, $S(\vec{x}_i)$ their states, Δ the distance between two points, N the number of points in the medium, and $R(\Delta)$ the total number of point pairs with distance Δ. For

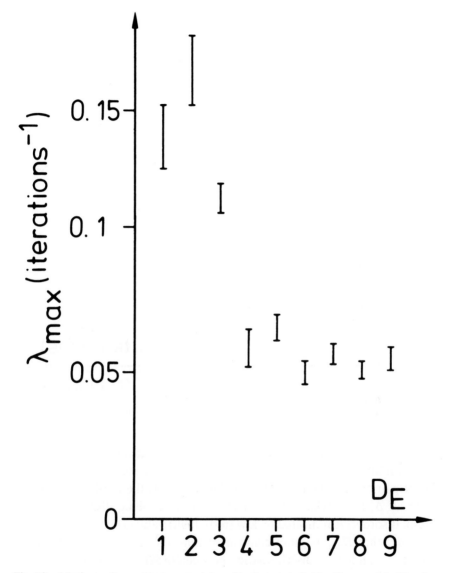

Fig. 10. Maximum Lyapunov exponent λ_{max} for the aperiodicities illustrated in Figs. 8c and 8d, versus the embedding dimension D_E.

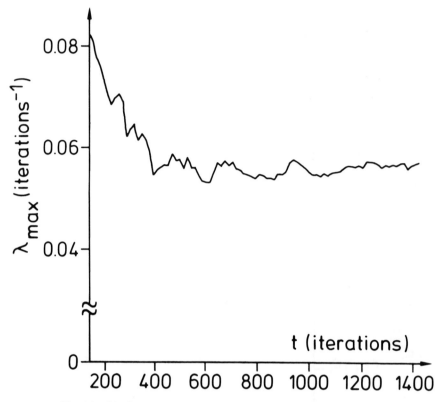

Fig. 11. Maximum Lyapunov exponent λ_{max} versus time for $D_E = 8$.

small Δ, R increases with increasing Δ. As Δ reaches the order of the medium size, $R(\Delta)$ starts to decrease because an increasing number of point pairs do not lie inside the medium. D_{min} is the minimum of $D(\Delta)$ over all Δ, the trivial minimum $D = 0$ at $\Delta = 0$ being discarded. Order (resp. disorder) is indicated by low (resp. high) values of D_{min}. It has been shown[24] that D_{min} is more convenient than the maximum autocorrelation function for the discrimination between order and disorder. Fig. 12a shows D_{min} as a function of time for the example illustrated in Fig. 8. The arrow marked "dis" indicates the time after which spatial periodicity of n is assumed in the calculations; the arrow "org" indicates the time at which n is reset to a constant value.

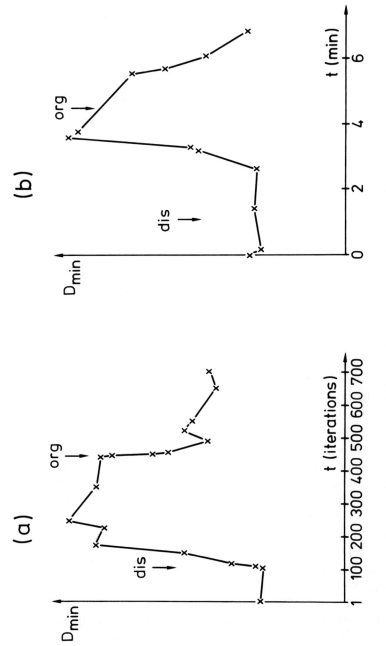

Fig. 12. (a) Minimum autodifference function D_{min}^{24} for the simulations in Fig. 8. (b) Same function from experiments with the Belousov-Zhabotinskii reaction. For both (a) and (b) the arrows marked "dis" and "org" indicate the times at which the systems are driven into disorder (aperiodicities) and order (periodicity). (b) is reprinted by permission from *Biological Cybernetics* **57** (1987) 187–195, courtesy of Springer-Verlag, Heidelberg

3. Experiments

Periodicity, disorganization, and self-organization of waves in an excitable medium, as simulated in the last section, can be observed in the Belousov-Zhabotinskii reaction.[24]

Periodic (ordered) waves obtained experimentally in the BZ reagent can be matched quantitatively with the automaton simulations by a proper choice of the parameters. In fact, setting $r = 5$, $n = 9$, $m_o = 2$, $p = 2$, and $S_{max} = 8$, one obtains a spiral period of 10.5 iterations (the spiral turns twice during 21 iterations) and a spiral propagation velocity of 3.8 cell lengths/iteration. Comparing these values with a measured spiral period (17.3 sec) and velocity (76 μm/sec)[48] yields: 1 cell length = 33 μm and 1 iteration = 1.65 sec. Using these unit transformations, other results of simulations can be compared with other measurements given in Ref. 48. For example, the velocity of planar waves resulting from these simulations is $c = 4.25$ cell lengths/iteration. This corresponds to 85 μm/sec, in excellent agreement with the experimental value 82.5 μm/sec. The critical radius ρ_{crit} (radius of the smallest circle capable of initiating a wave) obtained from these simulations is 0.8 cell lengths, corresponding to 26.4 μm, as compared with the experimental value of 23.1 μm. (Note: noninteger lengths obtained from the automaton are possible because averages are taken over the randomly distributed points.) The diffusion coefficient (obtained as the absolute value of the slope of the normal wavefront velocity versus curvature K, for $K \to 0^2$) resulting from these simulations is 3.4 (cell lengths)^2/iteration = 2.24×10^{-5} cm^2/sec, which compares well with the experimental value 2×10^{-5} cm^2/sec.[48]

Measurements of periodic waves in the BZ reagent (an example is the spiral wave shown in Fig. 13a) must be performed in such a way that convective motion is prevented. In a thin layer of reagent in a Petri dish this is accomplished by covering the dish with a glass plate (leaving an air gap of a few mm). In the experiments shown in Figs. 13b to 13d, convective motion is evoked by temperature gradients caused by evaporation after the dish is uncovered. Here, the coupling of periodic waves and periodic convective rolls yields aperiodic patterns. (This disorganization process was roughly simulated with the automaton yielding Figs. 8b to 8d.) When the dish is covered again after the aperiodicities have evolved, self-organization into spirals occurs

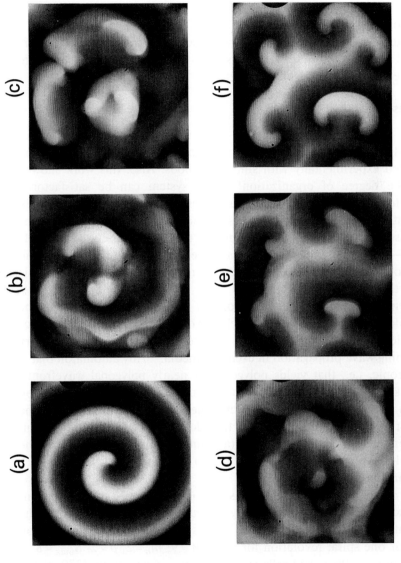

Fig. 13. Transitions from a spiral wave (a) to chaos (c,d) and autonomous reorganization into spirals (f) as measured in a thin layer of the Belousov-Zhabotinskii reagent.[24] Reprinted by permission from *Biological Cybernetics* **57** (1987) 187–195, courtesy of Springer-Verlag, Heidelberg.

(Figs. 13e and 13f), as in simulations when the homogeneity of n is restored (Figs. 8e and 8f). Note the "spiral buds" occurring at the onset of the organization process in Fig. 13e, similar to those in the experiments (Fig. 8e). Fig. 12b shows the minimum autodifference function D_{min} for the experiment shown in Fig. 13. The arrows "dis" (resp. "org") indicate the time at which the dish is uncovered (resp. covered) again. Fig. 12 shows that in experiments (b) the relaxation from order to disorder and vice versa takes longer than in simulations (a).

4. Discussion and Outlook

The present work has shown that the complex phenomena of spatio-temporal chaos and self-organization can be both simulated and observed experimentally with simple methods. Also, this work showed that the degree of organization of the system can be satisfactorily characterized by the minimum autodifference function D_{min}, both in computations and in measurements. It should be kept in mind that the processes investigated here are common to a large variety of excitable media—chemical, biological, physicochemical, epidemiological, and astronomical—in spite of the differences in their underlying microscopic properties.[1-3]

Solving the problem of anisotropy is essential for considering the dependence of normal wavefront-velocity on curvature. This dependence (so-called "eikonal relationship") is an essential ingredient of any theory describing wave propagation in excitable media.[22] Any automaton that is anisotropic, i.e., that yields polygonal or nearly polygonal wavefronts, will display a dependence of curvature on the direction of propagation and is thus not suitable to describe the eikonal relationship. In the present work, isotropy allowed the determination of a direction-independent diffusion coefficient (absolute value of the slope of the eikonal relationship at curvature $K \rightarrow O$) in excellent agreement with experiments.

The automaton presented here is not only in quantitative agreement with experiments using the BZ reagent, but is also much faster than the integration of the corresponding PDEs. As an example, the computation time for one spiral revolution for $r = 5$, $m = 9$, $m_o = 2$, $p = 2$, $S_{max} = 8$ in a medium of 100 × 100 cells is 5.2 min on a Concurrent 3230 (scalar microcomputer). A comparable PDE integration takes 160 min on the same computer with the two-dimensional version of the model and algorithm used in Ref. 49. Thus, the automaton yields a considerable computing acceleration, which is mainly due to the much shorter time steps required for PDE integration.

For three-dimensional systems the relative gain in computing time is even higher when using the automaton, as compared to PDEs. Therefore, the present model may be used in the future as an efficient tool for answering open questions regarding instabilities of different scroll waves. Also, the automaton may be helpful in finding optimal defibrillating oscillations to terminate reentrant disrhythmias in thick (three-dimensional) heart muscle. Furthermore, anisotropic media—occurring in a number of biological systems—may be easily simulated by introducing on purpose the corresponding anisotropy in the automaton.

Acknowledgements

I thank the Commission of the European Communities (Directorate-General for Science, Research and Development, Brussels) for financial support. Also, I thank Manfred Krafczyk for computing assistance and Gesine Schulte for the photographic work.

References

1. V. S. Zykov, *Simulations of Wave Processes in Excitable Media,* Manchester University Press (1988).
2. J. J. Tyson and J. P. Keener, Singular perturbation theory of travelling waves in excitable media, *Physica* **D32** (1988) 327–361.
3. A. V. Holden, M. Markus, and H. G. Othmer (eds.), *Nonlinear Wave Processes in Excitable Media,* Plenum Press, London (1991).
4. A. L. Hodgkin and A. F. Huxley, A quantitative description of membrane current and its application to conduction and excitation in nerve, *J. Physiol.* **117** (1952) 500–544.
5. N. A. Gorelova and J. Bures, Spiral waves of spreading depression in the isolated chicken retina, *J. Neurobiol.* **14** (1983) 353–363.
6. K. Hara, P. Tydeman, and M. Kirschner, Cytoplasmic clock with same period as division cycle in Xenopus eggs, *Proc. Natl. Acad. Sci. USA* **77** (1980) 462–466.
7. V. I. Koroleva and J. Bures, Circulation of cortical spreading around electrically stimulated areas and epileptic foci on the neocortex of rats, *Brain Res.* **173** (1979) 209–215.
8. N. Wiener and A. Rosenbluth, The mathematical formulation of the problem of conduction of impulses in a network of connected excitable elements, specifically in cardiac muscle, *Arch. Inst. Cardiol. Mex* **16** (1946) 205–265.

9. M. A. Allessie, F. I. M. Bonke, and F. J. G. Schopman, Circus movement in rabbit atrial muscle as a mechanism of tachycardia, *Circ. Res.* **33** (1973) 54–62; II. The role of nonuniform recovery of excitability, *Circ Res.* **39** (1976) 168–177; III. The "leading circle" concept, *Circ. Res.* **41** (1977) 9–18.
10. G. Gerisch, Periodische Signale steuern die Musterbildung in Zellverbänden, *Naturwissenschaften* **58** (1971) 430–438.
11. A. J. Durston, Dictyostelium discoideum aggregation fields as excitable media, *J. Theor. Biol.* **42** (1973) 483–504.
12. K. J. Tomchik and P. N. Devreotes, Adenosine 3′, 5′-monophosphate waves in Dictyostelium discoideum: a demonstration by isotope dilution-fluorography, *Science* **212** (1981) 443–446.
13. F. Siegert and C. Weijer, Digital image processing of optical density wave propagation in Dictyostelium discoideum and analysis of the effects of caffeine and ammonia, *J. Cell. Sci.* **93** (1989) 325–335.
14. P. Foerster, S. C. Müller, and B. Hess, Curvature and spiral geometry in aggregation patterns of Dictyostelium discoideum, *Development* **109** (1990) 11–16.
15. A. B. Carey, R. H. Giles, and R. G. MacLean, The landscape epidemiology of rabies in Virginia, *Am J. Trop. Med. Hyg.* **27** (1978) 573–580.
16. J. D. Murray, E. A. Stanley, and D. L. Brown, On the spatial spread of rabies among foxes, *Proc. Roy. Soc. Lond.* **B 229** (1986) 111–150.
17. J. D. Murray, Modeling the spread of rabies, *Am. Scientist* **75** (1987) 280–284.
18. A. W. M. Dress, M. Gerhardt, and H. Schuster, Cellular automata simulating the evolution of structure through the synchronization of Socillators, in *From Chemical to Biological Organization* (pp. 134–145), M. Markus, S. C. Müller, and G. Nicolis, eds. Springer, Heidelberg (1988).
19. K. I. Agladze and V. I. Krinsky, Multi-armed vortices in an active chemical medium, *Nature* **296** (1982) 424–426.
20. B. J. Welsh, J. Gomatam, and A. E. Burgess, Three-dimensional chemical waves in the Belousov-Zhabotinskii reaction, *Nature* **304** (1983) 611–614.
21. A. T. Winfree and S. H. Strogatz, Organizing centres for three-dimensional chemical wavefronts, *Nature* **311** (1984) 611–615.
22. J. P. Keener and J. J. Tyson, Spiral waves in the Belousov-Zhabotinskii reaction, *Physica* **21D** (1986) 307–324.
23. S. C. Müller, Th. Plesser, and B. Hess, Two-dimensional spectrophotometry of spiral wave formation in the Belousov-Zhabotinskii reaction, *Physica* **24D** (1987) 71–86.
24. M. Markus, S. C. Müller, Th. Plesser, and B. Hess, On the recognition of order and disorder, *Biol. Cybern.* **57** (1987) 187–195.
25. B. F. Madore and W. L. Freedman, Computer simulation of the Belousov-Zhabotinskii reaction, *Science* **222** (1983) 615–616.
26. L. S. Schulman and P. E. Seiden, Percolation and galaxies, *Science* **233** (1986) 425–431.
27. J. Feitzinger, Spiralarm waves in galaxies, in *Nonlinear Wave Processes in Excitable Media* (pp. 351–360), A. V. Holden, M. Markus, and H. G. Othmer, eds., Plenum Press, London (1991).

28. P. J. Nandapurkar and A. T. Winfree, A computational study of twisted linked scroll waves in excitable media, *Physica* **29D** (1987) 69–83.
29. P. J. Nandapurkar and A. T. Winfree, Dynamical stability of untwisted scroll rings in excitable media, *Physica* **D35** (1989) 277–288.
30. A. T. Winfree, E. M. Winfree, and H. Seifert, Organizing centers in a cellular excitable medium, *Physica* **17D** (1985) 109–115.
31. A. V. Panfilov and A. T. Winfree, Dynamical simulation of twisted scroll rings in three-dimensional excitable media, *Physica* **17D** (1985) 323–330.
32. V. S. Zykov and A. S. Mikhailov, Rotating spiral waves in a simple model of an excitable medium, *Sov. Phys. Doklady* **31** (1988) 51–52.
33. B. Dorjsurangiyn, *Autowave Processes in Active Media Formed by Cellular Automata*, Diploma Thesis, M. V. Lomonosov Moscow State Univ., Dept. of Physics (1987).
34. M. Gerhardt and H. Schuster, A cellular automaton describing the formation of spatially ordered structures in chemical systems, *Physica* **D36** (1989) 209–221.
35. M. Gerhardt, H. Schuster, and J. J. Tyson, A cellular automaton model of excitable media including curvature and dispersion, *Science* **247** (1990) 1563–1566.
36. R. Penrose, Pentaplexity, *Math. Intelligencer* **2** (1979) 32–37.
37. D. R. Nelson and B. I. Halperin, Pentagonal and icosahedral order in rapidly cooled metals, *Science* **229** (1985) 233–238.
38. M. Markus and B. Hess, An automaton with randomly distributed cells for the simulation of waves in excitable media, *Abstract-book of the 16th Aharon-Katchalsky Conf. on Dynamics in Molecular and Cellular Biology* pp. 37–38, Université Libre de Bruxelles, Brussels (1988).
39. M. Markus, Modelling morphogenetic processes in excitable tissues using novel cellular automata, *Biomedica Biochimica Acta*, **49** (1990) 681–696.
40. A. T. Winfree, Scroll-shaped waves of chemical activity in three dimensions, *Science* **181** (1973) 937–939.
41. A. V. Panfilov, Three dimensional vortices in active media, in *Nonlinear Wave Processes in Excitable Media* (pp. 361–381), A. V. Holden, M. Markus, and H. G. Othmer, eds. Plenum Press, London (1991).
42. A. T. Winfree, Wavefront geometry in excitable media, *Physica* **12D** (1984) 321–332.
43. A. B. Medvinsky, A. V. Panfilov, and A. M. Pertsov, Properties of rotating waves in three dimensions. Scroll waves in myocard, in *Self-Organization. Autowaves and Structures Far From Equilibrium* (pp. 195–199), V. I. Krinsky, ed. Springer-Verlag, Berlin (1984).
44. M. Markus and B. Hess, Transitions between oscillatory modes in a glycolytic model system, *Proc. Natl. Acad. Sci. USA* **81** (1984) 4394–4398.
45. M. Markus, D. Kuschmitz, and B. Hess, Properties of strange attractors in yeast glycolysis, *Biophys. Chem.* **22** (1985) 95–105.
46. A. Wolf, J. B. Swift, H. G. Swinney, and J. A. Vastano, Determining Lyapunov exponents from a time series, *Physica* **16D** (1985) 285–317.
47. H. G. Schuster, *Deterministic Chaos: An Introduction* (p. 107), Physik-Verlag, Weinheim (1984).

48. P. Foerster, S. C. Müller, and B. Hess, Critical size and curvature of wave formation in an excitable chemical medium, *Proc. Natl. Acad. Sci. USA* **86** (1989) 6831-6834.
49. Zs. Nagy-Ungvarai, J. J. Tyson, and B. Hess, Experimental study of the chemical waves in the cerium-catalyzed Belousov-Zhabotinskii reaction 1. Velocity of trigger waves, *J. Phys. Chem.* **93** (1989) 707-713.
50. P. Foerster, S. C. Müller, and B. Hess, Curvature and propagation velocity of chemical waves, *Science* **241** (1988) 685-687.

BROKEN SYMMETRY AND THE FORMATION OF SPIRAL PATTERNS IN FLUIDS

Ian Stewart

1. Introduction

The Couette-Taylor experiment is one of the most celebrated laboratory systems in fluid dynamics. The original apparatus consists of two coaxial cylinders; fluid is introduced between them, and the inner cylinder is rotated at speed Ω_i. Nowadays it is common to rotate the outer cylinder independently as well, at speed Ω_o (Fig. 1). The system was originally invented around 1890 by M. M. Couette[6] to study shear flows. In 1923, G. I. Taylor[26] discovered that if the inner cylinder is speeded up, then the fluid stratifies into vortex layers, rather like a stack of doughnuts or a pile of automobile tyres. In Tayor's original experiment the outer cylinder was not free to rotate, so $\Omega_o = 0$, and Ω_i was increased slowly from rest. In the initial state, *Couette flow* (Fig. 2a), the fluid rotates in a uniform shear flow without noticeable pattern. At a critical value of Ω_i this flow breaks up into vortex pairs known as *Taylor vortices* (Fig. 2b). At higher values of Ω_i the Taylor vortices develop a periodic oscillation to *wavy vortices* (Fig. 2c). The pattern is a rotating wave and it rotates as if rigid. Then a second frequency sets in, leading to *modulated wavy vortices* that behave rather like horses on a fairground carousel: the pattern itself varies in time. Beyond this, the flow becomes turbulent. This bifurcation scenario is the main sequence:

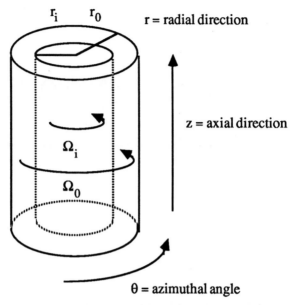

Fig. 1. The Taylor apparatus. Here Ω_i and Ω_o denote the angular velocities of the inner and outer cylinders, and r_i, r_o denote their radii.

$$\text{Couette flow} \\ \downarrow \\ \text{Taylor vortices} \\ \downarrow \\ \text{wavy vortices} \\ \downarrow \\ \text{modulated wavy vortices} \\ \downarrow \\ \text{chaos.}$$

In contrast, if the outer cylinder is initially counterrotated ($\Omega_i \ll 0$) then the first instability from Couette flow is to *spirals* (Fig. 2e), a helical pattern resembling a rotating barber's pole. This may be followed by *interpenetrating spirals* (Fig. 2f) and *wavy interpenetrating spirals*. Other states may occur depending on the precise dimensions of the apparatus. The Couette-Taylor system has been studied intensely ever since, and an enormous variety of different flow-patterns have been observed (see Di Prima and Swinney[10] and Andereck et al.[1]). In addition to the states mentioned so far, there are other kinds of helical spirals, wavy spirals (Fig. 2g), modulated wavy spirals, twisted vortices (Fig. 2d), braided vortices, wavy inflow and wavy outflow boundaries, and various forms of turbulent flow including at least two types (spiral turbulence and turbulent Taylor vortices) that have some kind of symmetry "on the average." Different flows occur for different combinations of the two speeds, and an experimental "phase diagram" of the appropriate parameter space, with the speeds nondimensionalized as Reynolds numbers R_i, R_o, is shown in Fig. 3.

This is already very rich behaviour, but a complete experimental picture is actually far more complicated, because several different flow-patterns may be stable at identical values of (R_o, R_i), and the pattern that occurs is selected by the system's past history and not just by the two speeds alone. For example, the order in which the two cylinders are speeded up can make a difference. There are thus hysteretic effects: the positions of the transition lines drawn in Fig. 3 would change if the apparatus were to follow a different route to the same pair of speeds.

The Couette-Taylor system poses an irresistible challenge to the theorist, a challenge that is not easy to meet, given the nonlinear nature of the Navier-Stokes equations of fluid dynamics and the complicated variety of flow-patterns known to occur. The more detail required, the harder the analysis becomes and much of it has to be numerical rather than analytical.

However, a considerable degree of unification has been achieved by first seeking an overall qualitative skeleton for the pattern-formation process, and then fleshing it out with quantitative detail. In this way it has become clear that virtually all of the nonturbulent states, and possibly even the "patterned" turbulent ones, are primarily controlled by the *symmetries* of the Couette-Taylor system (Chossat and Iooss,[5] Golubitsky and Stewart[14]). The physics of fluid flow is, in a sense, of secondary importance. Indeed, if we interpret "symmetry" as specifying not just a symmetry group but also a representation of that group on the appropriate space of modes, then the observed flow-patterns are the typical states that should be possible in *any* system with the same cylindrical symmetries. The detailed equations of fluid dynamics determine which of this range of patterns occurs under which conditions. Thus the symmetries determine a fixed "zoo" of possible patterns; the physics of a given system having those symmetries then selects the appropriate creature from the zoo.

This approach to the dynamics of symmetric systems is widely applicable: each type of symmetry has its own particular zoo, and an

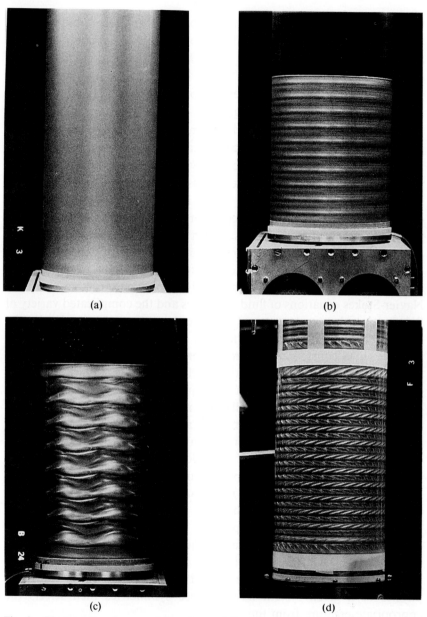

Fig. 2. Photographs of some of the observed flows in the Couette-Taylor experiment: (a) Couette flow, (b) Taylor vortices, (c) wavy vortices, (d) twisted vortices. (Courtesy of David Andereck and Harry Swinney.)

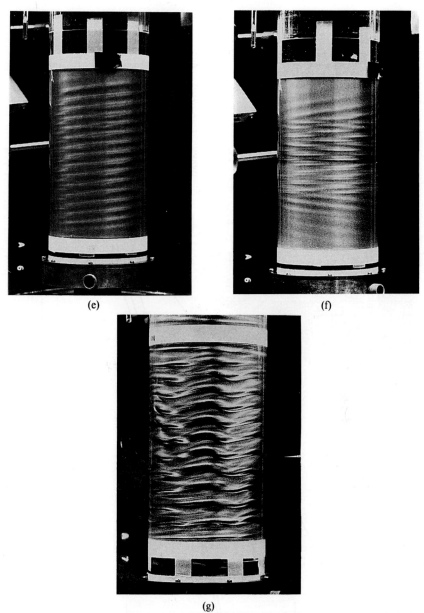

Fig. 2. (contd.) Photographs of some of the observed flows in the **Couette-Taylor** experiment: (e) helical spirals, (f) interpenetrating spirals, (g) wavy spirals. (Courtesy of David Andereck and Harry Swinney.)

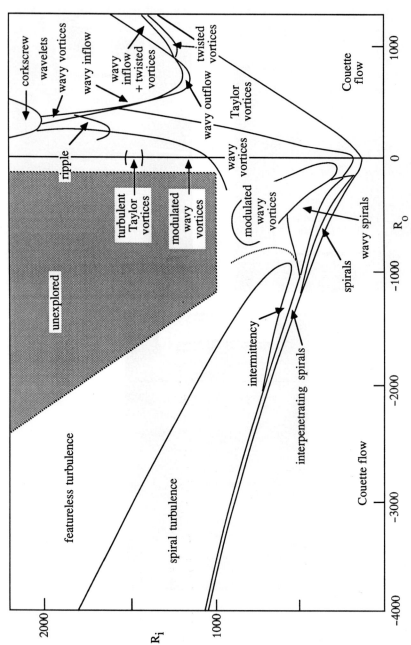

Fig. 3. Experimental "phase diagram" for flow-patterns in the Couette-Taylor experiment. (After Andreck et al.[1])

abstract understanding of the zoo often provides a head start in analysing any specific system's quantitative features. This point is developed below. In the case of Couette-Taylor flow, work that initially concentrated on abstract and qualitative questions about symmetry-breaking has led to new quantitative analysis of the Navier-Stokes equations and the successful experimental verification of new predictions.

It is of course the spiral flows of the Couette-Taylor system that are most appropriate for the present volume. However, as we shall see later, most of the other patterned states can be effectively represented as nonlinear combinations of spiral flows, together with Taylor vortices, which can be thought of as a rather degenerate type of spiral. In short, pattern-formation in Couette-Taylor flow is basically a matter of spirals.

Indeed, in a *linear* system with the appropriate symmetries, the possible flows are just superpositions of spiral waves. However, the Navier-Stokes equations for fluid flow are nonlinear. The surprising thing is that many traces of the linear picture remain, at least in appropriate ranges of parameters. Instead of superpositions of spirals, we find nonlinear interactions that approximate superpositions. However, not all such "nonlinear superpositions" are possible. Those that *are* possible are determined by the way that the symmetry of the system interacts with the nonlinear terms in the equations, that is, by invariant theory.

Two distinct mathematical ideas go into a symmetry-based analysis of the Couette-Taylor system. The first is *spontaneous symmetry-breaking*, a phenomenon whereby solutions to a system of equations may possess less symmetry than the equations themselves. The second is the concept of an *organizing centre*, a set of parameter values near which the interesting dynamics occurs "in microcosm." We discuss them in general terms before embarking on the resulting mathematical analysis.

2. Spontaneous Symmetry-Breaking

In everyday language the term "symmetry" is used in two distinct ways. The first is rather vague, something along the lines of "elegant proportions." The second is more specific, referring to a repetitive feature of a shape. It is this second meaning that is of mathematical interest. The symmetries of an object or system, in this sense, are determined by its *symmetry group*, the set of all transformations that leave its form unchanged. This paper will assume a general familiarity with group theory, although much of it can be followed without specialist knowledge.

Here is a nontechnological example, taken from Stewart.[22] In northern California huge trees, redwoods and sequoias, grow. The trunk of a tree is approximately cylindrical and it thus has a good approximation to cylindrical symmetry. The symmetries of a cylinder are of three kinds: rotations about the axis, translations along the axis, and reflections in planes through the axis, plus the mid-plane perpendicular to it (Fig. 4).

To be precise, translational symmetry holds only for an infinitely long cylinder; but it is valid to a good approximation for a sufficiently long one. It seems plausible that the pattern of bark on the tree should have similar symmetry to the tree itself. Now a pattern of bark with (a good approximation to) full cylindrical symmetry will have to look essentially the same (ignoring fine local detail) after any of those rotations, translations, and reflections. That means that the grooves in the bark should run roughly vertically, and "on average" every piece of bark should look like every other (Fig. 5a). Indeed, this is the common pattern.

However, the bark of some individual Californian trees has a *spiral* pattern (Fig. 5b). The spiral still has *some* symmetry, but of a different kind. If you rotate a helical spiral, *and* simultaneously translate it by an appropriate amount, it looks the same, so the symmetry of a spiral is a

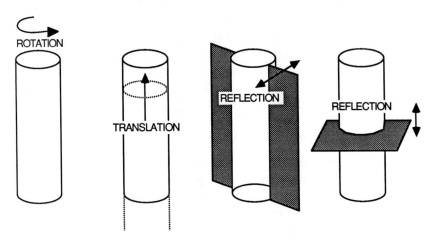

Fig. 4. The symmetries of an infinite cylinder, and ideal approximation to those of a sufficiently long tree trunk.

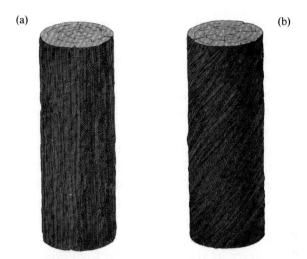

Fig. 5. Two distinct patterns of bark: (a) a pattern having the same symmetry "on the average" as the entire tree trunk, and (b) a spiral pattern that breaks the cylindrical symmetry but still has a great deal of symmetry left.

mixture of rotation and translation, known as a *screw*.

Did something strange happen to those trees that developed spiral bark? Pesticides, a bad winter, drought? Or should we *expect* spiral patterns as well as perfectly symmetric ones? To approach such problems we must answer a fundamental general question: how does the symmetry of a system affect its behaviour? In 1894, Pierre Curie[7] gave two logically equivalent statements of a general principle from the folklore of mathematical physics:

1. If certain causes produce certain effects, then the symmetries of the causes reappear in the effects produced.
2. If certain effects reveal a certain asymmetry, this asymmetry will be reflected in the causes which give rise to them.

Curie's principles are often assumed by scientists, usually unconsciously or at least subconsciously, in the course of their work. For example, in the Kensington Science Museum in London, there is an engineering model of a passenger jet, used in a wind tunnel to study the flow of air around the aircraft. Since the aircraft is bilaterally symmetric, the engineers only built half of the model, tacitly assuming that the air flow had to be bilaterally symmetric as well. Are such assumptions justified?

At first sight, Curie's principles are obviously true. If a planet in the shape of a perfect sphere acquires an ocean, we expect that ocean to be of uniform depth. If the planet rotates, then we expect the ocean to bulge at the equator, but retain circular symmetry about the axis of rotation. What else could the airflow be about a bilaterally symmetric aircraft except bilaterally symmetric?

Curie's principles may seem obvious, but they must be interpreted *very* carefully indeed, for in fact there are many symmetric systems whose behaviour has less symmetry than the system itself. The Couette-Taylor experiment is an example: the apparatus is cylindrical, but most flows fail to have full cylindrical symmetry. Similarly, if a perfect cylindrical shell is compressed by a sufficiently large axial force, then it will buckle into a shape that, whatever its symmetries, is far from cylindrical (Fig. 6). A computer picture of a spherical shell buckled by a spherically symmetric

Fig. 6. When a perfect cylinder (coke can) is subjected to a perfect uniform axial force (large hammer) it buckles into a pattern that breaks the cylindrical symmetry of the original can. More precise experiments also exhibit symmetry-breaking.

compressive force is shown in Fig. 7: the symmetry of the buckled state is circular rather than spherical.

The loss of symmetry in these systems is *not* merely a consequence of small imperfections: asymmetric solutions exist even in an idealized perfectly symmetric mathematical system. Indeed the perfect system largely controls *how* symmetries can break. However, imperfections play an important role in selecting exactly *where*. For example, when the sphere in Fig. 7 buckles, then for a perfect system the axis of circular symmetry can be *any* axis of the original sphere; for an imperfect system

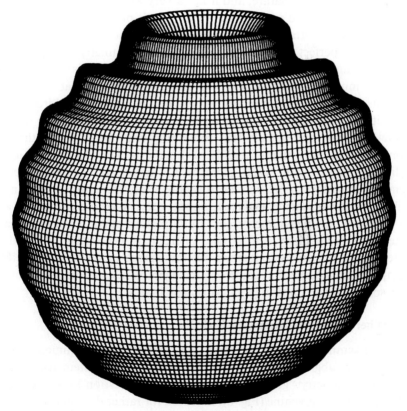

Fig. 7. Computer graphic of the deformation of a perfect sphere under a uniform radial compressive force. The spherical symmetry breaks but circular symmetry remains. (Graphic by R.J. Thompson. Reprinted with permission from J.M.T. Thompson, *Instabilities and Catastrophes in Science and Engineering*, Wiley, Chichester, 1982.)

some axes will be preferred, their positions being related to weaknesses in the spherical shell. The general form of the buckled sphere, however, will be the same in both cases. In this sense, Curie's principles are perhaps valid for an actual physical system (which is necessarily imperfect), but not for an idealized model. Rather than attempting to resurrect Curie's principles in this fashion, however, it seems preferable to understand the mechanism by which perfect idealized symmetric systems produce behaviour with less symmetry. This is called *spontaneous symmetry-breaking*. It seems to be responsible for many types of pattern-formation in nature, and it has a very well-defined mathematical structure that can be used to understand such processes. The adjective "spontaneous" distinguishes these effects from cases in which the whole system is imperfect and has only approximate symmetry to begin with.

What causes the symmetry to break? The answer is that natural systems must be *stable*. Curie was right in asserting that symmetric systems should have symmetric states, but he failed to address their stability. If a symmetric state becomes unstable, then the system will do something else, and usually that something else cannot retain the full symmetry. In particular, there is nothing especially surprising about cylindrical trees with spiral patterns to their bark. If the perfectly symmetric pattern represents an unstable state, then tiny disturbances will cause the symmetry to break. Spirals are one of the common ways to break cylindrical symmetry, so a spiral pattern might develop instead. The biology, chemistry, and physics of trees are necessarily involved in determining the precise outcome; but the change in symmetry does not compel us to assume that some outside agent is responsible. It might be just some strange quirk of that particular tree.

3. Where Does the Symmetry Go?

When symmetry breaks, we observe a solution whose symmetry group is a proper subgroup of the symmetry group of the system itself. How does the symmetry "get lost"? We answer this question by an example. The catastrophe machine (Fig. 8), invented for rather different reasons by E.C. Zeeman in 1969, shows that symmetry is not so much *broken* as *spread around*. You can make one and experiment with it. For an analysis, see Zeeman,[29] Poston and Woodcock,[21] Woodcock and Poston,[28] or Poston and Stewart.[20]

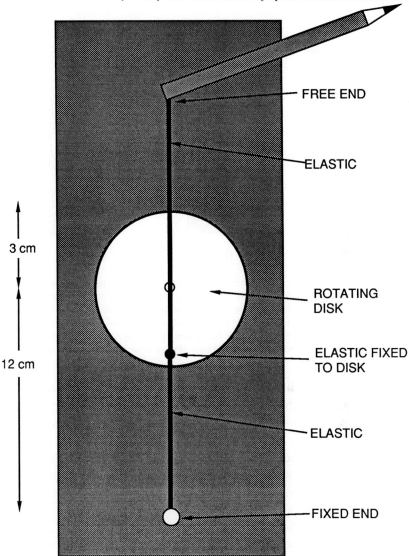

Fig. 8. The Zeeman catastrophe machine and its axis of symmetry. To construct the machine, attach a circular disk of thick card, of radius 3 cm, to a board using a thumb-tack and a paper washer. Fix another drawing-pin near the rim of the disk with its point *upward*. To this pin attach two elastic bands, of about 6 cm unstretched length. Fix one to a point 12 cm from the centre of the disk, and leave the end of the other free to move along the centre line as shown, for example by taping it to a pencil which can be moved by hand.

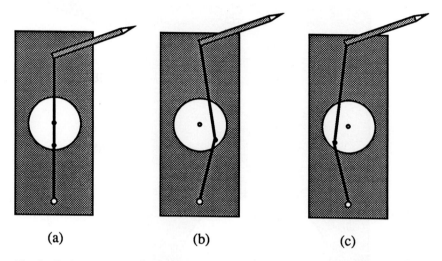

Fig. 9. Broken symmetry in the Zeeman machine. If the free elastic is stretched slightly then the catastrophe machine takes up a symmetric position, on the axis (a). If the elastic is stretched further, then despite the bilateral symmetry, the disc rotates off the axis (b). However, there is a second, symmetrically related state (c). The individual solutions (b, c) break the symmetry, but together they form a symmetric pair. Thus the symmetry is spread over several solutions rather than concentrated on one.

The entire system has reflectional symmetry about the centre-line. And if you begin to stretch the free elastic, keeping it on the centre-line, then you'll find that the system obeys Curie's principles and stays symmetric; that is, the disk does not rotate (Fig. 9a). But as you stretch the elastic further, the disk suddenly begins to turn—maybe clockwise, maybe anticlockwise (Fig. 9b). Now the state of the system fails to have reflectional symmetry. The symmetry has broken, and Curie's principles have failed.

Where has the missing symmetry gone?

Hold the elastic steady and rotate the disk to the symmetrically placed position on the other side (Fig. 9c). You will find that it will remain there. Instead of a single symmetric state we have two *symmetrically* related states. In general, a system with a given symmetry group breaks symmetry to a smaller group, or *subgroup*. Moreover, the system can then exist in several states, each obtainable from the others by one of the symmetries of the full system.

For example, the buckled spherical shell of Fig. 7 breaks symmetry from spherical to circular: from the group of all rotations in three-dimensional space to its subgroup of rotations with a given axis, here visible as the axis of symmetry of the buckled sphere. But the symmetry is not lost completely: if you rotate the buckled sphere in any way whatsoever you get another possible way for the sphere to buckle. In principle, infinitely many distinct axes can occur; in practice, imperfections in the shell select one of these.

Moving the elastic off the symmetry axis of the Zeeman machine breaks the symmetry of the entire system, leading to new phenomena best described by catastrophe/singularity theory: see Zeeman[29,31] and Poston and Stewart.[20]

4. Bifurcations

The changes in behaviour of the Zeeman machine are an example of *bifurcation*. This term describes any qualitative change in the state of a system that occurs as some parameter is varied. Here we observe *steady-state bifucation*: the change of the parameter describing the position of the elastic creates two new steady states in a *pitchfork bifurcation* (Fig. 10).

Another type of symmetry-breaking is of great importance for many

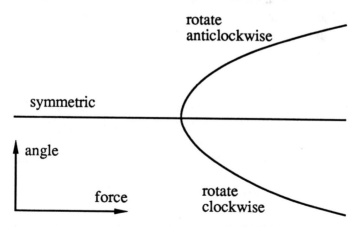

Fig. 10. The pitchfork bifurcation illustrated by means of a bifurcation diagram. As the position of the free end of the elastic in the Zeeman machine varies, two new branches of steady states bifurcate from the original branch.

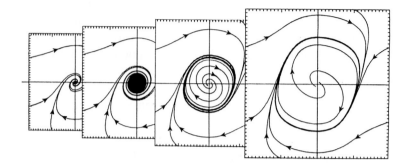

Fig. 11. Hopf bifurcation. The sequence of phase portraits shows how trajectories of a differential equation can create a periodic cycle from a steady state as a parameter varies continuously. Reprinted with permission from J.M.T. Thompson & H.B. Stewart, *Nonlinear Dynamics and Chaos*, Wiley, Chichester, 1986.

systems and is crucial to our understanding of the Couette-Taylor system. This is the breaking of symmetry in *time*. Steady states often bifurcate to periodic states, via a *Hopf bifurcation*. This is illustrated in Fig. 11. Any steady state looks exactly the same at all instants of time, but a periodic state does not. Thus Hopf bifurcation breaks the time-translation symmetry of a steady state. The time symmetry is not totally lost, however: periodic states look exactly the same when viewed at times that are integer multiples of the period. Thus in Hopf bifurcation the continuous temporal symmetry of a steady state breaks to give the discrete symmetry of a periodic one.

There are analogues of Hopf bifurcation in symmetric systems (see Golubitsky and Stewart[13] or Golubitsky et al.[15]). In most cases several distinct types of periodic solution bifurcate. For example, consider a hose of circular cross-section suspended vertically, nozzle downward, with water flowing steadily through it (Fig. 12). This system is circularly symmetric about an axis running vertically along the centre of the hose. And indeed, if the speed of the water is slow enough, the hose just remains in this vertical position, retaining its circular symmetry.

However, if the tap is turned on further, the hose can begin to wobble. In fact there are two distinct kinds of wobble. Whichever one occurs depends on the length and flexibility of the tubing. In one, it swings from side to side like a pendulum (Fig. 12a). In the other, it goes round and round, spraying water in a spiral (Fig. 12b). Similar effects are often observed when children wash the family car. These wobbles do not possess circular symmetry about a vertical axis: indeed they break it in two distinct ways.

The standing wave oscillation has spatial symmetry: if it is reflected in the plane of oscillation then it looks exactly the same. The rotating wave has a more interesting mixed spatio-temporal symmetry: if it is rotated

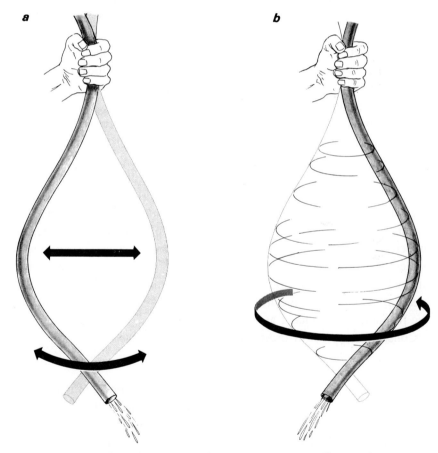

Fig. 12. Two ways for a circular hose to wobble: (a) a pendulum-like standing wave, (b) a rotating wave. Neither wobbles has the full cylindrical symmetry of the hose (from Stewart[22]).

in space the result is the same as a change of phase (shift of time-origin) in the periodic oscillation. That is, the combination of a spatial rotation and a temporal phase shift leaves it looking exactly the same. These *phase-shift symmetries* are common in the theory of symmetric Hopf bifurcation (see Golubitsky and Stewart[13]).

5. Mode Interactions and Organizing Centres

It is usually easier to analyse nonlinear systems in small regions near some interesting point, a method known as *local* analysis. In order to exploit this idea most effectively, the point should be chosen so that the greatest variety of dynamic behaviour occurs in its neighbourhood. Such a point is called an *organizing centre*, and it provides a local model that incorporates quasi-global information about dynamic bifurcations (Thom,[27] Zeeman[30]).

Our approach to the Couette-Taylor system involves choosing an organizing centre. The system has several variable parameters, of which the most significant are the Reynolds numbers R_i, R_o (nondimensionalised versions of the cylinder speeds Ω_i, Ω_o) and the radius ratio $\eta = r_i/r_o$. The idea is to vary these and to seek values at which the most degenerate singularities—the most complicated interactions—occur. The search for an organizing centre begins with experimental and numerical evidence for the presence of a *mode interaction*—a set of parameter values at which the system simultaneously becomes unstable to two distinct modes of behaviour. This is a "codimension 2" phenomenon, meaning that it can typically be found by varying just two parameters. We formalise the concept as follows.

Consider a system of ordinary differential equations

$$dx/dt + f(x,\lambda) = 0,$$

where $x = (x_1,...,x_n) \in \mathbb{R}^n$ and $\lambda \in \mathbb{R}$ is a bifurcation parameter. (For Couette-Taylor flow the traditional model is a partial differential equation, the Navier-Stokes equations, but standard reduction procedures lead to ordinary differential equations.) Two main types of local bifurcation can be distinguished, corresponding to distinct types of critical eigenvalue of the *Jacobian matrix*

$$(Df)x_0, \lambda_0 = \begin{pmatrix} \frac{\partial f_1}{\partial x_1} & \cdots & \frac{\partial f_1}{\partial x_n} \\ \frac{\partial f_n}{\partial x_1} & \cdots & \frac{\partial f_n}{\partial x_n} \end{pmatrix}_{x_0, \lambda_0}$$

A *critical eigenvalue* is one that lies on the imaginary axis, and it signals a change in stability of the trivial solution x_0 occurring at $\lambda = \lambda_0$. The two types of local bifurcation are:

(a) *Steady state.* $(Df)x_0, \lambda_0$ has a zero eigenvalue.
(b) *Hopf.* $(Df)x_0, \lambda_0$ has a conjugate complex pair of purely imaginary eigenvalues.

In steady-state bifurcation for asymmetric systems we obtain a new branch of steady-state solutions to the equations, bifurcating from the trivial branch $x = x_0$ at $\lambda = \lambda_0$. In the presence of symmetries, we may obtain several such branches (for example, in the pitchfork of Fig. 10, there are two.) In Hopf bifurcation a branch of periodic solutions bifurcates in the asymmetric case; in the presence of symmetry there may again be several such branches.

6. Multiple Eigenvalues Forced by Symmetry

We shall say that certain behaviour occurs *generically* if, first, it does occur, and second, any more complicated behaviour occurs only under extra unusual conditions. For example, if a point is chosen at random in a cartesian coordinate plane, then generically it will not lie on the x-axis. However, it can be forced to lie on the axis by specifying that it must be fixed by reflection in the x-axis, that is, that its y-coordinate must vanish. This is a simple example of how symmetry can cause nongeneric behaviour.

In systems without symmetry, critical eigenvalues are generically simple, that is, they have multiplicity 1 unless certain additional unusual conditions hold. But symmetry is itself unusual and when the system has symmetry — as is the case in Couette-Taylor flow — the eigenvalues may generically be forced to be multiple. In the world of symmetric systems, we must modify the notion of genericity appropriately: it should describe "typical behaviour *when symmetry is present.*"

The precise formulation of what is generic in symmetric steady-state or

Hopf bifurcation requires some ideas from representation theory. We sketch them here: see Golubitsky et al.[15] for the complete story. Let Γ be a compact Lie group acting linearly and orthogonally on a finite-dimensional real vector space **V**. Define the algebra of commuting mappings

$$\mathbf{D} = \{\text{linear maps } \varphi : \mathbf{V} \to \mathbf{V} | \varphi\gamma = \gamma\varphi \text{ for all } \gamma \in \Gamma\}.$$

If **V** is irreducible for Γ then the real version of Schur's Lemma shows that **D** is isomorphic to one of $\mathbb{R}$, $\mathbb{C}$, or $\mathbb{H}$, where these symbols denote the real numbers, complex numbers, and quaternions, respectively.

The analogue in a symmetric system of a simple eigenvalue is then as follows:

(a) *Steady-state.* The zero eigenspace E_0 of $(Df)x_0, \lambda_0$ is Γ-irreducible.
(b) *Hopf.* Let the imaginary eigenvalues be $\pm i\omega$. The real $\pm i\omega$ eigenspace $E_{i\omega}$ of $(Df)x_0, \lambda_0$ if Γ-*simple*, that is, either

$$E_{i\omega} \text{ is } \Gamma\text{-irreducible and the algebra } \mathbf{D} \simeq \mathbb{C} \text{ or } \mathbb{H},$$

or

$$E_{i\omega} \simeq W \oplus W \text{ where W is } \Gamma\text{-irreducible and } \mathbf{D} \simeq \mathbb{R}.$$

Here $\oplus$ is the direct sum symbol: $W \oplus W$ is the set of ordered pairs (w_1, w_2) with $w_1, w_2 \in W$, and more generally $U \oplus V$ is the set of ordered pairs (u, v) with $u \in U$ and $v \in V$. Associated to these eigenspaces are the corresponding solutions of the linearized equation $dx/dt + (Df)_{x_0, \lambda_0} x = 0$, which are called *eigenfunctions* or *modes*. The analogue of a multiple eigenvalue—a zero eigenspace that is not Γ-irreducible or an imaginary eigenspace this is not Γ-simple, or combinations of these—is thus a *mode interaction*.

6.1 *Example*

Suppose that $\Gamma = \mathbf{O}(2)$, the orthogonal group in $\mathbb{R}^2$ (the group of all symmetries of a circle). Consider the interaction of a steady-state mode and a Hopf mode. When $\Gamma = \mathbf{O}(2)$ we have $\mathbf{D} \simeq \mathbb{R}$ for all irreducible representations W. There are two distinct irreducibles W with dim $W = 1$ and all the rest have dim $W = 2$. The two-dimensional representations correspond to a choice of *mode number k* as follows. Identify $\mathbb{R}^2$ with the complex plane $\mathbb{C}$ and let $\mathbf{O}(2)$ act on $z \in \mathbb{C}$ by

$$\theta.z = e^{ki\theta}z, \qquad \kappa.z = \bar{z}.$$

Here θ is a rotation in $O(2)$ and κ is a reflection.

First, consider the standard representation U of $O(2)$ on $\mathbb{R}^2$, for which $k = 1$. The corresponding steady-state mode is U, the Hopf mode is $U \oplus U$, and the corresponding mode-interaction takes place on $U \oplus (U \oplus U) = U^3$. Thus there is a six-dimensional critical eigenspace.

A full catalogue of the distinct interactions between an $O(2)$ steady-state mode and an $O(2)$ Hopf mode must consider all possible mode numbers, that is all combinations $U \oplus (U' \oplus U')$, where U and U' are arbitrary irreducibles for $O(2)$. The total dimension can be anything between 3 and 6, depending on the choice of irreducible representations. The bifurcation geometry is to some extent sensitive to the choice of representation, and the interpretation of the dynamics even more so. See Golubitsky et al.[15] and Hill and Stewart.[16]

7. Symmetries of the Couette-Taylor System

Three distinct types of symmetry occur in the analysis of the Couette-Taylor system:

(a) physical symmetries of the apparatus,
(b) symmetries introduced by the choice of model, and
(c) mathematical symmetries introduced by the method of analysis.

We describe these in turn. For details see Golubitsky and Stewart.[14]

7.1 *Physical Symmetries*

The system is invariant under rotations θ about its axis, giving rise to a symmetry group $SO(2)$. It is also invariant under a reflection κ in the horizontal midplane. However, it is *not* invariant under reflection ρ in a vertical plane, because this would change the direction of rotation. There *is* a symmetry in which ρ is composed with time-reversal, but this does not seem to have been exploited.

7.2 *Symmetries in the Model*

The aim is to investigate spatially periodic phenomena occurring in the middle of a long cylinder. That is, end effects are ignored. The traditional model for this is to assume an infinitely long cylinder and to impose periodicity in the axial direction. In practice this means that we impose *periodic* boundary conditions on the ends of the cylinder. The advantage of this ansatz is that the equations become (moderately) tractable. But it does leave several major questions unanswered—including why periodic structures appear in the first place.

Effectively the assumption of periodic boundary conditions means that the vertical translational symmetry of an infinite cylinder is converted to a circle group action. When combined with the reflection κ this yields a symmetry group $O(2)$. It commutes with the rotations $SO(2)$ to yield a combined symmetry group $\Gamma = O(2) \times SO(2)$.

7.3 *Mathematical Symmetries*

In order to obtain an ordinary differential equation from the Navier-Stokes equations, we make use of one of several reduction procedures, such as centre manifold reduction, Liapunov-Schmidt reduction, or Birkhoff normal form reduction (see Golubitsky et al.[15]). For the present purposes it is sufficient to remark that after such reduction the problem reduces to finding the zero-set of a mapping defined on the sum of the critical eigenspaces. Moreover this reduction procedure introduces an extra symmetry. On a Hopf mode it introduces an action of the circle group S^1, which intuitively corresponds to phase shift to the corresponding time-periodic solutions, as dicussed in the example of an oscillating hosepipe.

As it happens, in the present system S^1 acts in exactly the same way as $SO(2)$. Thus we may either ignore S^1 or interpret $SO(2)$ as phase-shift. In the latter interpretation the spatial symmetries of the system are just $O(2)$, and $SO(2)$ becomes a group of *temporal phase-shift symmetries*. Because S^1 and $SO(2)$ coincide, all time-periodic solutions are rotating waves.

8. Numerical and Experimental Evidence for a Mode Interaction

In Taylor's original experiments, with $\Omega_o = 0$, the initial instability is to Taylor vortices. These represent a steady state of the velocity field of the fluid. However, if the outer cylinder is sufficiently counterrotated ($\Omega_o \ll 0$) then the initial instability is to spirals. This represents a time-periodic state of the velocity field.

Thus we expect to find, at some critical value of Ω_o, that the system becomes simultaneously unstable both to the Taylor vortex mode and the spiral mode. The simplest and most natural possibility is that this interaction should be an $O(2) \times SO(2)$ steady-state/Hopf mode interaction. This expectation is confirmed by numerical analysis of the eigenvalues (see Krueger et al.[18]). This shows that the bifurcation to Taylor vortices is associated with a double zero eigenvalue, and that to spirals is associated with a double imaginary eigenvalue.

In fact the circular symmetry of the Couette-Taylor apparatus implies that there is a direct mathematical analogy between the Hopf bifurcation to spirals and that of an oscillating hosepipe, with spirals in Couette-Taylor flow corresponding to rotating waves in a hosepipe. There must therefore be an analogue for Couette-Taylor flow of the other type of hosepipe oscillation, a standing wave. This flow is known as *ribbons*, but it is normally unstable and therefore not observed in experiments (see Demay and Iooss,[9] Chossat, Demay, and Iooss[3]).

As remarked above, the $SO(2)$-action can be interpreted as phase shift associated with periodic solutions, so effectively we are considering an $O(2)$-symmetric steady-state/Hopf mode interaction. The steady-state mode corresponds to a 2-dimensional eigenspace, the Hopf mode to a 4-dimensional eigenspace. Thus the entire mode interaction should take place on a 6-dimensional space.

Both the experiments and the numerical results demonstrate the existence of parameter values at which such a mode interaction occurs. We call such parameter values *bicritical points*.

9. The Reduced Model

The theoretical analysis begins with a study of the most general bifurcation problem, with the given symmetries, that can occur as a mode interaction of the appropriate type on a 6-dimensional space.

By looking at the linearized eigenfunctions it can be shown that the appropriate group action is that of Table 1. Here the z_i are complex coordinates, thought of as 2-dimensional real coordinates.

The next step is to find the *orbits* of $O(2) \times SO(2)$ on $\mathbb{C}^3$ for the group action of Table 1, where two points (z_0, z_1, z_2) lie in the same orbit if one can be transformed into the other by applying a group element. Having done that we calculate the *isotropy subgroups*, that is, the symmetry groups of orbit representatives. The results are shown in Table 2.

Table 1 Group Action for the 6-Dimensional Model

		2-dimensional steady state mode	4-dimensional Hopf mode	
		z_0	z_1	z_2
SO(2)	θ	z_0	$e^{i\theta}z_1$	$e^{i\theta}z_2$
O(2)	φ	$e^{i\varphi}z_0$	$e^{i\varphi}z_1$	$e^{-i\varphi}z_2$
	κ	$\bar{z}_0$	z_2	z_1

Table 2. Symmetry Data

Isotropy subgroup	State	Orbit representative
$O(2) \times SO(2)$	Couette flow	$z = 0$
$Z_2(\kappa) \times SO(2)$	Taylor vortices	$(x,0,0,)$ $x > 0$
$SO^-(2) = \{(\theta,\theta)\}$	Spirals	$(0,a,0)$ $a > 0$
$Z_2(\kappa) \oplus Z_2(\pi,\pi)$	Ribbons	$(0,a,a)$ $a > 0$
$Z_2(\kappa\pi,\pi)$	Wavy vortices	(iy,a,a) a, $y > 0$
$Z_2(\kappa)$	Twisted vortices	(x,a,a) a, $x > 0$

The *isotropy lattice*, which shows inclusions (up to conjugacy) between the isotropy subgroups and encapsulates the possibilities for symmetry-breaking, is shown in Fig. 13.

The reduced bifurcation equation takes the form $g(z, \mu)$ 0, where the mapping g commutes with the above group actions, that is

$$g(\gamma.z,\mu) = \gamma.g(z,\mu) \text{ for all } \gamma \in O(2) \times SO(2) \times SO(2), z \in \mathbb{C}^3, \mu \in \mathbb{R}.$$

Invariant theory implies that g can be written as

$$g(z,\mu) = (c^1 + i\delta c^2)\begin{bmatrix} z_0 \\ 0 \\ 0 \end{bmatrix} + (c^3 + i\delta c^4)\begin{bmatrix} \bar{z}_0 z_1 \bar{z}_2 \\ 0 \\ 0 \end{bmatrix} + (p^1 + iq^1)\begin{bmatrix} 0 \\ z_1 \\ z_2 \end{bmatrix} +$$

$$(p^2 + iq^2)\delta\begin{bmatrix} 0 \\ z_1 \\ -z_2 \end{bmatrix} + (p^3 + iq^3)\begin{bmatrix} 0 \\ z_0^2 z_2 \\ \bar{z}_0^2 z_1 \end{bmatrix} + (p^4 + iq^4)\delta\begin{bmatrix} 0 \\ z_0^2 z_2 \\ -\bar{z}_0^2 z_1 \end{bmatrix}.$$

The coefficients c^i, p^i, q^i are functions of the invariants ρ, N, Δ, Φ, Ψ, where

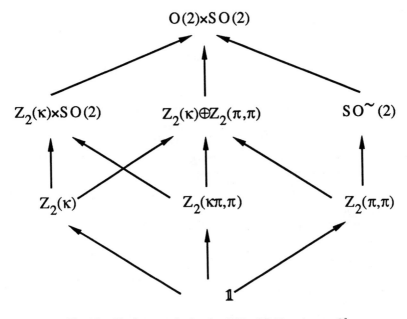

Fig. 13. The isotropy lattice for $O(2) \times SO(2)$ acting on $\mathbb{C}^3$.

$$\rho = |z_0|^2, \quad N = |z_1|^2 + |z_2|^2, \quad \delta = |z_1|^2 - |z_2|^2,$$
$$\Delta = \delta^2, \quad A = z_0^2 \bar{z}_1 z_2, \quad \Phi = \mathrm{Re}A, \quad \Psi = \mathrm{Im}A.$$

From the general form of g, specific information on branching directions and stabilities of the various solutions can be obtained. The results are complicated and will not be given here: see Golubitsky et al.[15] or Golubitsky and Langford.[12] They show in particular that open subsets of parameter space admit bifurcation scenarios consistent with the experimental results. For example, the main sequence is illustrated in Fig. 14.

10. The Eigenfunctions

The form of the flow can be approximated by an appropriate linear combination of linearized eigenfunctions. These take the form

$$\Phi_0 = e^{ikz} \begin{bmatrix} U_0(r) \\ V_0(r) \\ iW_0(r) \end{bmatrix}, \quad \Phi_1 = e^{i(m\theta + kz)} \begin{bmatrix} U_1(r) \\ V_1(r) \\ iW_1(r) \end{bmatrix}, \quad \Phi_2 = e^{i(m\theta - kz)} \begin{bmatrix} U_2(r) \\ V_1(r) \\ -iW_2(r) \end{bmatrix}$$

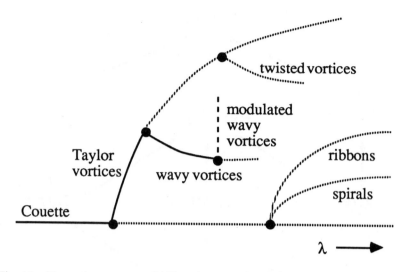

Fig. 14. The main sequence of bifurcations can be realised by a suitable choice of parameter values in g.

together with their complex conjugates. Here the U_i, V_i, W_i are certain vector functions of the radius r. These eigenfunctions have *azimuthal wave-number* $m = 0, 1, 2, 3, \ldots$ and *axial wave-number* k.

We can interpret the wave-functions Φ_1, Φ_2, Φ_3 as follows. The flow defined by Φ_1 is time-independent, and corresponds to Taylor vortex flow (Fig. 15a). The flows for Φ_2 and Φ_3 are mapped to each other by reflection $z \to -z$ in a horizontal plane. Because θ is identified with time, they are time-dependent travelling waves, and may be interpreted as spirals (Fig. 15b, c). Note that Fig. 15c is just Fig. 15b turned upside-down ($z \mapsto -z$). In this sense, every pattern that occurs in the 6-dimensional reduced model defined by the chosen mode-interaction is (approximated by) a superposition of Taylor vortices, left-handed spirals, and right-handed spirals. The appropriate coefficients are given in the third column of Table 2.

11. The Numerics

In order to apply the above analysis quantitatively, the parameter values for a bicritical point must be calculated, and then the coefficients c^i, p^i, q^i for the corresponding reduced bifurcation mapping $g(z,\mu)$ must

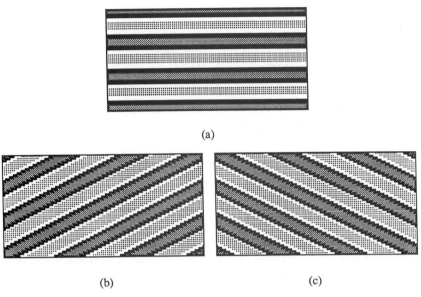

Fig. 15. (a) Taylor vortices, (b) right-handed spirals, (c) left-handed spirals. The shading shows the velocity component in the axial (z) direction at a typical radius intermediate between two cylinders. The coordinates are (θ, z).

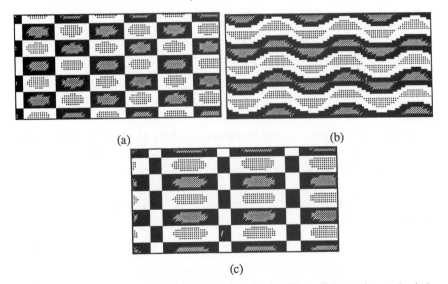

Fig. 16. Synthesizing possible states as nonlinear superpositions of the vortices and spirals of Fig. 15. (a) ribbons, (b) wavy vortices, (c) twisted vortices.

be found. Once this is done, we have detailed quantitative information on the bifurcations that occur near the bicritical point.

The first step is to perform the linear eigenvalue analysis, which has been tackled by several authors. Here we follow Golubitsky and Langford.[12] The starting point is the Navier-Stokes equations

$$\partial \mathbf{u}/\partial t = \nu \nabla^2 \mathbf{u} - (\mathbf{u}.\nabla)\mathbf{u} - \rho^- \nabla p$$
$$\nabla.\mathbf{u} = 0$$

where
- $\mathbf{u}(t,x)$ = velocity vector at time t, $x \in \mathbb{R}^3$
- ρ = pressure
- p = mass density
- ν = kinematic viscosity.

This equation is then nondimensionalised and coordinates are translated to make Couette-Taylor flow the "trivial" solution. The linear stability of the trivial solution is determined by the eigenvalues λ of an eigenvalue problem. This takes the form

$$\lambda u = \nabla^2 u - u/r^2 - 2v_\theta/r^2 - p_r - C(r)u_\theta + 2C(r)v$$
$$\lambda v = \nabla^2 v - v/r^2 + 2u_\theta/r^2 - p\theta/r - C(r)v_\theta + 2Du$$
$$\lambda w = \nabla^2 w - p_z - C(r)w\theta$$
$$\nabla.\mathbf{u} = 0.$$

Here $\mathbf{u} = (u,v,w,)$ in cylindrical coordinates (r,θ,z), and subscripts denote partial derivatives. The corresponding eigenfunctions are those defined above.

The *neutral stability curves* (parameter values at which critical eigenvalues occur) are shown in Fig. 17. Bicritical points—the desired organizing centres—occur where these curves intersect.

Numerical results, for a range of radius ratios η, have been found by Golubitsky and Langford[12] and Langford et al.[19] using no more computing power than an IBM PC clone. Sample values for $\eta = 0.800$ are:

$R_o = -99.29$, $R_i = 129.75$, $k = 3.573$, $\omega = 0.3450$, $\mu = -0.6130$.

Here $\eta = \Omega_i/\Omega_o$ and the relevant imaginary eigenvalues are $\pm i\omega$.

The corresponding coefficients in the mapping g (and those of their derivatives that are relevant to the bifurcation scenario) are:

$c_{R_1}^1 = .457$ $c_{R_2}^1 = .248$ $c_\rho^1 = -1.384$ $c_N^1 = -8.44$

$c^2 = -9.79$ $c^3 = -9.92$ $p_{R_1}^1 = .429$ $p_{R_2}^1 = .210$

$p_\rho^1 = -6.41$ $p_N^1 = -4.98$ $p^2 = -2.19$ $q^2 = -6.19$

$p^3 = -5.65$ $q^3 = -3.20$

From values such as these, the appropriate bifurcation diagram (near the bicritical point) can be plotted. It is convenient to represent the results as a *gyrant bifurcation diagram* (Fig. 18). This uses polar coordinates (r,θ), where r represents the "amplitude" of the relevant branch of solutions—in a schematic sense—and θ parametrises a small loop surrounding the bicritical point. The central circle represents zero amplitude (in other words, a constant has been added to the amplitude to prevent the Couette branch collapsing to a point). A number of "modulated states" also occur, formed by Hopf bifurcations from states modelled within the 6-dimensional reduced system.

12. Predictions

This method provides very precise quantitative predictions for the behaviour close to a bicritical point. Three specific predictions (visible in Fig. 18) are:

New states: Wavy vortex flow (WVF) can occur stably as a secondary bifurcation from Taylor vortices (TVF), extremely close to criticality.

Bistability: In suitable regions both spirals (SPI) and wavy vortices, or spirals and Taylor vortices, can both occur stably for the same parameter values.

Hysteresis: The transition between spirals and Taylor vortices occurs at different values of θ (the parameter along the small loop surrounding the bicritical point) if θ is changed clockwise or anticlockwise.

Experiments by Tagg et al.[25] have verified these predictions. Further work, both theoretical and experimental, is in progress.

13. Extensions

We briefly discuss two extensions of these ideas. The first is a 10-dimensional model that includes several other observed states; the second addresses the questions of patterned turbulence.

In Fig. 17 it can be seen that not only do the neutral stability curves

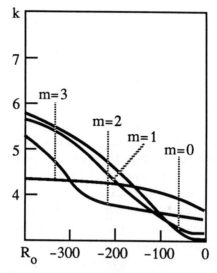

Fig. 17. Neutral stability curves for various wave-numbers. The relevant bicritical points are those for which the curve for $m = 0$ crosses that for $m = 1$. Here R_o, R_i are the Reynolds numbers corresponding to Ω_o, Ω_i. The radius ratio $\eta = 0.800$. (After Golubitsky and Langford[12] but with vertical coordinate magnified.)

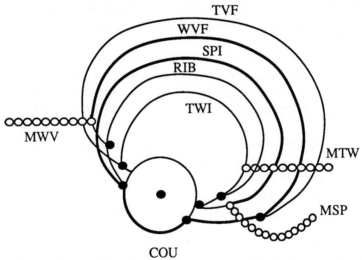

Fig. 18. Computed gyrant bifurcation diagram near the bicritical point when $\eta = 0.800$. Heavy curves represent stable solutions. COU = Couette flow; TVF = Taylor vortex flow; WVF = wavy vortex flow; SPI = spirals; RIB = ribbons; TWI = twisted vortices; MWV = modulated wavy vortices; MTW = modulated twisted vortices; MSP = modulated spirals.

$m = 0$ and $m = 1$ cross, giving an organizing centre for the 6-dimensional model, but they also cross very close to the neutral stability curve for $m = 2$. This represents another Hopf mode, corresponding to spirals with a different azimuthal wavenumber (Fig. 19). Chossat et al.[3] have combined the modes $m = 1$ and $m = -2$ into

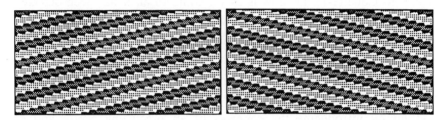

Fig. 19. Right- and left-handed spirals with azimuthal wavenumber 2.

an 8-dimensional model. By incorporating the mode $m = 0$ as well, we obtain a 10-dimensional model, which leads to a unified theory yielding many more states, determined as appropriate nonlinear superpositions or Hopf bifurcations away from these. An example is the state of *interpenetrating spirals* (Fig. 20). For details see Hill and Stewart.[17]

Chossat and Golubitsky[4] suggest that patterned turbulent states, such as turbulent vortices or spiral turbulence, may be modelled by *symmetric chaos*. The phenomenon of deterministic chaos is now widely known (see, for example, Gleick,[11] Stewart[23,24]). It explains how dynamical systems that have no explicit random terms can nevertheless possess solutions that appear to be random. This phenomenon is caused by the presence of *strange attractors*, long-term dynamics of a highly irregular

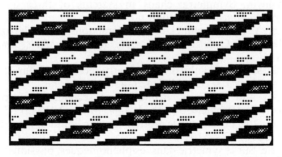

Fig. 20. Interpenetrating spirals formed by nonlinear superposition of Fig. 19a and Fig. 15b.

kind. When chaos occurs in a symmetric system the attractor can have a degree of symmetry; and this can be interpreted as a symmetry "on the average" of the corresponding irregular state.

A striking phenomenon in symmetric chaos is the occurrence of *symmetry-increasing* bifurcations. While the relevance of symmetric chaos to Couette-Taylor flow is as yet conjectural, it has been detected experimentally in systems of symmetrically coupled oscillators, together with symmetry-increasing bifurcation (see Ashwin[2]).

14. Conclusions

The idea of modelling the main qualitative features of a bifurcating system by introducing the concepts of an organizing centre and of symmetry-breaking can be applied successfully to real problems. Moreover, suitable numerical calculations, feasible on small microcomputers, can make the results quantitative. In the case of the Couette-Taylor system symmetries play a vital role in pattern-formation and can be exploited in a theoretical analysis. The method makes it possible to predict new states and transitions, and such predictions can be verified experimentally. Moreover, the techniques provide a paradigm for the analysis of other nonlinear systems in which mode-interactions are observed, especially systems with symmetry.

References

1. C. D. Andereck, S. S. Liu, and H. L. Swinney, Flow regimes in a circular Couette system with independently rotating cylinders, *J. Fluid Mech.* **164** (1986) 155–183.
2. P. B. Ashwin, Symmetric chaos in systems of three and four forced oscillators, *Nonlinearity* **3** (1990) 603–618.
3. P. Chossat, Y. Demay, and G. Iooss, Interactions des modes azimutaux dans le problème de Couette-Taylor, *Arch. Rational Mech. Anal.* **99** (1987) 213–248.
4. P. Chossat and M. Golubitsky, Symmetry-increasing bifurcation of chaotic attractors, *Physica D* **32** (1988) 423–436.
5. P. Chossat and G. Iooss, Primary and secondary bifurcations in the Couettte-Taylor problem, *Japan. J. Appl. Math.* **2** (1985) 37–68.
6. M. M. Couette, Études sur le frottement des liquides, *Ann. Chim. Phys.* **6** (ser. 21) (1890) 433–510.
7. M. P. Curie, Sur la symétrie dans les phénomèques, physiques symétrie d'un champ électrique at d'un champ magnétique, *Journal de Physique Théorique et Appliqueé* (3rd ser.) **3** (1894) 26–48.

8. A. Davey, R. C. DiPrima, and J. T. Stuart, On the instability of Taylor vortices, *J. Fluid Mech.* **31** (1968) 17–52.
9. Y. Demay and G. Iooss, Calcul des solutions bifurquées pour le problème de Couette-Taylor avec les deux cylindres en rotation, *J. de Mech. Theor. et Appl.* Numéro special (1984) 193–216.
10. R. C. DiPrima and H. L. Swinney, Instabilities and transition in flow between concentric rotating cylinders. In *Hydrodynamic Instabilities and the Transition to Turbulence*, H. L. Swinney and J. P. Gollub, eds., *Topics in Applied Physics* **45** (pp. 139–180), Springer, New York (1981).
11. J. Gleick, *Chaos—Making a New Science*, Viking, New York (1987)
12. M. Golubitsky and W. F. Langford, Pattern formation and bistability in flow between counterrotating cylinders, *Physica* **32D** (1988) 362–392.
13. M. Golubitsky and I. N. Stewart, Hopf bifurcation in the presence of symmetry, *Arch. Rational Mech. Anal.* **87** (1985) 107–165.
14. M. Golubitsky and I. N. Stewart, Symmetry and stability in Taylor-Couette flow, *SIAM J. Math, Anal.* **17** (1986) 249–288.
15. M. Golubitsky, I. N. Stewart, and D. G. Schaeffer, *Singularities and Groups in Bifurcation Theory: Vol. II*. Springer, New York (1988).
16. A. S. Hill and I. N. Stewart, *Hopf/Steady-State Mode Interactions with $O(2)$ Symmetry* (preprint). Mathematics Institute, Univ. of Warwick (1989).
17. A. S. Hill and I. N. Stewart, *Three-Mode Interactions with $O(2)$ Symmetry and a Model for Taylor-Couette Flow* (preprint). Mathematics Institute, Univ. of Warwick (1990).
18. E. R. Krueger, A. Gross, and R. D. DiPrima, On the relative importance of Taylor-vortex and nonaxisymmetric modes in flow between rotating cylinders, *J. Fluid Mech.* **24** (1966) 521–538.
19. W. F. Langford, R. Tagg, E. Kostelich, H. L. Swinney, and M. Golubitsky, Primary instabilities and bicriticality in flow between counter-rotating cylinders, *Phys. Fluids* **31** (1988) 776–785.
20. T. Poston and I. N. Stewart, *Catastrophe Theory and its Applications*. Pitman, Boston (1978).
21. T. Poston and A. E. R. Woodcock, On Zeeman's catastrophe machine, *Proc. Cambridge Philos. Soc.* **74** (1973) 217–226.
22. I. N. Stewart, La symétrie et ses brisures, *Pour La Science* **143** (September, 1989) 90–96.
23. I. N. Stewart, Chaos: Does God play dice? *1990 Yearbook of Science and the Future* (pp. 54–73). Encyclopedia Brittannica, Chicago (1989).
24. I. N. Stewart, *Does God Play Dice? The Mathematics of Chaos*. Basil Blackwell, Oxford (1989).
25. R. Tagg, D. Hirst, and H. L. Swinney, *Critical Dynamics Near the Spiral-Taylor Vortex Codimension Two Point* (in preparation).
26. G. I. Taylor, Stability of a viscous liquid contained between two rotating cylinders, *Phil. Trans. Roy. Soc. London* **A 223** (1923) 289–34.
27. R. Thom, *Structural Stability and Morphogenesis*. Benjamin, Reading, Mass.
28. A. E. R. Woodcock and T. Poston, A higher catastrophe machine, *Proc. Cambridge Philos. Soc.* **79** (1976) 343–350

29. E. C. Zeeman, A catastrophe machine. In *Towards a Theoretical Biology* **4** (pp. 276–282), C. H. Waddington, ed. Edinburgh Univ. Press, Edinburgh (1972).
30. E. C. Zeeman, Research ancient and modern, *Bull. Inst. Math. Appl.* **10** (1974) 272–281; reprinted in *Catastrophe Theory: Selected Papers 1972–77* (pp. 605–614). Addison-Wesley, Reading, Massachusetts (1977).
31. E. C. Zeeman, *Catastrophe Theory: Selected Papers (1972–1977)*. Addison-Wesley, Reading, Massachusetts (1977).
32. E. C. Zeeman, Bifurcation, catastrophe, and turbulence. In *New Directions in Applied Mathematics* (pp. 109–153) P. J. Hilton and G. S. Young, eds. Springer, New York (1981).

OSCILLATIONS, WAVES, AND SPIRALS IN CHEMICAL SYSTEMS

Endre Körös

1. Introduction

Temporal and spatial periodic phenomena occur at many levels of organization of matter. The temporal evolution of a system may show extremes, for example, in the number of species (ecology) or in the concentration of some chemical intermediates (chemistry, biochemistry). Spatial organizations may manifest themselves as layered structures, various types of waves of either physical or chemical activity, or as different patterns (mosaic, spotted, etc.). By proceeding from the highest level of organization (ecological) through the biological on down to the molecular level (chemistry), this paper gives a few examples to demonstrate the vital importance of such phenomena.

In *ecological systems* nonlinearity is well documented in population dynamics: oscillations of predator and prey populations have been found in many more-or-less isolated, geographical regions. When only two competing species are considered, and migration of the species is not allowed for, the Lotka model[12] provides an adequate description of the temporal changes in the numbers of the two species.

Nonlinear phenomena are especially well known in *biological systems*. Biological clocks and biorhythms are, in many cases, easily observable and detectable; to these belong the transmission of nerve impulses, propagation of brain waves, and circadian rhythms. Oscillations have also been found in growing cell populations, in tissues (e.g., muscle contraction, heart fibrillation), and during various cellular and intercellular organizations (e.g., cyclic-AMP oscillation in *Dictyostelium* cells and periodic cell communication in *Dictyostelium discoideum*, as well as in ion fluxes across the mitochondrial membrane).

At the *biochemical* level of organization, the most thoroughly studied and the best understood oscillations are the glycolytic oscillations, i.e. the concentration oscillations of chemical intermediates along the glycolytic pathway. Two other oscillatory biochemical reactions are also worth mentioning: the peroxidase-catalyzed oxidation, by O_2, of NADH, and the polymerization of fibrinogen.

Spatial structures of a large variety also abound in nature. They are exhibited in slime mold aggregation, in geological layering, in glycolyzing yeast extract, and during the formation of inorganic precipitates (Ag_2CrO_4, PbI_2), which are called Liesegang rings, under appropriate conditions. In addition, temporal and spatial periodicities occur in cold flames.

Over a very long period of time, most chemists suspected that in solutions in homogeneous phase, a chemical reaction might proceed in an oscillatory manner, that is, showing concentration extremes in time. The early observations of certain chemical oscillations ascribed these oscillations to heterogeneities, dust particles, the presence of bubbles, and so on. These observations include periodicities in the rate of carbon monoxide evolution during dehydration by concentrated sulfuric acid of formic acid (Morgan reaction)[1]

$$HCOOH \xrightarrow[-H_2O]{cc.H_2SO_4} CO$$

and the iodate-catalyzed decomposition of hydrogen peroxide in moderately acidic solution (Bray reaction)[2]

$$H_2O_2 \xrightarrow{IO_3^-, H^+} H_2O + 1/2\ O_2,$$

which showed periodicities in the concentration of iodine and iodide—both important intermediates of the reaction—and in the rate of oxygen evolution. Investigating the Bray reaction, Rice and Reiff[3] concluded that it is a heterogeneous reaction occurring on the surface of dust particles; later, Peard and Cullis[4] ascribed the oscillation to an uncommon combination of physical and chemical factors, including volatilization of iodine. Even as late as 1968, Shaw and Pritchard[5] argued against the possibility of a homogeneous oscillatory reaction.

Although, as described above, experimental observations of sustained oscillations in reacting chemical systems date back to the early part of this century, a sudden growth of interest in the subject has occurred about 20 years ago. One of the main reasons was the discovery, by Belousov, of a third reaction showing temporal periodicities, the ceriumion-catalyzed oxidation and bromination of an aliphatic polycarboxylic acid by acidic bromate. This discovery afforded an accessible and easily reproducible system. A detailed description of this reaction, referred to as the Belousov–Zhabotinsky (BZ) reaction, will be given below. Another reason has been the fundamental works of Prigogine and his school on the extension of thermodynamic theory to conditions far from equilibrium, in which they demonstrated that temporal and spatial organization could be sustained by evolving systems. Glansdorff and Prigogine[6] have shown that in systems with nonlinear dynamic laws there may be a critical distance from equilibrium beyond which new solutions to the dynamic laws of a chemical system may appear. These systems may show nonmonotic temporal and spatial behavior, that is, the concentration of reaction intermediates may oscillate and/or structures (patterns) form.

Within a very short period of time, much attention was turned to reveal the chemical mechanisms, which account for the chemical control, and to develop mathematically tractable models.

As late as 1980 only a few chemical oscillatory systems had been under investigation: the Bray reaction, the BZ reaction and its variants, the uncatalyzed (aromatic) bromate oscillators,[7,8] and the hybrid of the Bray and the BZ reactions (the Briggs–Rauscher reaction[9]). However, from 1980 on, due to a most successful activity of a Brandeis University research team, most of the chemical oscillators could be systematically designed after the conditions that were necessary for, or conductive to, oscillations[10,11] were established.

A reacting system at, or close to, equilibrium will not oscillate since it is a stable state. Therefore, one necessary condition for oscillation is that the system be *far from equilibrium*. This requirement could be met by having the reaction proceed under open conditions, in a continuously-fed stirred tank reactor (CSTR). The second condition is that some product of a step in the reaction series must exert influence on its own rate of formation. This mechanism is called *feedback*, commonly encountered both in chemical and in biological systems as autocatalysis: $X + Y \rightarrow 2Y$. In addition to autocatalysis, induced reactions and inhibition may manifest themselves in feedback mechanism. In 1920 Lotka presented a mechanism[12] in which two coupled autocatalytic reactions gave rise to oscillations. The mechanism he described is

$$\text{step 1: } A + X \rightarrow 2X$$
$$\text{step 2: } X + Y \rightarrow 2Y$$
$$\text{step 3: } \quad Y \rightarrow Q$$

The overall reaction $A \rightarrow Q$, and X and Y, are intermediates. Both step 1 and step 2 are autocatalytic. Assuming that the concentration of A is constant, Lotka found that the proposed mechanism led to sustained oscillations in the concentrations of X and Y. Although the Lotka mechanism has not proved to be a model for any actual chemical reaction, it certainly drew the attention of many chemists to the importance of autocatalysis in oscillatory processes.

The third condition, that we find often in conjunction with the oscillatory behavior of a chemically reacting system, is the existence of *bistability*, i.e., coexistence of two stable steady states for the same values of the externally controlled constraints. In a CSTR these constraints are the concentration of the inflow reactants, the rates of flow, and the temperature of the reacting system.

The questions of how a bistable system can be transformed to an oscillatory one, what close relationship exists between bistability and oscillations, and how a bistable system can be transformed to an oscillatory one are not our present concern, and we refer only to a review article on this subject.[13]

As a result of a systematic study on a large number of redox reactions in the halogen (Cl, Br, I), sulfur, nitrogen, and transition metal ion chemistry, numerous oscillatory chemical systems were discovered and characterized.

The taxonomy of chemical oscillators can be found in one of Epstein's recent papers.[14]

2. The Chemistry and Chemical Mechanism of the Belousov–Zhabotinsky Reaction

Numerous reacting chemical systems are known that show temporal oscillations and in which spatial structures emerge. It is the reacting BZ system that shows the greatest variety in nonlinear phenomena.

In a stirred BZ system, *temporal oscillation* occurs and the parameters of oscillation (frequency, amplitude) as well as its character change dramatically when even minor perturbations are applied. In an unstirred reagent, where diffusion also plays an important role, depending on the geometry of the reaction space, *ring-shaped oxidation zones* of chemical activity develop (in a test tube) or *chemical waves* propagate (in Petri dish), and in some cases mosaic or "pocky" patterns are discernible. If the waves are physically disrupted, *spiral structures* evolve, and in gels *scroll waves* form. On the surface of an ion-exchanger bead from a meandering center, *twisted oxidation zones* move along from pole to pole.

All the phenomena exhibited by a reacting BZ system have one thing intrinsically in common: the chemical mechanism. This will be discussed in detail and the occurrence of chemical oscillation explained. Then effects of different perturbing agents will be demonstrated, followed by an explanation of how waves and spiral structures develop.

In the early fifties, Belousov observed that during the cerium-ion-catalyzed oxidation of citric acid by acidic bromate, the color of the stirred reaction mixture turned periodically from yellow to colorless and back to yellow.[15] He gave a simple, but surprisingly correct qualitative explanation for that unusual behavior. The first important investigator of this reaction was Zhabotinsky, who, among others, found that citric acid can be substituted for other aliphatic polycarboxylic acids (e.g., malonic acid and malic acid) and cerium ion for manganese(II) ion or ferroin [(iron(II)-phenanthroline complex)].[16,17] At present, reaction mixtures composed of bromate, sulfuric acid, a catalyst, and an organic substrate, which show oscillatory behavior, are referred to as Belousov–Zhabotinsky (BZ) systems. The most thoroughly studied and by far the best characterized is the bromate, sulfuric acid, cerium(III), and malonic acid reacting system.

Zhabotinsky early during his studies found that the oxidation, by bromate, of cerium(III) was an autocatalytic reaction and bromide ion acted as an inhibitor.[18] This has been a most important observation that has helped us to reveal the chemical control mechanism in the BZ systems.

A detailed mechanism that could explain the rich array of dynamical behavior in the reacting BZ system was proposed by Field, Körös, and Noyes in 1972.[19,20]

The essence of the mechanism is that, depending on the actual bromide ion concentration, the reacting system can exist in one of the two kinetic states, and at a certain bromide ion concentration, referred to as the critical bromide concentration $[Br^-]_{crit}$, the one kinetic state switches to the other state. This switching on and off are repeated in time, and this can be monitored (mostly by electrochemical and optical methods) as a chemical oscillation.

The driving force for the BZ reaction is oxidation and bromination of malonic acid by bromate to carbon dioxide and bromomalonic acid as major products. The overall stoichiometry can be approximated by the following reaction:

$$2BrO_3^- + 3CH_2(COOH)_2 + 2H^+ \xrightarrow{catalyst} 2BrCH(COOH)_2 + 3CO_2 + 4H_2O.$$

The reaction is highly exothermic, and the heat of reaction is 650 ± 10 kJ/mol bromate.[21] The apparent activation energy of chemical oscillation in the BZ system is 65–70 kJ/mol.[22]

Notwithstanding that bromate is a strong oxidizing agent (the standard redox potential of the BrO_3^-/Br^- couple is $+1.5$ V), it reacts slowly with carboxylic acids. However, the presence of an appropriate one-electron redox couple (e.g., Ce(IV)/Ce(III); Mn(III)/Mn(II); ferriin [Fe(III)]/ferroin [Fe(II)]) greatly accelerates the reaction.

The sequence of the reactions in both of the kinetic states is as follows:

Kinetic state A (high bromide concentration state)

$BrO_3^- + Br^- + 2H^+ \rightarrow HBrO_2 + HOBr$	(R1)
$HBrO_2 + Br^- + H^+ \rightarrow 2HOBr$	(R2)
$HOBr + Br^- + H^+ \rightarrow Br_2 + H_2O$	(R3)
$Br_2 + CH_2(COOH)_2 \rightarrow BrCH(COOH)_2 + Br^- + H^+$	(R4)

$BrO_3^- + 2Br^- + 3CH_2(COOH) + 3H^+ \rightarrow 3BrCH(COOH)_2 + 3H_2O$ (A)

The net effect of kinetic state A is the removal of bromide ion from the system and bromine ends up organically bound.

Kinetic state B (low bromide concentration state)

$$BrO_3^- + HBrO_2 + H^+ \rightarrow 2BrO_2 + H_2O \qquad \text{(R5)}$$
$$BrO_2 + Ce^{3+} + H^+ \rightarrow HBrO_2 + Ce^{4+} \qquad \text{(R6)}$$
$$2HBrO_2 \rightarrow BrO_3^- + HOBr + H^+ \qquad \text{(R7)}$$

followed by (R3) and (R4) to give

$$BrO_3^- + 4Ce^{3+} + CH_2(COOH)_2 + 5H^+ \rightarrow 4Ce^{4+} + BrCH(COOH)_2 + 3H_2O. \qquad \text{(B)}$$

The heart of the mechanism is the autocatalytic formation of $HBrO_2$, since the consumption of each mole of $HBrO_2$ leads to the production of two more. However, (R7), the disproportionation of $HBrO_2$, limits the amount of this compound.

A third net process is the reduction of cerium(IV) to cerium(III) mainly by malonic acid.

^{82}Br tracer experiments disclosed[22] that bromide ions were produced partly as a consequence of the reaction of HOBr with the organic substrate, that is, directly from bromate (inorganic route) and in a much lesser extent in the Ce^{4+}, bromomalonic acid reaction by the rupture of the C—Br bond:

$$Ce^{4+} + Br-C- \rightarrow Ce^{3+} + Br^- + \text{organic oxidation products.}$$

As the bromide ion concentration decreases in kinetic state A, bromate starts to compete for $HBrO_2$ (R5) and, at a certain bromide ion concentration, the rates of R(2) and R(5) equal to each other. This leads to the following equation:

$$[Br^-]_{crit} = \frac{k_5}{k_2}[BrO_3^-],$$

where k_5 and k_2 are the respective rate constants. At $[BrO_3^-] = 0.05\ M$, using the rate constants by Field and Försterling,[24] $[Br^-]_{crit} = 7 \times 10^{-7}\ M$.

Inspecting an oscillatory system, we can classify the species into three distinct groups: (a) reactants (BrO_3^-, malonic acid), (b) recycling intermediates (e.g., Br^-, BrO_2, $HBrO_2$, Ce^{4+}), and (c) final products (e.g., CO_2, bromomalonic acid). The recycling intermediates exhibit temporal concentration oscillations that can be rather easily monitored, either potentiometrically or spectrophotometrically. With the reactants, the rates of consumption and with the final products, the rates of formation show extremes; the concentration–time curves develop in a stepwise manner. The same is valid for the heat evolution, which gives some information on the overall reaction. Typical examples are shown in Fig. 1 (a)–(d).

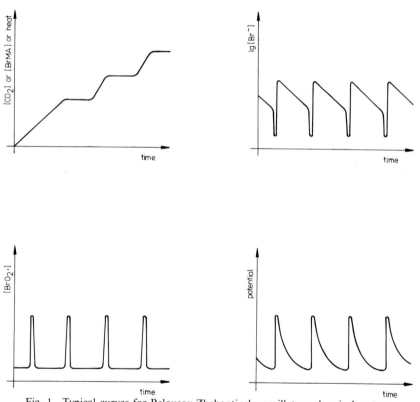

Fig. 1 Typical curves for Belousov-Zhabontinsky oscillatory chemical system.

The reacting BZ systems, as characteristic of most of the oscillatory reactions, are senstive to perturbations, i.e., to the addition of certain reagents in low concentrations. In Figs. 2–4, we demonstrate the effects of iodide ion[25] (opening new recycling routes), silver ion[26] (binding the control intermediate, the Br^-), and of ^{60}Co γ-irradiation[27] (generating in situ a species that influences the function of the catalyst).

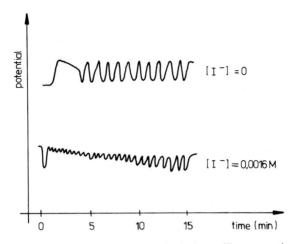

Fig. 2 The potential traces of the Belousov-Zhabotinsky oscillatory reaction. The reacting system is composed of 0.10 M $KBrO_3$, 0.20 M malonic acid, 0.001 M Mn^{2+}, and 1.0 M K_2SO_4. The lower trace shows the behavior of the, by iodide ions, perturbed system. The preoscillatory (or induction period) disappears and a high frequency, amplified oscillation is set on.

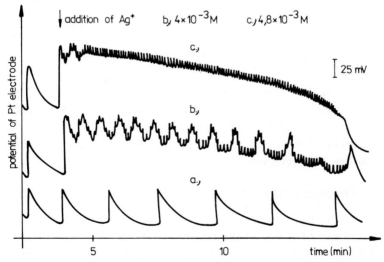

Fig. 3 The dynamic behavior of a bromate oscillator with the composition: 0.10 M $KBrO_3$. 0.05 M trihydroxibenzene and 2.25 M K_2SO_4. Trace: (a) without the addition of silver ions; traces (b) and (c) in the presence of silver ions. Silver ions induce a very high frequency oscillation, and it is noteworthy that a minor change in silver ion concentration, from 0.004 M to 0.0048 M, considerably alters the potential trace of the reacting system.

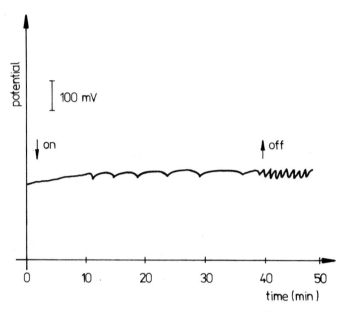

Fig. 4 The effect of ^{60}Co γ-radiation on a Belousov-Zhabotinsky system with the composition: 0.05 M $KBrO_3$, 0.20 M malonic acid, 0.0004 M Ce^{3+}, and 1.0 M H_2SO_4. Dose rate: 2.74×10^{18} eV/dm^3 s. The irradiation drastically increases the period time of the chemical oscillation. Compare the first part of the trace with the second one, when irradiation has been stopped.

In spite of the enormous experimental data accumulated during the last two decades on the BZ systems, there are still a few unclarified points, mostly regarding the organic reactions and the role of organic intermediates in the chemical mechanism. These are, however, inferior to our further considerations.

3. The Oregonator Model

The somewhat extended form[28] of the FKN mechanism, which considers some but not all the possible chemical reactions, is composed of 18 reactions that involve 21 independent species. This mechanism was simplified by Field and Noyes[29] to a five-step model including three intermediates, and they arrived at a model they labeled *Oregonator* (*Oregon* elaborated chemical oscill*ator*), which still retains the basic qualitative features of the FKN mechanism, the two kinetic states and the bromide regeneration step:

$$A + Y \to X + P \qquad (1)$$
$$X + Y \to P \qquad (2)$$
$$A + Y \to 2X + Z \qquad (3)$$
$$X + X \to A + P \qquad (4)$$
$$Z \to fY. \qquad (5)$$

The following identifications can be made $A \equiv BrO_3^-$, $Y \equiv Br^-$, $X \equiv HBrO_2$, $Z \equiv 2\ Ce^{4+}$ (or the oxidized form of the catalyst in general), $P \equiv$ product, a nonrecycling species. The organic chemistry is completely omitted and "appears" only in a most condensed form in reaction (5), since the regeneration of bromide ion is coupled, in a rather sophisticated way, to the bromate-malonic acid-cerium reaction.

The properties of the Oregonator have been extensively studied partly because it has been "derived" from a real reacting chemical system and partly because it combines a tractable level of mathematical difficulty with a wealth of dynamic behavior. The traces of $\lg[Ce^{4+}]$, $\lg[Br^-]$, and $\lg[HBrO_2]$ vs. time obtained by numerical integration of the mass action dynamic laws for the Oregonator have the similar form as those obtained experimentally (Fig. 5). These curves should be remembered at the discussion of chemical waves.

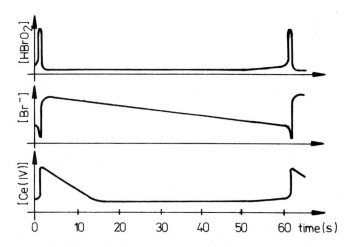

Fig. 5 The time evolution of three important intermediates of the Belousov-Zhabotinsky reaction obtained from the Oregonator model.

4. Waves

So far we have discussed the chemistry and the mechanism of the BZ reaction and the temporal behavior of the reacting system under stirred conditions. If the BZ systems are left unstirred they show a variety of spatial structures.

Busse[30] was the first to observe travelling waves in a BZ system, when mixing the reactants (except sulfuric acid) in a test tube and applying a vertical concentration gradient of sulfuric acid. Spatial bands of alternating oxidizing (blue) and reducing (pink) character propagated through the unstirred BZ solution containing ferroin as a catalyst. A more fascinating system, reported by Zaikin and Zhabotinsky,[31] and thoroughly investigated by Winfree,[32] was composed of malonic acid, bromomalonic acid, bromate, ferroin, and sulfuric acid. In this reagent, when spread in a thin layer, from accidentally appearing oxidizing (blue) centers, waves with sharp leading edges and blurred trailing edges propagate through the solution with velocities of about 5–8 mm/min and, after a certain period of time (in appr. 10–15 min), spectacular patterns develop.

Before discussing these phenomena and the underlying chemistry, the different types of travelling waves will be described.

4.1 Phase Wave

This type of wave occurs in reacting systems subject to temporal oscillation. When such a system is spread in a thin layer over a smooth surface, it may continue to oscillate. Phase waves may develop because it is possible that adjoining regions of the reaction mixture may get out of phase with each other. Phase waves that are not essentially dependent upon diffusion result from the existence of a monotonic phase gradient in an oscillatory reagent. Phase waves reflect only apparent movement; they pass through physical barriers and their velocity is inversely dependent upon the phase gradient in the medium. Because there is no limit to the shoalness of the phase gradient, there is essentially no upper limit to the speed of a phase wave. Fig. 6 shows how in a reagent, in its reducing state (red colored region), an oxidizing zone (blue) is passing along.

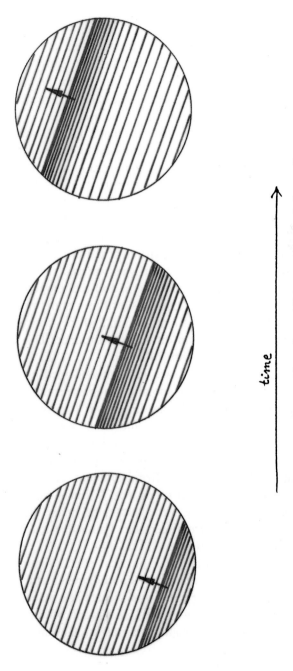

Fig. 6 A phase wave: the spreading of an oxidizing zone in a reducing medium.

4.2 Trigger Wave

The second mechanism of wave propagation is caused by the interaction of diffusion and chemical reaction that leads to the so-called trigger waves. These waves may occur in either an oscillatory or nonoscillatory reacting system. The nonoscillatory reagent must be excitable if trigger waves are to manifest themselves.

Propagation of trigger waves can eventuate only if, in the reacting system, the rates of formation or consumption of some crucial intermediates are very high. Large changes in concentrations occur as a result of a perturbation in an excitable system. Perturbation can be induced, for example, by immersing a hot Pt-wire into the reagent or by solid particles. The rate by which the trigger waves propagate through a reaction mixture is determined both by the rates of the chemical reactions and the diffusive characteristics of the crucial intermediates.

Unlike phase waves, the velocity of trigger waves is strictly determined by the composition of the reacting system and the temperature. Trigger waves can travel macroscopic distances within a short period of time (typical rates are between 3 and 8 mm/min), and they move much more rapidly than intermediates could diffuse that distance. This is the result of the interaction of reaction with diffusion.

The sources of trigger waves are randomly formed centers (in the ferroin-catalyzed BZ system blue spots) that represent the oxidized state of the system. From these, oxidizing waves start to propagate, resulting in the appearance of concentric rings, and finally a pattern of rings fills the reaction vessel. Two important observations should be mentioned: (a) consecutive waves do not overtake those in front of them, and (b) advancing bands—originating from different centers (called also "pace-makers")—annihilate each other. Fig. 7 shows the temporal evolution of the pattern, and in Fig. 8 the "close ups" make the annihilation of contacting wave fronts more discernible. Bands that reach the walls of the reaction vessel also disappear.

If the bands are "ruptured" by a physical effect, about the sites of discontinuity, spiral bands start to develop; the broken ends curl around the refractory zone to produce spirals that rotate at a velocity characteristic of the chemical composition of the reagent. If the layer of the reacting solution is deep enough, rotating scrolls (three-dimensional structures) are produced.

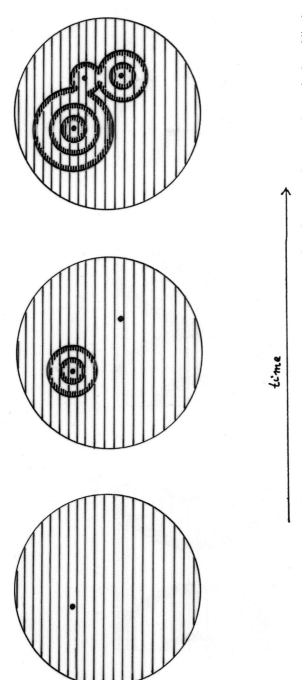

Fig. 7 The development of a ring pattern from trigger waves: (a) the appearance of the first pacemaker center (a spot in the oxidized state of the reagent); (b) the evolution of concentric rings and the formation of the second center; (c) the ring pattern in a more developed state.

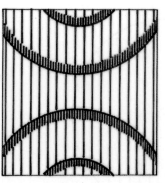
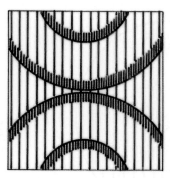
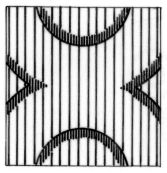

Fig. 8 The collision of two wave fronts. The contacting waves, both in oxidized state (blue), annihiliate each other.

4.2a *The Chemical Mechanism of Trigger Wave Propagation*

The appearance of trigger waves and their "fine structure" (the concentration of the key intermediates between consecutive wave fronts) can be interpreted in terms of the FKN mechanism.[33,34] The key intermediates of the reaction are bromide ion, bromous acid ($HBrO_2$), and the catalyst (ferroin, in most of the experiments). These intermediates are considered also in the Oregonator model as Y, X, and Z. By taking the temporal concentration changes of the intermediates, Field and Noyes[33] suggested concentration cross-sections for the propagating trigger waves in the cerium-catalyzed BZ system, which is shown in Fig. 9. Here the concentrations of the three intermediates are plotted against distance; at the origo the leading edge of the consecutive wave should be considered. The direction of the wave movement is from left to right, and as the wave propagates it consumes bromide ions and leaves

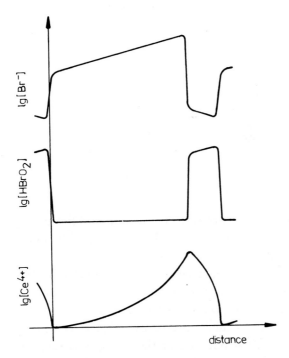

Fig. 9 Concentration cross-sections for a trigger wave. Three important intermediates of the BZ reaction (Br^-, $HBrO_2$, and Ce^{4+}) are plotted against distance.

behind it a much higher bromide ion concentration than that in front of it. This means that, in the leading edge, the reacting system is in kinetic state B and behind it in kinetic state A. In the trailing region the reduction of the catalyst, $Ce^{4+} \rightarrow Ce^{3+}$ or iron(III)-complex→iron(II)-complex, occurs and, as a result, the redox potential of the system continuously decreases till the leading edge of the consecutive wave (see the bottom plot in Fig. 9).

The region of high bromide ion concentration behind the front of the wave is unfavorable for another wave front to move into it; that is, this region is temporarily refractory to the penetration of another wave. This is the reason a moving band will never catch up with the one in front of it and will not be reflected by any physical barrier.

Field and Noyes[34] were the first to measure the rate of propagation of trigger waves in an uncreated reagent, i.e., where the initial chemical conditions exist, as a function of the concentration of reactants (bromate, malonic acid, bromomalonic acid, and ferroin). They found that the rate can be approximated by the $v = k\sqrt{[H^+][BrO_3^-]}$ equation. This shows that, as regards the rate of wave propagation, the concentrations of the other reactants are irrelevant. They and others[35] could rationalize this result on the basis of the Oregonator reaction diffusion Eqs. (1) and (2)

$$\left(\frac{\partial X}{\partial t}\right)_l = D_X\left(\frac{\partial^2 X}{\partial l^2}\right) + k_1[Y] - k_2[X][Y] + k_3[A][X] - 2k_4[X]^2 \quad (1)$$

$$\left(\frac{\partial Y}{\partial t}\right)_l = D_Y\left(\frac{\partial^2 Y}{\partial l^2}\right) - k_1[Y] - k_2[X][Y] + fk_5[Z] \quad (2)$$

by making some simplifications: suppression of diffusion terms for a few variables and suppression of some kinetic terms. The rate constants are for the respective Oregonator reactions (see above).

The temperature dependence of k, studied between 284 and 318 K, gave an apparent activation energy of about 35 kJ/mol.

It is a much-debated point how trigger waves, which appear in a thin (1–2 mm) layer of an oscillatory or excitable reagent, originate. There exist two possibilities: one is that they arise as a result of symmetry-breaking fluctuations (e.g., in concentration or in temperature), and the other is that they are produced deterministically from a heterogeneous center (e.g., a gas bubble, a mote). Although it is an experimental fact that in ultrafiltered BZ solutions the formation of trigger waves can be suppressed, this can be achieved only in excitable systems and not in systems exhibiting temporal oscillations. At present more evidence support the first possibility; the problem, however, is far from being solved.

There exist an analogy between the propagation of trigger waves and that of nerve impulses. This analogy is not coincidental but can be traced to similarities in the mathematical description of these phenomena: the threshold excitation in the Oregonator equations and the threshold phenomenon in the Hodgkin–Huxley model.[35] The latter model is a complex system of nonlinear partial differential equations that describes the propagation of an electric signal along a nerve axon. Chemical waves propagate because diffusion of a reactive species ahead of the wave front triggers the autocatalytic production of that species in the adjacent region. The nerve impulses spread by passive diffusion of membrane potential ahead of the wave, which triggers a kind of autocatalytic increase in membrane potential.

By physical influence chemical waves—the concentric moving reaction-diffusion systems—can be broken and a variety of geometric forms can be produced. In spite of the wealth in geometric forms, the underlying chemistry is preserved. In order to account for the emerging geometries, the prevailing physical conditions should be considered.

4.3 Sustained Chemical Waves

The usual way to investigate spatial self-organization in chemical systems (e.g., the formation of concentric ring patterns, etc.) is to conduct the experiments under batch configuration (e.g., in a Petri dish or in a test tube). Due to a continuous decrease in the concentrations of the reactants, the evolution of the spatial structures proceeds in an ever-changing reaction medium. Thus the appearing spatial structures undergo, to a certain extent, an uncontrollable change and finally, as the reacting system approaches the thermodynamic equilibrium, they cease to exist. In order to overcome this problem, Noszticzius et al.[36] constructed a novel type of open reactor in which they fed the reagents at the inner and outer rims of an annular inert gel. They studied the BZ reaction and replenished the reducing components of the BZ system (malonic acid and ferroin) from the center of the reactor and the oxidizing agent (acidic bromate) from the outside. After a couple of hours, pacemaker centers form and waves propagate both clockwise and counterclockwise, eliminating each other when they "collide." The authors used temporary concentration gradients and could block waves travelling in one direction. They thus select waves that can circulate around the ring one direction and provide evidence that in a circular geometry the nonlinear chemical dynamics may lead to the development of rotating wave structures.

4.4 *Spiral Waves*

When an expanding wave front is disrupted, for example, by applying a gentle blast of air from a pipette onto the surface of the reacting solution, spiral-shaped waves start to develop. This type of wave was first reported by Winfree,[32] who found that when trigger waves propagating in a BZ solution are broken, spiral waves appear and hinder the formation of concentric ring waves. These waves have received special attention ever since their discovery in 1972, and, owing to the activity of a research group led by Hess in Dortmund, even the finest details of the spiral structures have been unraveled.

At sites where the circular wave is broken, i.e., at the open ends, a rather high-speed evolution starts and a structure composed of a pair of counterrotating spiral waves with greatly regular geometry emerges. The tips of the spirals turn inward, whereas the fronts move in the outward direction. The rotation period of the tips of the spiral shaped vortices is in the order of a few tenths of seconds.

Three phases of the spontaneous development of a spiral structure can be seen in Fig. 10.

The relation between the curvature and the propagation velocity of a spiral wave front has also been studied and, by modeling rotating waves in an excitable BZ system, Keener and Tyson[37] predicted that curvature effect would prevent a circular front propagating away from a center if its initial radius is smaller than a critical size of typically 20 μm. Keener[38] developed an approximate theory of wave propagation and found that the geometric theory he proposed for waves travelling in an excitable reacting medium strongly resembles the geometrical diffraction theory of

Oscillations, Waves, and Spirals in Chemical Systems 241

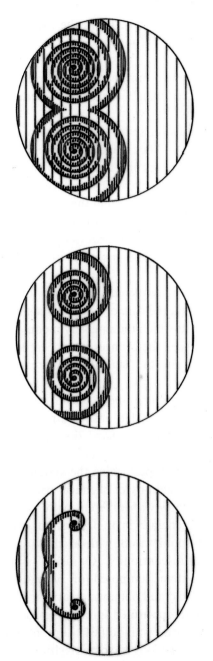

Fig. 10 The evolution of a spiral ring pattern in a reacting BZ system.

high-frequency waves in hyperbolic systems. From the theory he could determine the wavelength and frequency for spiral structures in excitable media.

The computer simulation of the two-dimensional reacting BZ system was performed by Madore and Freedman.[39] They could generate, among others, spiral structures as well.

The most thorough investigations on spiral waves have been carried out by Müller, Plesser, and Hess.[40–43] They could characterize the spatial patterns quantitatively by measuring the concentration distribution of an essential compound applying nonintrusive space-resolved observation techniques of high resolution. With the BZ system the color photographs they obtain give information on the ion-concentration distribution of the oxidized form (ferriin, blue) and the reduced form (ferroin, pink) of the catalyst in two dimensions and show the esthetic order in the developing structures.

The technique Müller et al. applied rendered a close view of the tip of the spiral possible. It was shown that the core of a spiral is a singular site not exceeding 30 μm in diameter at which the intensity modulations due to the ferroin-ferriin distribution (i.e., the changes in concentrations) are at least ten times smaller than in the surrounding area of spiral propagation.

Tam et al.[44] could generate *sustained spiral waves* when the BZ reaction was performed in a continuously fed unstirred reactor (CFUR) and the system have been maintained at a fixed distance away from equilibrium. Spirals were formed above a certain bromate concentration (> 0.018 M) and could be initiated by perturbing the system. CFUR is a suitable tool for investigating the stability of chemical pattern and the transitions between well-defined states with different patterns.

Agladze and Krinsky[45] could generate in a BZ system *multiarmed spirals* by breaking the core of spiral waves with a silver needle and adding chloride ions at the site of discontinuity. Rovinsky[46] simulated a two-dimensional BZ reacting system composed of bromate ferroin, bromomalonic acid, and sulfuric acid and obtained solutions in the form of propagating single-, double-, and four-armed spiral waves, as shown in Fig. 11. For the simulation, he used a modified form of the FKN mechanism in which protonation reactions of some oxybromine species

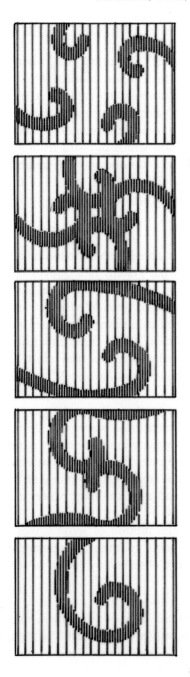

Fig. 11 The different phases of development of a multiarmed spiral in a reacting BZ system.

and the oxidation, by ferriin, of bromomalonic acid and its bromination were included into the mechanism.

Recently, Maselko and Showalter[47] reported that when a cation-exchange bead (~ 1.4 mm in diameter) was loaded with ferroin and bathed in BZ reaction mixture containing no catalyst, a wave, on the *surface* of the bead, started to develop to form a *rotating spiral*. The waves rotate from pole to pole in a single direction. If we inspect the whole process we find that the spiral winds outward from a wandering site at one pole, crosses the equator, and undergoes self-annihilation as it winds into itself at the other pole. The different phases of this peculiar wave movement and the transient structures are shown in Fig. 12.

4.5 *Scroll-Shaped Waves*

Since even a thin layer of the excitable BZ solution has finite thickness, the spiral wave we observe in a BZ system (see above) is actually a cross-section of a three-dimensional wave. This wave has a scroll shape that persists over a rather long period of time. Winfree[48] was the first to look into the third dimension when examining two-dimensional trigger waves. He gave a detailed description of the three-dimensional (scroll-shaped) wave forms and discussed this wave type most extensively, among others, in his thought-provoking book[49] and in his monographies.[50,51]

One of the possible wave forms in three dimensions is the involute spiral shown in Fig. 13. Three-dimensional waves can be made easily visible by allowing the wave to propagate in an inert matrix provided by stacks of filter paper. After a few minutes the developed wave pattern can be "frozen in" by suitable chemical treatment (by plunging the stack in a cold perchloric acid solution), and then the filter papers examined in cross-section.

Keener and Tyson[52] analyzed the motion of scroll waves in the BZ reagent theoretically and found good quantitative agreement between their model and the experimentally observed evolution of scroll waves.

5. Epilogue

The BZ reacting system is not the only one in which, under certain chemical and physical conditions, waves of chemical activity appear and propagate, and structures exhibit themselves. Neither of the other

Oscillations, Waves, and Spirals in Chemical Systems   245

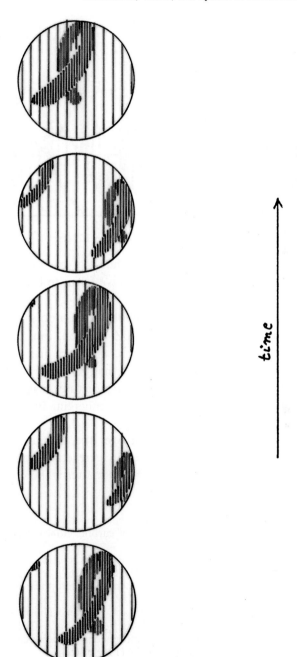

Fig. 12   The movement of an oxidizing zone on the surface of a sphere (ion exchanger bead). The bead is soaked with the BZ reagent.

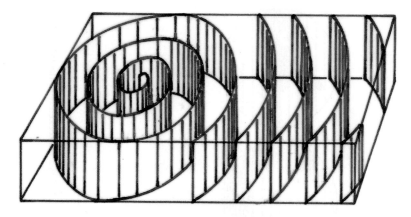

Fig. 13 Scrolled-shaped wave in a BZ system.

systems, however, can express such a richness of phenomena as the BZ reaction, and it has therefore been conducive for us to highlight this very system.

We demonstrated that a relatively simple redox process, proceeding in a homogeneous aqueous solution, the BZ reaction, can, under stirring, show temporal concentration oscillation, and can, in unstirred state, develop chemical waves. When it is either physically or chemically disturbed, various spiral- and scroll-shaped structures may appear. For all these phenomena, the peculiar kinetics inherent in the system is primarily responsible.

It is no wonder that, when leaning over the laboratory bench and watching the shaping of the structures in a reaction vessel, Johannes Kepler's idea emerges in our mind: "Where matter is there is geometry."

References

1. J. S. Morgan, The periodic evolution of carbon monoxide, *J. Chem. Soc. Trans.*, **109** (1916) 274–278.
2. W. C. Bray, A periodic reaction in homogeneous solution and its relation to catalysis, *J. Am. Chem. Soc.*, **43** (1921) 1262–1267.
3. F. O. Rice and O. M. Reiff, The thermal decomposition of hydrogen peroxide, *J. Phys. Chem.* **31** (1927) 1352–1356.
4. M. G. Peard and C. F. Cullis, A periodic reaction. The reaction between hydrogen peroxide and iodic acid, *Trans. Faraday Soc.*, **47** (1951) 616–630.
5. D. H. Shaw and H. O. Pritchard, Homogeneous periodic reactions, *J. Phys. Chem.*, **72** (1968) 1403–1404; *ibid,* **72** (1968) 2693.

6. P. Glansdorff and I. Prigogine, *Thermodynamic Theory of Structure, Stability and Fluctuations*, Wiley, New York (1971).
7. E. Körös and M. Orbán, Uncatalyzed oscillatory chemical reactions, *Nature*, **273** (1978) 371–372.
8. M. Orbán and E. Körös, Chemical oscillations during the uncatalyzed reaction of aromatic compounds with bromate, *J. Phys. Chem.*, **82** (1978) 1672–1673.
9. T. S. Briggs and W. C. Rauscher, An oscillating iodine clock, *J. Chem. Educ.*, **50** (1973) 496–497.
10. I. R. Epstein, K. Kustin, P. De Kepper, and M. Orbán, Oscillating chemical reactions, *Sci. Am.* **248** (1983) 96–108.
11. I. R. Epstein and K. Kustin, Design of inorganic chemical oscillators, *Structure and Bonding*, **56** (1984) 1–33.
12. A. J. Lotka, Undamped oscillations derived from the law of mass action, *J. Am. Chem. Soc.*, **42** (1920) 1595–1599.
13. P. De Kepper and J. Boissonade, From bistability to sustained oscillations, in *Oscillations and Traveling Waves in Chemical Systems*, (Chap. 7, pp. 223–256), R. J. Field and M. Burger, eds. Wiley, New York (1975).
14. I. R. Epstein, Patterns in time and space, *Chem. Eng. News* (1987) 24–36.
15. B. P. Belousov, *A Periodic Reaction and its Mechanism, Sbornik Referatov po Radiatsionni Medicine* (p. 145–147). Medgiz, Moskow (1958).
16. A. M. Zhabotinsky, Periodic processes of the oxidation of malonic acid in solution, *Biofizika*, **9** (1964) 306–310.
17. A. M. Zhabotinsky et al., *Oscillating Processes in Biological and Chemical Systems* (p. 149, 223, 252, 269, 279, 288). Science Publ., Moscow.
18. V. A. Vavilin and A. M. Zhabotinsky, Autocatalytic oxidation of cerium (III) by bromate, *Kinet. Katal.*, **10** (1969) 83–88.
19. R. M. Noyes, R. J. Field, and E. Körös, Oscillations in chemical systems. Part 1. Detailed mechanism in a system showing temporal oscillations, *J. Am. Chem. Soc.*, **94** (1972) 1394–1395.
20. R. J. Field, E. Körös, and R. M. Noyes, Oscillations in chemical systems. Part 2. Thorough analysis of temporal oscillations in the Ce-BrO_3^--malonic acid system, *J. Am. Chem. Soc.*, **94** (1972) 8649–8664.
21. E. Körös, M. Orbán, and Zs. Nagy, Periodic heat evolution during temporal chemical oscillations, *Nature*, **242** (1973) 30–31; *Acta Chim. Hung.*, **100** (1979) 449–461.
22. E. Körös, Monomolecular treatment of chemical oscillation, *Nature*, **251** (1974) 703–704.
23. M. Varga, L. Györgyi, and E. Körös, Bromate oscillators: Elucidation of the source of bromide ion and modification of the chemical mechanism, *J. Am. Chem. Soc.*, **107** (1985) 4780–4781.
24. R. J. Field and H.-D. Försterling, On the oxybromine chemistry rate constants with cerium ions in the Field-Körös-Noyes mechanism of the Belousov-Zhabotinsky reaction, *J. Phys. Chem.*, **90** (1986) 5400–5407.
25. E. Körös and M. Varga, A quantitative study of the iodide-induced high frequency oscillation in bromate-malonic acid — catalyst systems, *J. Phys. Chem.*, **86** (1982) 4839–4843.

26. M. Varga and E. Körös, Various dynamic behavior of Ag$^+$-induced oscillations in uncatalyzed bromate oscillators, *React. Kin. Catal. Lett.,* **28** (1985) 259–268.
27. E. Körös, G. Putirskaya, and M. Varga, Perturbation of bromate oscillators, Part I. Perturbation by gamma radiation, *Acta Chim. Hung.,* **110** (1982) 295–303.
28. D. Edelson, R. J. Field, and R. M. Noyes, Mechanistic details of the Belousov-Zhabotinsky reaction, *Int. J. Chem. Kin.,* **7** (1975) 417–423.
29. R. J. Field and R. M. Noyes, Oscillations in chemical systems, Part 4. Limit cycle behavior in a model of a real chemical reaction, *J. Chem. Phys.,* **60** (1974) 1877–1884.
30. H. G. Busse, A spatial periodic homogeneous chemical reaction, *J. Phys. Chem.,* **73** (1969) 750.
31. A. N. Zaikin and A. M. Zhabotinsky, Concentration wave propagation in a two-dimensional liquid-phase self-oscillating system, *Nature,* **225** (1970) 535–536.
32. A. T. Winfree, Spiral waves of chemical activity, *Science,* **175** (1972) 634–636.
33. R. J. Field and R. M. Noyes, Explanation of spatial band propagation in the Belousov reaction, *Nature,* **237** (1972) 390–392.
34. R. J. Field and R. M. Noyes, Oscillations in chemical systems. Part 5. Quantitative explanation of band migration in the Belousov-Zhabotinsky reaction, *J. Am. Chem. Soc.,* **96** (1979) 2001–2006.
35. W. C. Troy, Mathematical modeling of excitable media in neurobiology and chemistry, in *Theoretical Chemistry, Vol 4: Periodicities in Chemistry and Biology* (pp. 133–157, and references therein), H. Eyring and D. Henderson eds., Academic Press, New York (1978).
36. Z. Noszticiusz, W. Horsthemke, W. D. McCormick, H. L. Swinney, and W. Y. Tam, Sustained chemical waves in an annual reactor: a chemical pinwheel, *Nature,* **329** (1987) 619–620.
37. J. P. Keener and J. J. Tyson, Spiral waves in the Belousov-Zhabotinsky reaction, *Physica D,* **21** (1986) 307–324.
38. J. P. Keener, A geometrical theory for spiral waves in excitable media, *SIAM J. Appl. Math.,* **46** (1986) 1039–1056.
39. B. F. Madore and W. L. Freedman, Computer simulations of the Belousov-Zhabotinsky reaction, *Science,* **222** (1983) 437–438.
40. S. C. Müller, T. Plesser, and B. Hess, Two-dimensional spectro-photometry and pseudo-color representation of chemical reaction patterns, *Naturwissenschaften,* **73** (1986) 165–179.
41. S. C. Müller, T. Plesser, and B. Hess, Three-dimensional representation of chemical gradients, *Biophys. Chem.,* **26** (1987) 357–365.
42. S. C. Müller, T. Plesser, and B. Hess, Two-dimensional spectro-photometry of spiral wave propagation in the BZ reaction, I., II. *Physica D,* **24** (1987) 71–86; ibid **24** (1987) 87–96.
43. S. C. Müller, T. Plesser, and B. Hess, Distinctive sites in chemical waves: the spiral core and the collision area of two annuli, *J. Stat. Phys.,* **48** (1987) 991–1004.

44. M. Y. Tam, W. Horsthemke, Z. Noszticiusz, and H. L. Swinney, Sustained spiral waves in a continuously fed unstirred chemical reactor, *J. Chem. Phys.*, **88** (1988) 3395–3396.
45. K. I. Agladze and V. I. Krinsky, Multi-armed vortices in an active chemical medium, *Nature*, **296** (1982) 424–426.
46. A. B. Rovinsky, Spiral waves in a model of the ferroin catalyzed Belousov-Zhabotinsky reaction, *J. Phys. Chem.*, **90** (1986) 217–219.
47. J. Maselko and K. Showalter, Chemical waves on spherical surfaces, *Nature*, **339** (1989) 609–611.
48. A. T. Winfree, Scroll-shaped waves of chemical activity in three dimension, *Science*, **181** (1973) 937–939.
49. A. T. Winfree, *The Geometry of Biological Time*. Springer, New York (1980).
50. A. T. Winfree, Stable rotating patterns of reaction and diffusion, in *Theoretical Chemistry, Vol. 4: Periodicities in Chemistry and Biology* (pp. 2–51), H. Eyring and D. Henderson, eds. Academic Press, New York (1978).
51. A. T. Winfree, Organizing centers for chemical waves in two and three dimensions, in *Oscillations and Traveling Waves in Chemical Systems*, (pp. 441–472), R. J. Field and M. Burger, eds. Wiley, New York (1985).
52. J. P. Keener and J. J. Tyson, The motion of untwisted untorted scroll waves in Belousov-Zhabotinsky reagent, *Science*, **239** (1988) 1284–1286.

DETERMINATION OF SPIRAL SYMMETRY IN PLANTS AND POLYMERS

Dawn Friedman

1. Introduction

Almost every plant that produces shoots from a growth point creates a pattern.[1] These patterns of shoot arrangement or phyllotaxy fall into several classes, with spiral symmetry only the most mathematically and esthetically intriguing of a continuum of possibilities. There is one observation, however, that is suggested by every common form of phyllotaxy: new shoots distance themselves from existing ones. For example, as a stem grows and produces leaves, new leaves will appear at positions displaced from older leaves by rotation or by both translation and rotation. The impression given is that existing shoots somehow repel new ones, that they have some power that forces them away. This is not a naive observation. Most current theories invoke something very much like a repulsive force—an *acropetal influence*— to explain the patterns of phyllotaxy. The position of a new primordium, the structure that becomes a leaf, petal, or other organ, is determined by the inhibiting effect produced by existing primordia.

There are interesting analogies between plant phyllotaxy and molecular conformation, particularly in the case of polymer chains. Some biopolymers, organic polymers, and inorganic polymers exhibit a helical backbone conformation similar to the spiral arrangement of primordia around a stem, while others assume nonspiral configurations. A few may display more than one type of symmetry within a single natural sample, recalling certain plants that exhibit transitions between phyllotactic systems. The acropetal influence that drives plant phyllotaxy has a chemical counterpart: substituents on a backbone tend to maximize their distance from one another due to the repulsive forces between pairs of electrons. These interactions determine which conformational system a polymer chain will adopt in the same way that interactions between primordia determine a plant's phyllotaxy. In both cases, the patterns found in natural systems can be reproduced by modeling the significant interactions.

A model of phyllotaxy presented by Schwabe and Clewer[2] will be reviewed here, and its results will be compared with those of an orbital interaction model of polymer conformation. It is helpful to begin by sorting the rich natural variety of phyllotactic patterns into a few categories. The model that will be discussed attempts to reproduce these categories and the transitions between them.

2. Phyllotaxies Taxonomized

Phyllotactic arrangements can be described in terms of two factors: the number of primordia initiated at each stem level, and the spatial relationship of the organs at one level to those in the next. In most plants, shoots are produced along a stem as it grows. One or more shoots may be initiated at each level on the stem. If several shoot primordia are initiated at the same level, they form a ring or *whorl*. This arrangement of similar units around a center generally exhibits point symmetry;[3] the individual primordia are spaced as far apart as possible. Whorls often appear in groups of organs at the end of a stem, such as the parts of a flower. The sepals and petals of a strawberry blossom (Fig. 1) are arranged in whorls of five, and the whole flower has C_{5v} symmetry, showing its rose family pedigree. Woodruff (*Asperula odorata*), among other plants, produces its leaves in whorls: each stem consists of many-leaved whorls alternating with bare stem (Fig. 2).

The fewer the members in each whorl, the easier it is to distinguish a significant feature of whorled phyllotaxis: screw rotation. The five petals of the strawberry flower fall between each pair of sepals. Oleander produces whorls of three leaves, and each successive whorl is rotated by 60°, a pattern designated *tricussate* (Fig. 3). The intriguing resemblance of this system to the staggered conformation of the ethane molecule is not a coincidence, as later discussion will show.

Fig. 1. A strawberry flower has five petals and five sepals, arranged with C_{5v} symmetry. Adapted figure from *A Student's Atlas of Flowering Plants* by Carroll E. Wood and Harvard University. Copyright 1974 by the president and fellows of Harvard College. Reprinted by permission of Harper Collins Publishers.

Plants in the mint family produce pairs of leaves, two-membered whorls (Fig. 4). Each successive whorl is rotated by 90°; this boxy arrangement, which in the mints is emphasized by a four-sided stem, is called *decussate* phyllotaxy. The decussate pattern is significant in many dicotyledonous plants, because the pair of seed leaves in the young plant is often succeeded by a pair of new leaves rotated 90°. This is true even in some cases where later growth shows quite different phyllotaxis.

A "whorl" can consist of a single leaf. In this case, the next whorl will be rotated by 180°, so that leaves appear above and opposite their predecessors. This zig-zag pattern in a single plane is called alternate or *distichous* phyllotaxis. Many monocotyledonous plants, such as corn and onion, show this pattern; Fig. 5 shows a woodland relative of the lily, twisted-stalk (*Streptopus roseus*). The planar zig-zag pattern with successive units rotated by 180° is also a common polymer conformation.

The angle rotation between successive organs is called the *divergence angle*. When organs are produced at different levels on a stem, the divergence angle is the dihedral angle between them. It characterizes all

Fig. 2. Sweet woodruff (Asperula odorata) exhibits whorled phyllotaxy (adapted with permission from E. M. Felsko, *Blumen-Fibel,* F. A. Herbig Verlagsbuchhandlung, Berlin, 1956.)

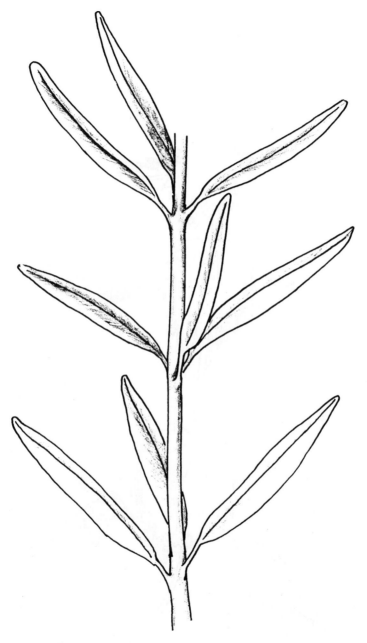

Fig. 3. Nerium oleander, showing tricussate phyllotaxy. Each whorl of three leaves is perfectly staggered with respect to its neighbors.

Fig. 4. Urtica urens exhibits the decussate phyllotaxy that characterizes the Mint family. (Adapted from E. M. Felsko.)

Determination of Spiral Symmetry in Plants and Polymers 257

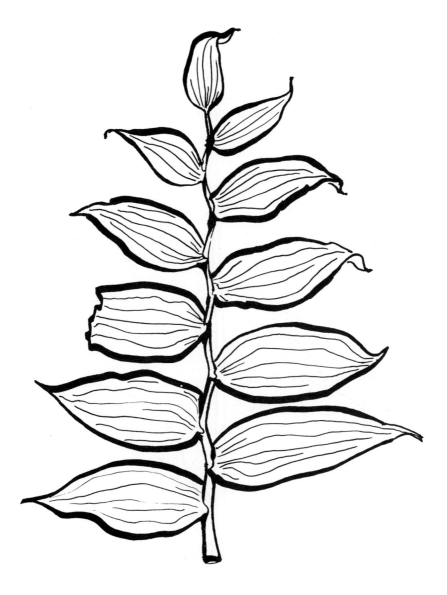

Fig. 5. Twisted-stalk (Streptopus roseus) shows alternate or distichous phyllotaxy. (Adapted from G. L. Walton, *The Flower-Finder* (3rd Edn.) J.B. Lippincott Company, Philadelphia, 1930.)

the arrangements in which a single primordium is initiated at each level. When the divergence angle is less than 180°, the result is some form of *spiral phyllotaxis*. For example, in Fig. 6 a series of leaves spaced along a stem forms a single prominent spiral.

Where many primordia crowd into a short vertical span and form a broad cone or disc, it is possible to count sets of spirals, clockwise and

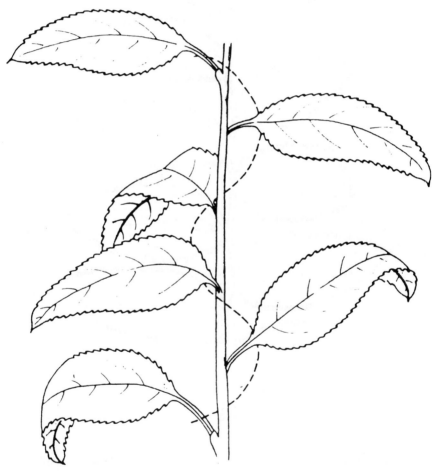

Fig. 6. Leaves on a cherry twig form a single spiral. The phyllotactic fraction is 2/5: five leaves complete two rotations. (From A. Fahn, *Plant Anatomy*. Pergamon Press, New York, 1969.)

counterclockwise. Because these spirals occur in numbers that are members of the Fibonacci series, expressed in general terms as

$$n_i = n_{i-2} + n_{i-1},$$

spiral phyllotaxy of this kind has attracted great attention from biologists, mathematicians, and computer scientists. The Fibonacci numbers approximate the evocatively named golden ratio, $r = 1.618...$, with interesting mathematical and esthetic properties, and the convergence of many spirals on a single point is the most visually arresting of plant phyllotaxies.

In this connection, it may be conceded that no discussion of phyllotaxy is complete without a stunning example of vertically crowded spiral symmetry. D'Arcy Thompson reproduces a many-spiraled cauliflower as Fig. 449 in the second edition of *On Growth and Form*.[4] Close relatives of this cauliflower appear as broccoli cultivar "Minaret" in Johnny's Selected Seeds catalog (Fig. 7), and as "Romanesco" elsewhere.

Readers who garden may be interested to know that these varieties are recommended for their flavor as well as for their unusual shape, which the catalogs seem to have a little trouble describing. Thompson distinguishes first- and second-order spirals in his cauliflower; three orders of self-similar spirals are visible in Fig. 7. Oddly enough, this example does not appear in *The Fractal Geometry of Nature*, even in a speculation on the genetic difference in fractal dimension between broccoli and cauliflower.[5] The genetic regulation of the difference between Romanesco-type and ordinary broccoli poses questions at least equally interesting.

Even a multileveled splendor like Minaret, however, need not daunt the researcher or computer modeler. Each of these spirals, whether single or multiple, is the result of a specific divergence angle between successive units. Fibonacci-numbered spirals occur when the divergence angle is close to 137.5°, which equals 2π divided by the golden ratio squared. Larger and smaller divergence angles produce less symmetrical spirals. A single model should be capable of reproducing spiral, whorled, and distichous phyllotaxis. It should also allow for variations within these types and transitions between them. Such a model, presented by Schwabe and Clewer, is summarized below.

Fig. 7. Broccoli cultivar Minaret shows multiple orders of vertically condensed spiral phyllotaxis. (Reproduced by permission of Johnny's Selected Seeds.)

3. A Model of Phyllotaxy Based on Inhibitor Diffusion

Almost all theories of phyllotaxis begin with the assumption that existing organs help determine the site of initiation for new organs. For example, new leaves form in a region below the tip of the growing shoot and above existing leaves, but within this region (the *anneau initial*) new leaves are initiated at particular points determined by the repelling influence of existing leaves. The nature of the acropetal influence, chemical or physical, has not been demonstrated, and there are those who postulate biophysical changes in the growing point due to existing primordia.[1,6] However, most current theories assume that the influence exerted by existing primordia is biochemical. They concern themselves with modeling the uptake of a promoter substance[7] or the diffusion of an inhibitor compound.[2] As it turns out, it is entirely possible to reproduce phyllotactic patterns without identifying the presumed "phyllotaxy compound" or compounds. A few parameters, based on assumptions about growth and inhibitory influence, are all that is needed.

Schwabe and Clewer base their computer model on a set of assumptions that may be summarized as follows. Primordia form on a ring just beneath the apex of the growing point. As growth occurs, existing primordia move downward relative to this ring, which under steady-state conditions is of constant radius. Existing primordia produce an inhibitor that prevents the initiation of new primordia. Inhibitor diffuses sideways and downward, but only minimally upward, from each existing primordium, so that new primordia form above, not below, the level of the newest existing primordium. Growth is slow enough compared to inhibitor diffusion that the concentration of inhibitor at a point on the ring due to existing primordia can be calculated by

$$C = s \Sigma e^{-x(i)},$$

where $x(i)$ is the distance of the point from the ith primordium and s is the inhibitor source strength of each primordium. A new primordium is initiated when the concentration of inhibitor at some point falls below a certain threshold value.

Only four parameters are used in the model. Two are geometric: the radius of the ring is represented by r, the angle of the growth cone by α. The values $r = 0.5$ and $\alpha = 0$ represent a cylinder. The remaining parameters describe the inhibitory effect of existing primordia on new ones: g, the polar transport constant, represents the preferential diffusion of inhibitor downward; s/t is the ratio of the inhibitor source strength of each primordium to the threshold concentration of inhibitor.

Growth and formation of new primordia can be played out in the following way. The distance from a potential growth site on the ring to the newest existing primordium depends on the vertical distance of the ring and angular separation of that point from the existing primordium. The vertical distance reflects the time that the stem has had to grow since the last primordium was formed. If the inhibitor concentration at all points on the ring is above the threshold value, growth must occur before a new primordium can be initiated. When the vertical distance reaches a certain value, designated d, the concentration falls below the threshold value at some point on the ring, and a new primordium is formed, with angular separation θ from its predecessor. This angle θ is the divergence angle.

The model begins with one primordium, P1, already present, its position due to chance or the nature of the seed embryo. The minimal concentration of inhibitor will be 180° from P1. The second primordium, P2, will therefore form directly opposite P1, with or without a vertical distance between them. P3 will be initiated at some distance d and angle θ from the second, depending on the inhibitory effects of P1 and P2. As successive primordia are formed, d and θ each tend to converge on a stable value, and a particular phyllotactic pattern emerges. The pattern that forms depends on the parameters that describe the behavior of the inhibitory effect due to existing primordia.

4. Results of the Model

The simplest system, alternate or distichous phyllotaxis, occurs when only one existing primordium, the latest formed, determines the placement of the next. For the cylindrical model, parameter values of $g = 1.0$ and $s/t = 3.0$ represent a system where each primordium is a strong source of inhibitor compared to the threshold value. The diffusion of inhibitor from the initial primordium, P1, forces vertical growth before P2 can be formed, above and 180° away from P1. Further vertical growth is now needed to drop concentration of inhibitor due to P2 below the threshold. By the time P3 can be initiated, only P2 can influence its position—P1 is too far way. P3 forms above and 180° away from P2, directly above P1. Each new leaf solely determines the position of its successor; this is the meaning of distichous phyllotaxis.

Decussate (two-membered whorl) systems can be reproduced with $g = 3.5$, $s/t = 1.1$. In this case, the first two leaf primordia, P1 and P2, are formed at the same level, but growth must take place before inhibitor concentration drops sufficiently for more leaves to be initiated. The third primordium, P3, will be initiated above and between the first pair, at a point 90° (or 270°) from P2; but where will P4 appear? If it is initiated simultaneously with P3, then it must appear 180° from P3, because this is the other point where concentration of inhibitor produced by P1 and P2 is lowest. This will ensure a decussate system. But simultaneous initiation is not necessary for decussate phyllotaxis; a relatively large value of g will produce this system even when successive primordia are initiated independently.

This result follows because a large value of g corresponds to a rapid fall-off in inhibitor concentration due to older primordia. If the effects of P1 and P2 have fallen off sufficiently, the point opposite P3 and at the same level will be below the threshold concentration, and P4 will appear there without further vertical growth. P3 and P4 are now a second whorl of two, 180° apart but at the same level. Vertical growth is now required before a P5 and P6 can be initiated. When they are, the large value of g will mean that the effects of P1 and P2 are negligible. P5 will appear at 90° from P4, and P6 opposite P5, even though this means that they will be directly above the pair P1 and P2. This is the decussate system.

Whorled systems with n leaves per whorl can be generated with similar parameters. The first n primordia form at the same level to create the first whorl. If g is sufficiently large, after vertical growth, all of the next n primordia can be initiated at the same level as primordium $n + 1$, forming the second whorl. The second will be rotated relative to the first whorl, since primordium $n + 1$ will be initiated as far as possible from the n existing primordia.

In distichous, decussate, and other whorled systems, placement of primordia at vertical level L is affected only by existing primordia at level $L - 1$. The falloff in source strength with vertical growth, g, is too high to allow any but the immediately preceding whorl to influence the new one. In order to achieve spiral phyllotaxis, the position of new primordia must depend on at least two levels or generations of existing primordia.

A set of parameters for a cylindrical model that simulates influence from more than one existing level of primordia is $g = 1.4$, $s/t = 1.1$. Again, the model starts with a single primordium, P1. P2 forms 180° from P1, and because s/t is small, it forms at the same level. Now the concentration of inhibitor is above the threshold level at all points on the ring, and growth must occur. When P1 and P2 are a certain distance below the ring, P3 forms, 90° or 270° from P2. So far, these parameters yield a decussate pattern. But because g is so much lower, inhibitor concentration due to P1 and P2 is higher, and the initiation of P3 brings the inhibitor concentration above the threshold all around the growth ring. P4 cannot form at the same level as P3; it forms above and opposite P3.

There are now existing primordia at three levels: the pair P1 and P2 at the oldest level, then P3, and finally P4. Inhibitor concentration is once again above the threshold level all around the ring, and vertical growth follows. If s/t were high enough, the growth ring would be carried so high that only the influence of P4 could affect the position of P5, and a distichous system would result. But with $s/t = 1.1$ and $g = 1.4$, both P3 and P4 influence the position of P5. P5 appears at a divergence angle of 152.7° from P4. Due to the effects of P4 and P5, P6 is initiated at divergence angle $\theta = 131.4°$. A spiral pattern begins to appear. Eventually, after about P16, a steady state is reached, with the Fibonacci divergence angle of 137.5° and a constant distance between levels. This set of parameters is one of many that produce spiral phyllotaxis.

What about transitions between systems? Almost every possible transition can be simulated by starting with the correct set of parameters and decreasing or increasing g or s/t. For example, a high value of g and a moderate ratio s/t generates a decussate system; if the value of g is decreased, new primordia will be initiated in a pattern that soon becomes spiral. A model with low g and moderate s/t will start with a spiral and switch to a distichous system (divergence angle increasing to 180°) as s/t is increased. Changes in the shape and size, α and r, of the growth area also influence the pattern created, because they change the distance between primordia. As the growth area is made flatter and broader by increasing α and r, preceding generations of primordia remain closer to the growth ring. The influence of many generations of primordia ensures spiral phyllotaxis over a wider range of parameters.

Is there any naturally occurring phyllotactic system that this model cannot reproduce? In nature there always seems to be an exception to the rule. In about 13% of families of flowering plants, species occur in which members of floral whorls are superposed, member directly on top of member, rather than alternating or staggered; in very few cases, this superposition has been reported in vegetative growth. Explaining these exceptions in terms of inhibitor diffusion might involve mechanisms for lowering the inhibitor concentration near an existing primordium. For the time being, these "eclipsed" whorls remain a challenge to most theories of phyllotaxy.[8]

Schwabe and Clewer conclude that rather simple mathematics can produce almost the entire range of systems and transitions found in nature. The basic principle is independent of the biological mechanisms involved: pattern is determined by the behavior of the influence due to existing primordia. The robustness of this model, they suggest, may explain the ubiquity of the patterns it generates.

5. The "Acropetal Influence" in Chemistry: Orbital Interactions

There is more than one similarity between the placement of primordia on a plant stem and the arrangement of chemical substituents around a polymer backbone. Both problems require the researcher to explain a wealth of natural variation. Both are deeply connected with symmetry. Both have been addressed by mathematical treatments and models that may or may not make claims concerning underlying mechanism; and in both cases these models have often yielded useful results. Both continue to provoke debates about the interpretation, not only of experimental data, but also of theories and models. But the most compelling similarity involves a generally accepted conclusion: many polymers assume either zig-zag or helical conformations depending on the behavior of the interaction between electron pairs, just as plants show distichous, whorled, or spiral phyllotaxy according to the behavior of the "repulsion" between primordia.

The basic questions of molecular conformation are old ones, but theories and their interpretation are still being debated. One of the classic problems is the rotational barrier of ethane. The two CH_3 groups that make up ethane, C_2H_6, can rotate relative to each other around the C-C bond. But all possible positions—rotational *conformers*—are not equally likely. Certain conformations are favored, lower in energy, while others are unfavorable, higher energy states. The three hydrogen atoms on one carbon will usually be *staggered* with respect to those on the other—rotated 60°, like the leaves in tricussate whorls (Fig. 8a). An *eclipsed* conformation, with each hydrogen atom directly "above" its counterpart, is unfavorable (Fig. 8b). In order for the CH_3 groups to rotate from one staggered conformation to another, the ethane molecule must pass through the eclipsed state. It is the energy price that must be paid to do so that constitutes the rotational "barrier."

Where does the energy barrier come from? Why is the eclipsed state un-

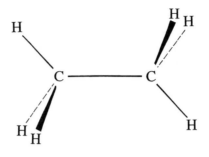
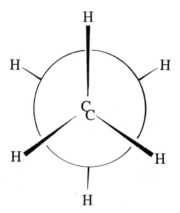

Fig. 8a. Ethane in the staggered conformation. The dihedral angle between hydrogen atoms is 60°. The lower sketch is a projection of the upper.

favorable? As in phyllotaxy, the simplest observation is that the hydrogens bonded to each carbon somehow repel each other. And as in phyllotaxy, many years of debate on mechanism have not made this observation less true or less useful. Many molecular conformations can be predicted based on the principle of electron pair, or filled orbital, repulsion.

The Pauli electron exclusion principle allows no more than two electrons to have the same spatial wave function; very roughly, this means that no more than two electrons can occupy the same set of positions with the same probabilities. These position probabilities define an *orbital*. Orbitals may correspond to electron positions that are closely bound to the atomic nucleus, loosely bound, or shared between two atoms. Each orbital may be filled—occupied by its full complement of two electrons—half filled, or empty.

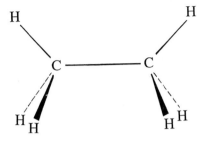

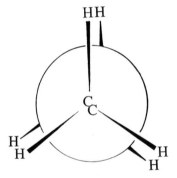

Fig. 8b. Ethane in the eclipsed conformation.

An orbital that is tightly bound to the nucleus of the atom has a lower energy than one that is less tightly bound. There is always a tendency for lower energy states to be occupied. For this reason, all the low-energy orbitals belonging to an atom will usually be filled with that atom's electrons. But if the atom has only enough electrons to half fill some of its low-energy orbitals, these orbitals may be shared by electrons from other atoms. An orbital with electrons from two different atoms is a bonding orbital. Bonding orbitals represent the electronic positions that bind atoms into molecules.

An orbital filled only with electrons from its own atom is a nonbonding orbital. These orbitals do not hold molecules together, but they affect molecular conformation. Because no more than two electrons can occupy the same orbital, orbitals that are already filled, whether they are bonding or nonbonding, repel the electrons in other filled orbitals. Filled orbitals avoid each other, seek positions as far as possible from each other, much as leaf primordia do.

There are always some orbitals in a molecule that are higher in energy than the others, and so are left completely unoccupied. Empty orbitals have no electrons with which to repel other orbitals. They usually have no direct effect on filled orbitals or molecular conformation. But a few empty orbitals are relatively low-energy, not much higher than a high-energy filled orbital. If a filled and an empty orbital are of similar energy, there may be a favorable interaction between them. The two orbitals will seek positions that promote this interaction. Such situations occur much more rarely than the repulsive interaction between two filled orbitals. It is the filled–filled orbital interaction that defines most molecular conformations.

To a certain order of approximation, then, the hydrogen atoms in ethane avoid each other because each of the carbon–hydrogen bonds comprises a filled bond orbital. Within the ethane molecule, in each CH_3 group, the four bond orbitals representing the C-C and the three C-H bonds are directed to the corners of a tetrahedron, as far away from each other as they can get. The filled C-H bond orbitals on each carbon constitute a "whorl," spaced evenly around the carbon center. The elements of each "whorl" of filled orbitals also exert a repulsive influence on the filled orbitals in the neighboring whorl. This explains the resemblance of staggered ethane to tricussate phyllotaxy.

These four orbitals pointing to the corners of a tetrahedron are a common feature of many molecular structures. Each element has atoms with a different number of electrons, and therefore a different number of filled orbitals. But it is favorable for most atoms to fill a certain set of four *valence* orbitals. If they cannot fill the valence orbitals with eight of their own electrons, they will share electrons from other atoms. This means that some of the valence orbitals will be bond orbitals while others may be lone pairs, depending on how many of its own electrons the atom had originally. A carbon atom has only four of its own electrons to fill valence orbitals. It is somewhat more favorable to fill bond orbitals than lone pair orbitals, so a carbon atom usually shares four electrons belonging to other atoms to make four bond orbitals. An oxygen atom has six electrons in valence orbitals; by sharing two electrons belonging to another atom, it fills a total of two bond orbitals and two lone pairs.

All these filled orbitals, bonding or lone pair, repel each other. The arrangement that allows four structural elements around a center to get as far from each other as possible is a tetrahedron, with each element separated from the others by an angle of 109.5°. If an atom has four filled valence orbitals, with no two of them representing bonds to the same neighboring atom, they will point to the corners of a tetrahedron.

In the case of ethane, each carbon atom is surrounded by a tetrahedral arrangement of filled bonding orbitals. Each carbon atom is bonded to three hydrogen atoms and to the other carbon atom. The ethane molecule consists of two tetrahedral units joined through the C-C bond. If a hydrogen atom on each carbon were replaced by another CH_3 unit, the result would be butane, C_4H_{10}. As more identical units are added to any molecule in this way, a longer and longer chain is created, until the molecule is considered a polymer. What conformations might a polymer molecule assume?

Consider a polymer in which each unit is based on an atom with four filled valence orbitals. Each atom uses two valence orbitals to bond to two neighbor atoms, one on each side, forming a backbone. As long as the two remaining valence orbitals are not used to bond to the same atom, the four valence orbitals will form a tetrahedral arrangement. However, these tetrahedral units will be more or less free to rotate around the bonds between backbone atoms, just as the CH_3 groups in ethane could rotate around the C-C bond. As each unit rotates with respect to the next, a huge number of possible conformations may be generated.

As with ethane, it is possible to determine which conformations are more favorable than others, based on filled-orbital repulsions. Just as in whorled phyllotaxis, there are two orders of positional influences. The repulsion effect between elements on the same center creates the whorl in one case, the tetrahedral arrangement in the other. In both cases, the repulsive effect between elements on neighboring centers determines a staggered conformation between centers. The polymer will consist of a series of tetrahedral centers, each staggered with respect to its neighbors.

In the case of ethane, this description was sufficient to define the conformation of the molecule. That was because each of the six "whorl" orbitals, the C-H bonds, was identical. Any staggered arrangement was equivalent to any other. Here, however, the six orbitals rotating with respect to each other are not identical; two of them represent backbone bonds to neighboring units. The dihedral angle—"divergence angle"—between the backbone bonds on any two neighbor atoms may be 60°, 180°, or 300°. Each of these possibilities produces a different result.

Dihedral angles near 60° or 300° represent *gauche* conformations; the 180° conformation is called *trans*. Assume that since all units are identical, each one will favor the same rotation with respect to its neighbors as every other. If each unit is rotated by 60°, the backbone will form a right-handed helix (Fig. 9a).

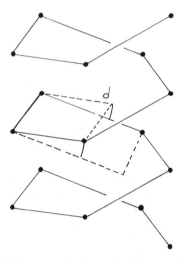

Fig. 9a. Polymer in the all-gauche helical conformation.

A rotation of 300° at every unit will give rise to the left-handed variant. These all-gauche conformations produce the chemical equivalent of spiral phyllotaxy. But a "divergence angle" of 180°, whether in plant or in polymer, cannot produce a spiral. The result is a zig-zag in the plane, the arrangement called alternate or distichous in plant phyllotaxy, all-trans in chemistry (Fig. 9b).

Both the all-gauche and the all-trans conformations satisfy the requirements analyzed so far: the orbitals on each unit are tetrahedrally oriented with respect to each other, staggered with respect to other units. In order to decide between systems of polymer "phyllotaxy," a more detailed model of interorbital repulsions is required.

Up till now, this discussion has assumed that all filled orbitals repel

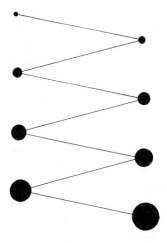

Fig. 9b. Polymer in the all-trans zig-zag conformation.

each other to the same degree. This is not the case. To model the natural range of phyllotactic systems, it was necessary to assume that leaf primordia on different plants might behave differently as inhibitor sources. Similarly, filled orbitals of different types interact differently with other orbitals. The repulsion between two lone pair orbitals on the same atom is very strong. Two bond orbitals repel each other less strongly; the exact strength of the repulsion depends on which atoms the bond connects. The repulsion between a lone pair and a bond orbital on the same atom also varies depending on the atoms involved, but usually falls between the other two types.

This difference in repulsive strength means that four filled valence orbitals only form a perfectly regular tetrahedron if they are all identical, for example, if they are all bonds to the same type of atom. Otherwise, some orbitals will require more space than others. For example, the methane molecule, CH_4, has four C-H bond orbitals pointing to the corners of a perfect tetrahedron, because they all repel each other equally. A water molecule, H_2O, also contains a central atom with four filled valence orbitals surrounding it. But two of those orbitals are oxygen lone pairs, and they repel each other more strongly than the O-H bond orbitals do. The lone pairs end up further apart, the O-H bonds closer together, than they would be if they were all identical.

If differences in filled-orbital repulsions only affected orbitals on the same atom, they might only change the shape of a molecule locally, around each center. But different orbitals also interact differently with orbitals on neighboring atoms. Two lone pair orbitals on different atoms still repel each other strongly, usually more strongly than two bond orbitals do, depending on the atoms in the bond. The strength of this repulsion will affect the preferred dihedral angle between the orbitals. Between filled bonding orbitals on neighboring atoms, the trans dihedral angle of 180° may be only moderately favored relative to the smaller gauche angle. In the case of neighboring lone pairs, the trans preference may be much stronger and is likely to override the conformational preference of bond orbitals on the same atoms.

A lone pair orbital may exhibit unusual conformational behavior with respect to a bond on a neighboring atom. The reason for this is that the space between two bonded atoms contains both filled bonding orbitals and empty orbitals called *antibonding* orbitals. Antibonding orbitals tend to be relatively low-energy as empty orbitals go. The antibonding orbitals that accompany certain bonds are not much higher in energy than a lone pair orbital, since lone pairs tend to be high-energy for filled orbitals. When an antibonding orbital and a lone pair on a neighboring atom are close in energy, a favorable filled–empty orbital interaction is possible. This interaction is *maximized* when the lone pair orbital is trans to the bond. The filled–filled interaction between the lone pair and the bonding orbital remains unfavorable, but it is minimized in the trans position. The result is an unusually strong preference for a dihedral angle of 180° between lone pair and bond.[9] The possibility that this effect will come into play makes the interaction between a lone pair orbital and a neighboring bond the most variable "parameter" in the modeling of polymer conformation.

6. Results of the Orbital Interaction Model: Polymer Conformations

In a recent study, Cui and Kertesz[10] modeled a series of polymers based on backbone atoms with four filled valence orbitals. Their mathematical treatment, based on modified extended Hückel theory, describes the electron wave functions in terms of continuous bands instead of individual orbitals. The results, however, can still be explained in terms of individual orbital interactions.

One of the simplest systems studied was the polysulfur chain, S_n. Like oxygen, sulfur has six electrons of its own available for its valence orbitals. To fill up its valence orbitals, a sulfur atom usually forms bonds with two other atoms; the other two valence orbitals are lone pairs. Sulfur atoms can form a polymeric chain in which each unit consists of a single sulfur atom with its two lone pair orbitals, linked to two neighbors by S-S bond orbitals. The four valence orbitals around each sulfur atom will be oriented approximately to the corners of a tetrahedron, but not precisely, because the two lone pairs will take up more room than S-S bonds. Each of these tetrahedra will be staggered with respect to the next, as in ethane. To determine whether the overall conformation will be zig-zag or spiral, it is necessary to examine individual orbital interactions.

In Fig. 10, four linked sulfur atoms are shown, first in the trans position, then in the gauche. In the trans conformation (Fig. 10a), the S-S

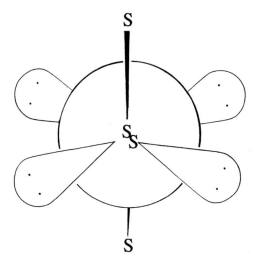

Fig. 10a. Polysulfur chain segment of four sulfur atoms in the trans conformation.

bonds are rotated 180° with respect to each other, and any repulsive interaction between them is minimized. However, the overall trans conformation places the sulfur lone pairs on neighboring atoms gauche to each other. The lone pair–lone pair interaction is much stronger than the bond–bond interaction, and it is not minimized by this conformation. The repulsive force between neighboring lone pairs would tend to make this conformation unfavorable.

In the gauche conformation (Fig. 10b), the dihedral angle between one pair of lone pair orbitals has increased, but the angle between the other two has decreased. This might add up to a less favorable situation than the two gauche interactions in the trans conformation. But this effect is overshadowed by the interaction between the lone pairs and the S-S bonds. The sulfur–sulfur antibonding orbitals that accompany these bonds lie close in energy to the sulfur lone pairs. The favorable filled orbital–empty orbital interaction between them is maximized when the lone pair and the S-S bond are trans to each other. In the overall trans

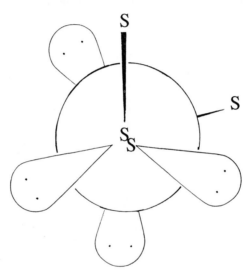

Fig. 10b. Polysulfur chain segment, gauche conformation.

conformation, all four sulfur lone pairs are gauche to the S-S bonds. In the overall gauche conformation, each S-S bond is approximately trans to a lone pair on the neighboring sulfur atom.

Detailed calculations confirm that it is this lone pair–bond interaction that determines the result. The optimum conformation has spiral symmetry: an all-gauche helix with a dihedral angle between S-S bonds of about 85°. The all-trans zig-zag form is less favored by a significant margin.

What would happen in a case where there were no unusual lone pair–antibonding orbital interactions? A polyethylene molecule, $(CH_2)_x$, is a chain of units based on carbon atoms. Each carbon is surrounded by four bond orbitals: two bonds to neighbor carbons, two to hydrogen atoms. (The ethane molecule can be considered a "chain" of two such units, with the loose ends tied with extra hydrogens.) The units in a polyethylene chain will be staggered with respect to their neighbors, so that the possible overall conformations are the all-trans zig-zag and the all-gauche spiral. Calculations show that the all-trans conformation is the most favorable. But a gauche spiral form with a dihedral angle of about 100° is also a stable conformation for polyethylene. Experiments suggest that polyethylene exists in both conformations. The interactions between valence orbitals must explain why polyethylene doesn't "see" the difference between the conformations as well as polysulfur does.

Fig. 11 shows the geometric relationships between valence orbitals in the two conformations. In the trans case (Fig.11a), all the C-H bonds are

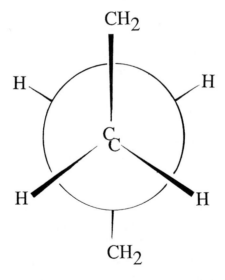

Fig. 11a. Polyethylene chain segment, trans conformation.

gauche to another C-H bond and to a C-C bond. In the gauche case (Fig. 11b), the C-H bonds are still all gauche to each other or to C-C bonds, and of course the two C-C bonds are gauche to each other. C-H bonds repel each other moderately; they also repel C-C bonds moderately, and C-C bonds repel each other moderately; in fact, there is not

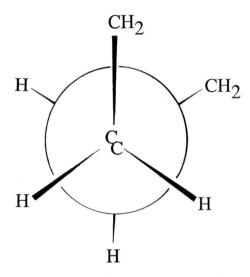

Fig. 11b. Polyethylene chain segment, gauche conformation.

much difference between the three interactions. The two C-C bonds seem to repel each other just a little more strongly than the others, making the trans conformation, where they are further apart, the more stable of the two. But the difference is slight enough to allow both conformations to exist in nature.

A more complex example is poly(oxymethylene), $(CH_2O)_n$. This polymer is composed of alternating CH_2 groups and oxygen atoms. The potential orbital interactions are those between C-H and C-O bonds and the oxygen lone pairs. If these interactions differ sufficiently in strength, then the conformation of poly(oxymethylene), spiral or alternate, will be the one that maximizes the relatively favorable interactions or minimizes the unfavorable. The trans and gauche conformations for a segment of two units, $(CH_2O)_2$, are shown in Figs. 12a and 12b, respectively.

In this case, there are no powerful repulsive interactions between lone pairs on neighboring atoms. All of the gauche interactions, in both conformations, are either moderate repulsions between two bonds, or slightly stronger repulsions between a bond and a lone pair. In the trans conformation, the two C-O bonds are separated by 180°; whatever repulsion exists between them is already minimized. The two C-H bonds lie gauche to the C-O bond on the neighboring oxygen; this interaction is

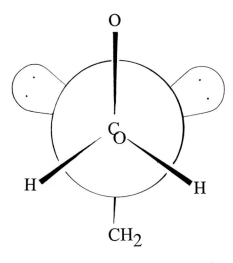

Fig. 12a. Poly(oxymethylene) segment, trans conformation.

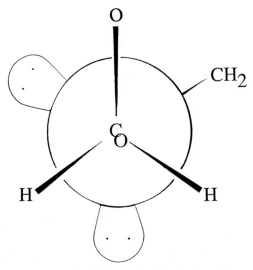

Fig. 12b. Poly(oxymethylene) segment, gauche conformation.

moderately repulsive, not much different from the C-H–C-C interaction in polyethylene. In the gauche conformation, the C-O bond is gauche to one C-H bond and to the other C-O bond. These interactions are unlikely to create a decisive difference between the conformations.

This leaves the oxygen lone pairs. In the trans conformation, they are each trans to a C-H bond, gauche to the other C-H bond and to the C-O bond. In the gauche conformation, one lone pair is still trans to C-H, gauche to the other C-H and the C-O bond; the second lone pair is gauche to both C-H bonds, trans to C-O. This is a complex set of interactions, but it can be simplified. The antibonding orbital of a C-H bond is too high in energy to allow a favorable filled–empty interaction with an oxygen lone pair. Therefore, the repulsive effect of the filled C-H bond orbital is not counteracted; the overall interaction between bond and lone pair is moderately repulsive. The trans dihedral between O lone pair and C-H bond may be favored over the gauche, but not strongly; the C-H bond has relatively weak repulsive influence, even on a lone pair. In this case, the gauche position is likely to provide sufficient separation.

The C-O antibonding orbital, however, lies much lower in energy than its C-H counterpart, close enough to the oxygen lone pair energy to allow a strong favorable interaction. This results in a powerful preference for conformations in which an oxygen lone pair is trans to a C-O bond. The gauche conformation of the $(CH_2O)_2$ unit allows one such trans relationship. In the case of polysulfur, the all-gauche conformation allowed a lone pair on every backbone atom to interact with a low-energy antibonding orbital. In all-gauche poly (oxymethylene) there is such an interaction only on every second backbone atom, but this is sufficient to decide the issue. The dihedral angle determined by experiment is 77°, not far from the polysulfur value of 85°. Cui and Kertesz conclude that similar orbital interactions in the two polymer chains have resulted in this conformational resemblance.

7. Conclusion

The factors governing the existence of spiral symmetry in plants and polymers are similar in many respects. In both cases, a relatively simple model has considerable predictive power. A description based on variations in the "acropetal influence" between elements can account for conformation without invoking biochemical mechanisms or explicit quantum chemical calculations. In one case, variations in this influence are described by parameters that specify the inhibitory effect of existing primordia on new ones. In the other, orbital interactions of varying favorability are maximized or minimized through geometrical changes.

Spiral phyllotaxy in plants occurs when primordia have inhibitory influence over a broader span of generations than in other systems. Helical symmetry in the polymers described is favored by the unusually strong interaction between a lone pair and a low-energy antibonding orbital. It may seem from these requirements that spirals should be rare and transient, existing only under conditions infrequently met. But the ease with which they are generated by models both simple and complex matches their natural ubiquity. Spiral symmetry is robust and resilient—not a fragile flower, but a willing weed.

Acknowledgement

The author would like to thank W. N. Lipscomb for his support of research on orbital interactions in fluoromethanol.

References

1. P. B. Green, Inheritance of pattern: Analysis from phenotype to gene, *Amer. Zool.* **27** (1987) 657–673.
2. W. W. Schwabe and A. G. Clewer, Phyllotaxis—a simple computer model based on the theory of a polarly-translocated inhibitor, *J. Theor. Biol.* **109** (1984) 595–619.
3. R. Dixon, Spiral phyllotaxis, *Computers Math. Applic.* 17 (4–6) (1989) 535–538, and in *Symmetry 2: Unifying Human Understanding,* I. Hargittai, ed. Pergamon Press, Oxford (1989).
4. D. W. Thompson, *On Growth and Form* (2nd. Edn.). Cambridge University Press, U.K. (1942).
5. B. B. Mandelbrot, *The Fractal Geometry of Nature.* Freeman, San Francisco (1982).
6. S. Jesuthasan and P. B. Green, On the mechanism of decussate phyllotaxis: biophysical studies on the tunica layer of Vinca major, *Amer. J. Bot.* **76**(8) (1989) 1152–1166.
7. J. M. Chapman and R. Perry, A diffusion model of phyllotaxis, *Ann. Bot.* **60** (1987) 377–389.
8. C. Lacroix and R. Sattler, Phyllotaxis theories and tepal-stamen superposition in Basella rubra, *Amer. J. Bot* **75**(6) (1988) 906–917.
9. A. J. Kirby, *The Anomeric Effect and Related Stereoelectronic Effects at Oxygen.* Springer-Verlag, New York (1983).
10. C. X. Cui and Miklos Kertesz, Conformation study of helical main-group polymers: Organic and inorganic, trans and gauche, *J. Am. Chem. Soc.* **111** (1989) 4216–4224.

ELECTROMAGNETIC THEORY FOR CHIRAL MEDIA

Akhlesh Lakhtakia

1. Introduction and Preliminaries

The commonest examples of chiral bodies are spirals and helices; the lovely photographs of the chamfered nautilus on wall calenders come to mind. Undoubtedly, the most readily available chiral objects are the reader's two hands. Before the advent of unisex fashions, handedness could be used to easily distinguish a woman's shirt from a man's. Golf clubs and field hockey sticks are also handed, not to mention screws and nuts. Spiral ear-rings are displayed in the museum of antiquities at the University of Uppsala and are merchandised in great numbers even today in jewellery catalogues. And one has only to look at the ubiquity of spiral motifs on Celtic artifacts all over Europe to realize humankind's fascination with handedness.

At the beginning of the last century, the phenomenon of *natural* optical activity in certain kinds of biological substances was discovered independently by Biot;[1] see Buchwald[2] for a fascinating account of the machinations engendered in the transactions of the French Academy at that time. Yet even before the proposal of the tetravalent carbon atom,[3,4] Pasteur[5] interpreted Biot's observations by imagining that the arrangement of atoms within an optically active material is asymmetric in having a noncongruent mirror image; that is, this arrangement is handed or chiral. Since Pasteur's time, the field of stereochemistry has expanded radically and has continually provided insights into the structure of molecules.

A chiral medium is characterized by either a left-handedness or a right-handedness in its microstructure. As a result, left- and right-circularly polarized (LCP and RCP) waves propagate through it with differing phase velocities: the field with the latter polarization propagating through a right-handed medium faster than the left-circularly polarized field, and vice versa. Natural optical activity, exhibited at optical frequencies by many organic molecules, is a manifestation of the innate chirality of these molecules.[6] The twin phenomena of circular dichroism (CD) and optical rotatory dispersion (ORD) are observed when a linearly polarized plane wave travels through a suspension of chiral molecules. ORD refers to the rotation of the vibration ellipse of the transmitted planewave with respect to that of the incident plane wave while CD is the differential absorption of the left- and right-circularly polarized plane waves inside the chiral suspension. Since these measures of optical activity are material-specific, in the last century and a half they have been extensively used by physical chemists to characterize molecular structure.

Helices are about the most common chiral objects found in nature. They are found as seashells,[7] levo- and dextro-molecules, and the α-helix model of many organic molecules that has been much studied by chemists.[8] By conducting microwave experiments around 1 GHz frequency, Lindman[9] found that the wavelength dependence of the ORD of optically active molecules had the same form as that of 9-cm copper helices. Although this claim was disputed by Winkler,[10] Tinoco and Freeman[11] soon demonstrated the truth of Lindman's results. With advances in digital computers, simulation of scattering by helices has become a very prominent area of research in the physical chemistry community, as is testified by the large number of workers in the field (e.g., Refs. 12–15). Particular emphasis has been laid on the calculation of the circular intensity differential scattering (CIDS), which is the normalized difference between the scattering patterns for incident LCP and RCP plane waves, averaged over all possible orientations of the helix at a given frequency.

Research on the electromagnetic theory of chiral media may have some impact on our understanding of vision as well. Structural anisotropy was used by deVries et al.[16] to explain the difference in the sensitivity of eyes to left- and right-circularly polarized light. However, it has recently been shown that when a rod is illuminated end-on with linearly polarized light, dichroism is not observed because the chromophore dipole of rhodopsin rotates; both vertically and horizontally polarized light components are absorbed equally.[17] Also, the retinylidene chromophore of rhodopsin has two distinct circular dichroism bands at 335 and 487 nm,[18] and measurements suggest that the visual pigments contain a high α-helical contant.[17] Furthermore, the individual collagen molecule has chirality,[19] and the primary stroma of the cornea is composed of collagen; the triple helix formed from three collagen α-chains is right-handed. Hence, fibrils and most tissues, layers, and bundles must be handed. However, the issue is probably even more complex, with the handedness possibly alternating from left to right in consecutive layers.[20] Thus, not only should any treatment of the physics of the eye include anisotropy due to the microstructure — rods, cones, ganglions, etc. — but it should also consider the nanostructure, for example, the helicities of the component molecules whose dimensions may be significant fractions of the optical wavelengths.

In this mini-review, emphasis has been laid on recent developments in electromagnetic field theory pertinent to linear, homogeneous, isotropic chiral media. As is customary since the discovery of the electron, phenomenology is incorporated in the constitutive equations. The independent electric and magnetic fields, respectively, are taken to be **E** and **H**, which can exist even in free space (i.e., vacuum). The presence of matter gives rise to intrinsic electric and magnetic polarizabilities, which are then incorporated into the material electric and magnetic fields, **D** and **B**. The material fields are related to **E** and **H** by the constitutive equations. On the other hands, Maxwell's equations at a given frequency also provides a connection, as per

$$\nabla \cdot \mathbf{D} = \rho_e ; \qquad \nabla \cdot \mathbf{B} = 0 ,$$
$$\nabla \times \mathbf{E} = i\omega \mathbf{B} ; \qquad \nabla \times \mathbf{H} = -i\omega \mathbf{D} + \mathbf{J} ,$$

all fields being functions of space **r**, and have the $exp[-i\omega t]$ time-dependence with ω being the circular frequency. Since charge can neither be created nor destroyed, the electric charge density ρ_e and the electric current density **J** are also related by a continuity equation,

$$\nabla \cdot \mathbf{J} - i\omega \rho_e = 0.$$

It must be noted that only electric sources will be considered here for simplicity; magnetic sources can be incorporated quite easily if need be. Furthermore, Maxwell's equations are sacrosanct, as also are the boundary conditions on the continuity of tangential **E** and **H** fields at a bimaterial interface. The nature of matter must enter solely through the (phenomenological) constitutive equations. Finally, the charge and current densities will be assumed to be impressed sources, and will not relate to Ohmic currents; hence, the chiral materials treated here will be completely lossless.

The plan of this review is as follows: In Sec 2, the various constitutive equations proposed and used for isotropic chiral media are discussed, and the pertinent field equations and their solutions are given in Sec. 3. Next, in Sec. 4 issues pertaining to chiral volumes of finite extent and embedded in free space are reviewed. Finally, in Sec. 5 are mentioned issues pertaining to unbounded chiral media. Only essential details have been provided; for a more exhaustive treatment, the interested reader is referred to Lakhtakia et al.[21]

2. Constitutive Equations

As stated earlier, chirality was first observed as optical activity, which is the rotation of the vibration ellipse. Drude[22] indicated that the rotation of the vibration ellipse in an isotropic medium can be predicted using Maxwell's equations, provided the intrinsic electric polarizability has an additional term proportional to $\nabla \times \mathbf{E}$, which premise appears to have been accepted by Born.[23] The consequent proposal for isotropic chirality was modified by Fedorov to

$$\mathbf{D} = \varepsilon[\mathbf{E} + \beta \nabla \times \mathbf{E}] ; \quad \mathbf{B} = \mu[\mathbf{H} + \beta \nabla \times \mathbf{H}], \qquad (1a,b)$$

this set being symmetric under time-reversality and duality transformations. Its validity has been affirmed by studies carried on optically active molecules[24] as well as from the examination of light propagation in optically active crystals.[25-27] The nonlocal character of Eqs. 1 needs to be noticed, because the material field **D** (resp. **B**) depends not only on **E** (resp. **H**) but also on the *circulation* of **E** (resp. **H**). Not very rigorously, one may even observe that **D** (resp. **B**) has a component due to the time-rate of change of **H** (resp. **E**), *vide* Faraday and Ampere-Maxwell laws.[28] It is assumed here, and hereafter, that the chiral medium is intrinsically lossless.

In addition to the Drude-Born-Fedorov equations (Eqs. 1), several other constitutive equations have been proposed and used. Condon[29] used

$$\mathbf{D} = \varepsilon_C \mathbf{E} - \chi \partial \mathbf{H}/\partial t\,; \qquad \mathbf{B} = \mu_C \mathbf{H} + \chi \partial \mathbf{E}/\partial t\,; \qquad (2\text{a,b})$$

while a take-off from Tellegen's formulation for the gyrator[30] gives

$$\mathbf{D} = \varepsilon_T \mathbf{E} + \zeta \mathbf{H}\,; \qquad \mathbf{B} = \mu_T \mathbf{H} - \zeta \mathbf{E}\,. \qquad (3\text{a,b})$$

A remarkable set of equations

$$\mathbf{D} = \varepsilon_p \mathbf{E} + i\xi \mathbf{B}\,; \qquad \mathbf{B} = \mu_p[\mathbf{H} - i\xi \mathbf{E}]\,, \qquad (4\text{a,b})$$

was deduced by Post,[31] who did not consider the medium microstructure at all, but simply required all equations to be generally covariant. The merits of the various constitutive equations given above are debatable and, using Maxwell's equations, they have been shown to be equivalent to each other for time-harmonic fields.[32] For example, the correspondence between Eqs. 1a and 1b and Eqs. 4a and 4b is given as: $\varepsilon_p = \varepsilon$, $\mu_p = \mu/(1 - \omega^2 \varepsilon \mu \beta^2)$ and $\xi = \omega \varepsilon \beta$. However, *mirror-asymmetry is immediately apparent in Eqs. (1) as opposed to the other constitutive equations:* the curl is not a vector under a reflection of coordinate systems. Hence, the Drude-Born-Fedorov constitutive equations (Eqs. 1) have been used in the sequel.

3. Chiral Field Equations

The prescription of the constitutive equations has to be followed by their integration into Maxwell's equations. It is necessary to realize that a chiral volume will exhibit its handedness only while interacting with time-varying fields, though chirality can be introduced into an otherwise achiral medium simply by subjecting it to a static magnetic field.[33] In addition, it should be noted also that ε, μ, and β are frequency-dependent.

Use of the first two source-free monochromatic Maxwell's equations, $\nabla \cdot \mathbf{D} = 0$ and $\nabla \cdot \mathbf{B} = 0$, in conjunction with Eqs. 1, easily yields the fact that $\mathbf{E}$ and $\mathbf{H}$ are also divergenceless at a source-free point.

The source-incorporated Maxwell's equations,

$$\nabla \times \mathbf{E}(\mathbf{r}) = i\omega \mathbf{B}(\mathbf{r})\,; \qquad \nabla \times \mathbf{H}(\mathbf{r}) = -i\omega \mathbf{D}(\mathbf{r}) + \mathbf{J}(\mathbf{r})\,, \qquad (5\text{a,b})$$

in which **J** is the volume electric current density, along with the constitutive equations (Eqs. 1), can be manipulated to yield the differential equations:

$$\nabla \times \nabla \times \mathbf{E} - 2\gamma^2 \beta \nabla \times \mathbf{E} - \gamma^2 \mathbf{E} = i\omega\mu(\gamma/k)^2[\mathbf{J} + \beta\nabla \times \mathbf{J}], \quad (6a)$$

$$\nabla \times \nabla \times \mathbf{H} - 2\gamma^2 \beta \nabla \times \mathbf{H} - \gamma^2 \mathbf{H} = (\gamma/k)^2 \nabla \times \mathbf{J}. \quad (6b)$$

where

$$\gamma^2 = k^2[1 - k^2\beta^2]^{-1}; \qquad k^2 = \omega^2 \varepsilon\mu. \quad (7a,b)$$

The solution of the source-incorporated governing differential equations (Eqs. 6) for **E** and **H** fields is given by

$$(k/\gamma)^2 \mathbf{E}(\mathbf{r}) = i\omega\mu \int dv' [\mathfrak{I} + \beta\nabla \times \mathfrak{I}] \cdot \mathfrak{B}(\mathbf{r}, \mathbf{r}') \cdot \mathbf{J}(\mathbf{r}'), \quad (8a)$$

$$(k/\gamma)^2 \mathbf{H}(\mathbf{r}) = \int dv' [\nabla \times \mathfrak{I}] \cdot \mathfrak{B}(\mathbf{r}, \mathbf{r}') \cdot \mathbf{J}(\mathbf{r}'), \quad (8b)$$

where the integrals extend over the source volume, **r** is the field point, and **r**′ is the source point. In these equations, $\mathfrak{I}$ is the unit dyadic,[34] and the infinite-medium Green's dyadic $\mathfrak{B}(\mathbf{r}, \mathbf{r}')$ can be split into two parts[32,35]:

$$\mathfrak{B}(\mathbf{r}, \mathbf{r}') = \mathfrak{B}_1(\mathbf{r}, \mathbf{r}') + \mathfrak{B}_2(\mathbf{r}, \mathbf{r}'), \quad (9a)$$

$$\mathfrak{B}_1(\mathbf{r}, \mathbf{r}') = (k/8\pi\gamma^2)[\gamma_1 \mathfrak{I} + \gamma_1^{-1} \nabla\nabla + \nabla \times \mathfrak{I}] g(\gamma_1; \mathbf{R}), \quad (9b)$$

$$\mathfrak{B}_2(\mathbf{r}, \mathbf{r}') = (k/8\pi\gamma^2)[\gamma_2 \mathfrak{I} + \gamma_2^{-1} \nabla\nabla - \nabla \times \mathfrak{I}] g(\gamma_2; \mathbf{R}), \quad (9c)$$

where

$$g(\sigma; \mathbf{R}) = exp[i\sigma R]/R. \quad (10)$$

is a scalar Green's function, $\mathbf{R} = \mathbf{r} - \mathbf{r}'$ denotes the spatial invariance of the chiral medium, and the wavenumbers

$$\gamma_1 = k/(1 - k\beta), \qquad \gamma_2 = k/(1 + k\beta). \quad (11a,b)$$

Only those values of β are considered in the sequel, which cause both γ_1 and γ_2 to be finite and positive real.

The reciprocal nature of the chiral media lead to the transpose properties,

$$\mathcal{B}(\mathbf{r}, \mathbf{r}') = [\mathcal{B}(\mathbf{r}', \mathbf{r})]^{(tr)}; \quad \nabla \times \mathcal{B}(\mathbf{r}, \mathbf{r}') = [\nabla' \times \mathcal{B}(\mathbf{r}', \mathbf{r})]^{(tr)}, \quad (12a,b)$$

in which the superscripted qualifier $^{(tr)}$ denotes transpose. Furthermore, the rotational properties

$$\nabla \times \mathcal{B}_1(\mathbf{r}, \mathbf{r}') = \gamma_1 \mathcal{B}_1(\mathbf{r}, \mathbf{r}'); \quad \nabla \times \mathcal{B}_2(\mathbf{r}, \mathbf{r}') = -\gamma_2 \mathcal{B}_2(\mathbf{r}, \mathbf{r}'), \quad (13a,b)$$

show that the isotropic chiral media are circularly birefringent. Circular birefringence in chiral media has been obtained here from the Green's function: this implies that all kinds of fields — near as well as far fields, fields with spherical, cylindrical, planar, or any other wavefronts — are circularly birefringent.

Maxwell's equations, $\nabla \times \mathbf{E} = i\omega \mathbf{B}$ and $\nabla \times \mathbf{H} = -i\omega \mathbf{D}$, can be seen to be completely satisfied by a *vector potential* $\mathbf{A}$ and a *scalar potential* V, given by the relations

$$\mathbf{H} = \mu^{-1} \nabla \times \mathbf{A}, \quad (14a)$$

$$\mathbf{E} = i\omega[\mathbf{A} + \beta \nabla \times \mathbf{A}], \quad (14b)$$

$$\mathbf{D} = i\omega\varepsilon[\mathbf{A} + 2\beta \nabla \times \mathbf{A} + \beta^2 \nabla \times \nabla \times \mathbf{A}] - \varepsilon \nabla V, \quad (14c)$$

$$\mathbf{B} = \nabla \times [\mathbf{A} + \beta \nabla \times \mathbf{A}], \quad (14d)$$

subject to specified gauge conditions. Whereas the electromagnetic field vectors simply exhibit birefringence, it turns out that the vector potential $\mathbf{A}$ is trirefringent and the scalar potential V is unirefringent.[21]

For time-harmonic fields, the conservation of energy principle can be obtained[21] in two different fashions. Either this principle can be stated as

$$\nabla \cdot \mathbf{P} + (1/2)Re\{i\omega(\mathbf{E} \cdot \mathbf{D}^* - \mathbf{H} \cdot \mathbf{B}^*)\} + (1/2)Re\{\mathbf{E} \cdot \mathbf{J}^*\} = 0, \quad (15a)$$

or, equivalently,

$$\nabla \cdot \mathbf{P} + (1/2)Re\{i\omega(\mathbf{D} \cdot \mathbf{D}^*/\varepsilon - \mathbf{B} \cdot \mathbf{B}^*/\mu)\} + (1/2)Re\{\mathbf{D} \cdot \mathbf{J}^*/\varepsilon\} = 0, \quad (15b)$$

in which

$$\mathbf{P} = (1/2)Re\{\mathbf{E} \times \mathbf{H}^*\} \quad (15c)$$

is the usual time-averaged time-harmonic Poynting vector, with the asterisk denoting the complex conjugate, and Re meaning "the real part of." The principle of conservation of momentum in these media has also been derived.[32]

It is often possible that a field problem is simplified by replacing electric sources with equivalent magnetic sources, and vice versa. To that end, for homogeneous, chiral, isotropic media, the necessary source-equivalence theorems and field duality theorems have also become available.[36]

4. Bounded Chiral Volumes

General speaking, only finite volumes may be endowed with chirality; and until the development of composite media, there was no specific need to develop a full-fledged electromagnetic theory pertinent to chiral media. Furthermore, the consequences of native chirality are exhibited by organic substances only around optical frequencies, where for most purposes ray-tracing techniques suffice. Thus, it was only in the 1970s that the need arose for more general treatments, when chiral polymers, active at lower frequencies, became feasible.[37-40]

In this context, Bohren[41] devised a transformation that permits the solution of such problems as scattering of electromagnetic waves by bounded three-dimensional chiral objects embedded in achiral media. The essence of Bohren's transformation is the prescription of a LCP field $\mathbf{Q}_1$ and a RCP field $\mathbf{Q}_2$, defined as per the following relations:

$$\mathbf{E} = \mathbf{Q}_1 + a_R \mathbf{Q}_2, \tag{16a}$$

$$\mathbf{H} = a_L \mathbf{Q}_1 + \mathbf{Q}_2, \tag{16b}$$

with

$$a_R = -i\sqrt{(\mu/\varepsilon)} = -1/a_L. \tag{17}$$

Substitution of Eqs. 16 into source-free Maxwell's equations leads to the wave equations

$$\nabla^2 \mathbf{Q}_1 + \gamma_1^2 \mathbf{Q}_1 = 0; \quad \nabla^2 \mathbf{Q}_2 + \gamma_2^2 \mathbf{Q}_2 = 0. \tag{18a,b}$$

The polarization characteristics of the **Q** fields are reaffirmed via the circulation equations

$$\nabla \times \mathbf{Q}_1 = \gamma_1 \mathbf{Q}_1; \quad \nabla \times \mathbf{Q}_2 = -\gamma_2 \mathbf{Q}_2; \tag{19a,b}$$

both the **Q** fields are divergenceless. Solutions of Eqs. 18, subject to the conditions of Eqs. 19, are easily obtainable in the cartesian coordinate system, and were given in the cylindrical and the spherical coordinate systems by Bohren.[41,42]

Layered structures are often the easiest to analyze, and for that purpose the reflection and transmission characteristics of a planar achiral/chiral interface serve as an important building block. In modern times this problem was systematically studied by Ramachandran and Ramaseshan,[43] and more recently by Lakhtakia et al.[44] Extension to the absorption characteristics of a metal-backed chiral slab was made by Varadan et al.[45] It is to be noted that the use of an imaging theory for chiral media is very complicated for scattering problems in general: not only do the sources get imaged, the medium also does.[21]

Bohren[42] examined the scattering response of an infinitely long, right circular, chiral cylinder. The solution of the scattering problem involving a chiral sphere is due to Bohren,[41] who also studied the scattering response of spherical chiral shells[46] using the presented formalism. The T-matrix method was extended to consider scattering by a three-dimensional chiral object of arbitrary shape embedded in free space.[47]

5. Unbounded Chiral Volumes

Bohren's decomposition shows that all fields in a sourceless chiral volume are either LCP or RCP. The admissible LCP and RCP planewave solutions of the homogeneous parts of Eqs. 6a and 6b are, respectively, given as

$$\mathbf{Q}_1 = (\mathbf{e}_\parallel + i\mathbf{e}_\perp)\exp[i\gamma_1 \mathbf{e}_p \cdot \mathbf{r}]; \quad \mathbf{Q}_2 = (\mathbf{e}_\parallel - i\mathbf{e}_\perp)\exp[i\gamma_2 \mathbf{e}_p \cdot \mathbf{r}], \quad (20\text{a},\text{b})$$

in which $\mathbf{e}_p$ is the direction of propagation; $\mathbf{e}_p$, $\mathbf{e}_\parallel$, $\mathbf{e}_\perp$ are mutually orthogonal unit vectors forming a cartesian coordinate system. If β is assumed to be positive real (right-handed medium), then $\mathbf{Q}_1$ propagates with a slower phase velocity; or else, $\mathbf{Q}_2$ is the slower of the two waves. In addition, in unbounded, isotropic chiral media these plane waves are transverse electromagnetic (TEM). It must, however, be emphasized here that combining $\mathbf{Q}_1$ and $\mathbf{Q}_2$ in any fashion cannot lead to a linearly polarized planewave, because the LCP and the RCP plane waves travel with different phase velocities.

By substituting Eqs. 16 into Eqs. 8, it can be seen that

$$(k/\gamma)^2 \mathbf{Q}_1(\mathbf{r}) = (i\omega\mu/k)\gamma_1 \int dv' \mathcal{B}_1(\mathbf{r}, \mathbf{r}') \cdot \mathbf{J}(\mathbf{r}'), \qquad (21a)$$

$$(k/\gamma)^2 \mathbf{Q}_2(\mathbf{r}) = -\gamma_2 \int dv' \mathcal{B}_2(\mathbf{r}, \mathbf{r}') \cdot \mathbf{J}(\mathbf{r}'); \qquad (21b)$$

these equations demonstrate that the radiation of $\mathbf{Q}_1$ is independent from that of $\mathbf{Q}_2$. The coupling between the LCP and the RCP fields takes place only at bimaterial boundaries[47,48] where conditions on the tangential components of $\mathbf{E}$ and $\mathbf{H}$ must be satisfied; that is, the boundary conditions are specified not on the $\mathbf{Q}$s singly, but on the tangential components of the combinations $[\mathbf{Q}_1 - (i\omega\mu/k)\mathbf{Q}_2]$ and $[\mathbf{Q}_2 - (i\omega\varepsilon/k)\mathbf{Q}_1]$.

The time-averaged Poynting vectors for the fields $\mathbf{Q}_1$ and $\mathbf{Q}_2$ can be easily derived using the decompositions (Eq. 16) in the definition (Eq. 15). Consequently,

$$\mathbf{P}_1 = (1/2) \, Re\{\mathbf{Q}_1 \times a_L^* \mathbf{Q}_1^*\}, \qquad (22a)$$

$$\mathbf{P}_2 = (1/2) \, Re\{a_R \mathbf{Q}_2 \times \mathbf{Q}_2^*\}, \qquad (22b)$$

while the cross-coupling of $\mathbf{Q}_1$ and $\mathbf{Q}_2$ cannot give rise to any energy since $\gamma_1 \neq \gamma_2$.

In order to investigate the scattering response of an obstacle embedded in a chiral medium, it is necessary to formulate Huygens' principle for such media.[21] Therefore, upon considering the sourceless external chiral volume V_e bounded by the surfaces S of Fig. 1, it can be shown that Huygens' principle for chiral media can be simply stated *via* the twin relations

$$\mathbf{E}(\mathbf{r}') = -2\gamma^2\beta \int_S ds \mathcal{B}(\mathbf{r}', \mathbf{r}) \cdot [\mathbf{e}_n \times \mathbf{E}(\mathbf{r})]$$

$$+ \int_S ds \, \{\mathcal{B}(\mathbf{r}', \mathbf{r}) \cdot (\mathbf{e}_n \times [\nabla \times \mathbf{E}(\mathbf{r})])$$

$$+ [\nabla' \times \mathcal{B}(\mathbf{r}', \mathbf{r})] \cdot (\mathbf{e}_n \times \mathbf{E}(\mathbf{r}))\}, \, \mathbf{r}' \in V_e \qquad (23a)$$

$$0 = -2\gamma^2\beta \int_S ds \, \mathcal{B}(\mathbf{r}', \mathbf{r}) \cdot [\mathbf{e}_n \times \mathbf{E}(\mathbf{r})]$$

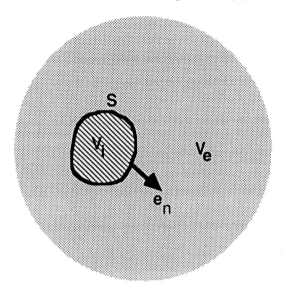

Fig. 1. Relevant to Huygens' principle for isotropic chiral media.

$$+ \int_S ds \, \{\mathfrak{B}(\mathbf{r}', \mathbf{r}) \cdot (\mathbf{e}_n \times [\nabla \times \mathbf{E}(\mathbf{r})])$$
$$+ [\nabla' \times \mathfrak{B}(\mathbf{r}', \mathbf{r})] \cdot (\mathbf{e}_n \times \mathbf{E}(\mathbf{r}))\}, \mathbf{r}' \in V_i, \tag{23b}$$

in which $\mathbf{e}_n$ is the unit normal to the surface S shown in Fig. 1. Thus, it is possible to compute the field $\mathbf{E}$ in any source-free region provided the tangential components of $\mathbf{E}$ and $\nabla \times \mathbf{E}$ are known on the surfaces enclosing the volume of interest; a similar principle can be derived on replacing $\mathbf{E}$ by $\mathbf{H}$ everywhere in Eqs. 23a and 23b. The development of Huygens' principle contains the seeds of a scattering formalism pertinent to isotropic chiral media; in this context, the properties of the plane wave scattering matrices have been explored.[21]

The foregoing developments have adequately set the foundations for an electromagnetic field theory relevant to isotropic chiral media. Several researchers have contributed to these developments, though they have not always used the Drude-Born-Fedorov constitutive equations, and for which reason they have not necessarily been identified in the preceding text. For interested readers, a selected list of relevant papers is included in Ref. 21, and is also available in Refs. 49 and 50.

References

1. J. B. Biot, Mémoire sur les rotations que certains substances impriment aux axes de polarisation des rayons lumineux, *Mémoires de l'Académie royale des sciences de l'Institut de France* **2** (1817) 41–136.
2. J. Z. Buchwald, The battle between Arago and Biot over Fresnel, *J. Optics (Paris)* **20** (1989) 109–117.
3. J. H. van't Hoff, Sur les formules de structure dans l'espace, *Archives néerlandaises des sciences exactes et naturelles* **9** (1874) 445–454.
4. J. A. Le Bel, Sur les relations qui existent entre les formules atomiques des corps organiques et le pouvoir rotatoire de leures dissolutions, *Bull. Soc. chimique Paris* **22** (1874) 337–347.
5. L. Pasteur, Recherces sur les propriétés spécifiques de deux acides qui composent l'acide racémique, *Annales de chimie et de physique* **28** (1850) 56–99.
6. E. Charney, *The Molecular Basis of Optical Activity*. Krieger, Malabar (Florida) (1979).
7. C. Illert, Formulation and solution of the classical seashell problem. I. Seashell geometry, *Nuovo Cimento D* **9** (1987) 791–811.
8. R. B. Setlow and E. C. Pollard, *Molecular Biophysics*. Addison-Wesley, Reading (Mass.) (1964).
9. K. Lindman, Über die durch ein aktives Raumgitter erzeugte Rotationspolarisation der elektromagnetischen Wellen, *Ann. Physik.* **69** (1822) 270–284.
10. M. H. Winkler, An experimental investigation of some models for optical activity, *J. Phys. Chem.* **60** (1956) 1665–1668.
11. I. Tinoco, Jr. and M. P. Freeman, The optical activity of oriented copper helices. I. Experimental, *J. Phys. Chem.* **61** (1957) 1196–1200.
12. C. Bustamante, I. Tinoco, Jr., and M. F. Maestre, Circular intensity differential scattering of light. IV. Randomly oriented species, *J. Chem. Phys.* **76** (1982) 3440–3446.
13. A. S. Belmont, S. Zietz, and C. Nicolini, Differential scattering of circularly polarized light by chromatin modeled as a helical array of dielectric ellipsoids within the Born approximation, *Biopolymers* **24** (1985) 1301–1321.
14. D. Keller, C. Bustamente, M. F. Maestre, and I. Tinoco, Jr., Model computations on the differential scattering of circularly polarized light by dense macromolecular particles, *Biopolymers* **24** (1985) 783–795.
15. C. W. Patterson, S. B. Singham, G. C. Salzman, and C. Bustamante, Circular intensity differential scattering of light by hierarchial molecular structures, *J. Chem. Phys.* **84** (1986) 1916–1921.
16. H. L. deVries, A. Spoor, and R. Jielof, Properties of the eye with respect to polarised light, *Physica* **19** (1953) 419–432.
17. H. Shichi, *Biochemistry of Vision*. Academic Press, New York (1983).
18. T. I. Shaw, The circular dichroism and optical rotatory dispersion of visual pigments, in *Handbook of Sensory Physiology, Vol 7, Part 1*, H. J. A. Dartnall, ed. Springer-Verlag, Berlin (1972).
19. K. A. Piez, Molecular and aggregate structures of the collagens in *Extracellular Matrix Biochemistry*, K. A. Piez and A. H. Reddi, eds. Elsevier, New York (1984).

20. R. L. Trelstad, Multistep assembly of Type I collagen fibrils, *Cell* **28** (1982) 197–198.
21. A. Lakhtakia, V. K. Varadan, and V. V. Varadan, *Time-Harmonic Electromagnetic Fields in Chiral Media*. Springer-Verlag, Berlin (1989).
22. P. Drude, *Lehrbuch der Optik*. S. Hirzel, Leipzig (1900).
23. M. P. Silverman, Reflection and refraction at the surface of a chiral medium: comparison of gyrotropic constitutive relations invariant or noninvariant under a duality transformation, *J. Opt. Soc. Am. A* **3** (1986) 831–837; *ibid* **4** (1987) 740.
24. C. F. Bohren, Angular dependence of the scattering contribution to circular dichroism, *Chem. Phys. Lett.* **40** (1976) 391–396.
25. F. I. Fedorov, On the theory of optical activity in crystals. I. The law of conservation of energy and the optical activity tensors, *Opt. Spectrosc. (USSR)* **6** (1959) 49–53.
26. F. I. Fedorov, On the theory of optical activity in crystals. II. Crystals of cubic symmetry and plane classes of central symmetry, *Opt. Spectrosc. (USSR)* **6** (1959) 237–240.
27. B. V. Bokut' and F. I. Fedorov, On the theory of optical activity in crystals. III. General equations of normals, *Opt. Spectrosc. (USSR)* **6** (1959) 342–344.
28. J. van Bladel, *Electromagnetic Fields*. Hemisphere, New York (1985).
29. E. U. Condon, Theories of optical rotatory power, *Rev. Mod. Phys.* **9** (1937) 432–457.
30. B. D. H. Tellegen, The gyrator: A new electric network element, *Phillips Res. Repts.* **3** (1948) 81–101.
31. E. J. Post, *Formal Structure of Electromagnetics*. North-Holland, Amsterdam (1962).
32. A. Lakhtakia, V. V. Varadan, and V. K. Varadan, Field equations, Huygens' principle, integral equations, and theorems for radiation and scattering of electromagnetic waves in isotropic chiral media, *J. Opt. Soc. Amer. A* **5** (1988) 175–184.
33. L. D. Barron, *Molecular Light Scattering and Optical Activity*. Cambridge University Press, Cambridge (UK) (1982).
34. H. C. Chen, *Theory of Electromagnetic Waves*. McGraw-Hill, New York (1983).
35. S. Bassiri, N. Engheta, and C. H. Papas, Dyadic Green's function and dipole radiation in chiral media, *Alta Freq.* **55** (1986) 83–88.
36. V. V. Varadan, A. Lakhtakia, and V. K. Varadan, On the equivalence of sources and duality of fields in isotropic chiral media, *J. Phys. A* **20** (1987) 6259–6265.
37. L. S. Corley and O. Vogl, Optically active polychloral, *Polymer Bull.* **3** (1980) 211–217.
38. W. J. Harris and O. Vogl, Synthesis of optically active polymers, *Polymer Preprints* **22** (1981) 309–310.
39. M. Vacatello and P. J. Flory, Helical conformations of isotactic poly(methyl methacrylate). Energies computed with bond angle relaxation, *Polym. Commun.* **25** (1984) 258–262.

40. G. Heppke, D. Lötzsch, and F. Oestreicher, Chirale dotierstoffe mit außerwöhnlich hohem verdrillungsvermögen, *Z. Naturforsch. (Leipzig)* **41A** (1986) 1214–1218.
41. C. F. Bohren, Light scattering by an optically active sphere, *Chem. Phys. Lett.* **29** (1974) 458–462.
42. C. F. Bohren, Scattering of electromagnetic waves by an optically active cylinder, *J. Colloid Interface Sci.* **66** (1978) 105–109.
43. G. N. Ramachandran and S. Ramaseshan, *Encyclopedia of Physics XXV/1*. Springer-Verlag, Berlin (1961).
44. A. Lakhtakia, V. V. Varadan, and V. K. Varadan, A parametric study of microwave reflection characteristics of a planar achiral-chiral interface, *IEEE Trans. Electromag. Compat.* **28** (1986) 90–95.
45. V. K. Varadan, V. V. Varadan, and A. Lakhtakia, On the possibility of designing anti-reflection coatings using chiral composites, *J. Wave-Mater. Interact.* **2**, (1987) 71–81.
46. C. F. Bohren, Scattering of electromagnetic waves by an optically active spherical shell, *J. Chem. Phys.* **62** (1975) 1566–1571.
47. A. Lakhtakia, V. K. Varadan, and V. V. Varadan, Scattering and absorption characteristics of lossy dielectric, chiral, nonspherical objects, *Appl. Opt.* **24** (1985) 4146–4154.
48. A. Lakhtakia, V. K. Varadan, and V. V. Varadan, Regarding the sources of radiation fields in an isotropic chiral medium, *J. Wave-Mater. Interact.* **2** (1987) 183–189.
49. A. Lakhtakia (ed.) *Selected Papers on Natural Optical Activity*. SPIE Opt. Engg. Press, Bellingham, Wash. (1990).
50. A. Lakhtakia, Recent contributions to classical electromagnectic theory of chiral media: What next? *Speculat. Sci. Technol.* **14** (1991) 2–17.

SUNFLOWER QUASICRYSTALLOGRAPHY

L. A. Bursill, J. L. Rouse, and Alun Needham

1. Introduction

Phyllotaxis is the study of plant growth with emphasis on the geometrical arrangement of the parts. The history of such studies is long and fascinating. Jean[1] provides a recent, extensive, and critical review of the literature, including botanical observations as well as mathematical modelling and other theoretical analyses. We will not attempt a further review of the literature at this stage. Members of the *Compositae* family, including sunflowers and daisies, possess a multiple set of florets which after fertilization result in a seed packing. At various stages of growth of the capitulum (documented by Marc and Palmer[2]) the florets, ovaries, and seeds are arranged in visible spirals, known as *parastichies* (Bravais and Bravais[3]). The curving of the parastichies, from the rim to the centre, either in a clockwise or counterclockwise direction with respect to the direction of upward growth of the plant, implies that the spiral packings have a handedness or chirality. Church[4] and Cook[5] provide historical accounts of attempts to define a chirality convention, including references to many observation studies. It is only fair to say that, despite the best intentions of these authors, neither proposed an unambiguous chirality convention; they were mutually inconsistent, as well as self-contradictory! Gardner[6] has provided a survey of the ubiquitous nature of chirality, cutting across all the natural sciences. However, he made no attempt to define a chirality convention.

As physicists, we are interested in the development of new models of ordered aperiodic structures (quasicrystals), which cannot be described by traditional crystallographic methods.[7] In order to understand the growth mechanism of spiral lattice structures, which may lead to synthesis of such structures on macroscopic or microscopic scales, a thorough study of the natural growth of some sunflower varieties has been made, as reported below.

Our interest in chirality began as an attempt to define the chirality and phyllotaxis of the floret packing of the multiheaded sunflower species *H. tuberosus*.[8,9] This led to the present study of the single-headed species *H. annuus*, when we were confronted with providing a chirality convention that would correctly account for the handedness of *both* the leaf and seed-packing phyllotaxis. Extensive observations were made of the growth and development of *whole* sunflowers. The need for a chirality convention is introduced in Sec. 2. Experimental methods are described in Sec. 3, followed in Sec. 4 by observations of the initial seedling development, stem and leaf arrangement, and capitulum development. An appropriate mathematical definition of chirality is then developed (Sec. 5) and applied to *H. annuus* varieties (Sec. 6). Some recent results concerning the mathematical description of *H. annus*, obtained by analysis of digitized coordinates and computer modelling, are presented in Sec. 7. We conclude with a discussion in Sec. 8.

2. Need for a Chirality Convention

It is possible to gain an understanding of the nature of chirality for two-dimensional spiral lattices by noting the relationship between successive visible spirals (i.e., parastichies) and a common generating spiral, as shown in Fig. 1. An inwardly directed generating spiral may be described using a power law

$$r_n = A(H - n)^k \qquad (1)$$

or a logarithmic spiral

$$r_n = A' k^{H-n}, \qquad (2)$$

where $n = 1, 2, 3 \ldots H$, and H is the total number of seeds (or cells, using crystallographic terminology). The appropriate angular coordinate is

$$\theta = (\phi_d)\, n, \tag{3}$$

where the divergence angle ϕ_d is

$$\phi_d = 2\pi/\left(t + \frac{1}{\tau}\right). \tag{4}$$

This choice of ϕ_d generates parastichy numbers given by

$$S_n(t) = 1, t, t+1, 2t+1, \ldots, \tag{5}$$

where

$$\tau = \lim_{n \to \infty} \frac{S_{n+1}(t)}{S_n(t)} \tag{6}$$

is the golden mean $(1 + \sqrt{5})/2$. Table 1 lists some useful properties of sequences whose general term is $S_n(t)$, for small integral values of t. The most frequently occurring parastichy numbers are called the Fibonacci ($t = 2$)

$$F_n = 1,2,3,5,8,13,21\ldots \tag{7}$$

and Lucas ($t = 3$)

$$L_n = 1,3,4,7,11,18,29\ldots \tag{8}$$

sequences. Note that relatively rare series also occur, namely

$$A_n = 2,5,7,12,19,31\ldots \tag{9}$$

(known as a higher accessory series[10])

$$B_n = 2,4,6,10,16,26\ldots \tag{10}$$

(known as bijugate[10]), and

$$T_n = 3,6,9,15,24,39\ldots \tag{11}$$

(known as trijugate[10]). Earlier[7] it was noted that the details of spiral lattices may be very sensitive to small changes in divergence angle. Therefore it is always advisable to confirm the divergence angle by direct measurement, and not rely solely upon the occurrence of specific parastichy numbers.

Note (Figs. 1a, b) that consecutive parastichy numbers (see Eqs. 7–11) give visible spirals that sweep alternatively clockwise and anticlockwise, for example, starting from the rim in Fig. 1. Clearly, there are always two

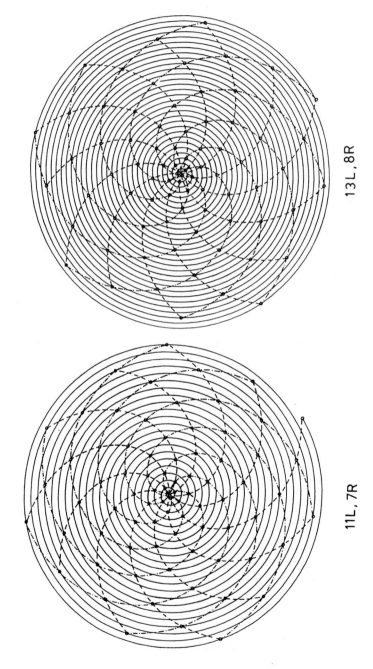

Fig. 1. (a) Left-handed ($\phi_d = -99.502°$) and (b) right-handed ($\phi_d = +137.508°$) spiral lattices. Note appearance of Lucas and Fibonacci numbers of visible spirals (parastichies).

distinct structures (in two dimensions) depending upon the handedness of the underlying generating spiral (cf. Figs. 1a and b). Thus ϕ_d may take absolute values $\pm|\phi_d|$. It is necessary to impose a chirality convention in order to label left- or right-handed structures, with an associated unambiguous practical algorithm.

Table 1. (a)   Properties of the General Fibonacci Sequence $S_n(t)$

t	$S_n(t)$	α	ϕ_d
1	1,1,2,3,5,8,13,...	1.618	222.49°
2	1,2,3,5,8,13,21,...	2.618	137.51°
3	1,3,4,7,11,18,29,...	3.618	99.50°
4	1,4,5,9,14,25,39,...	4.618	77.96°

Table 1. (b)   Convention for Labelling Chirality

Name	ϕ_d	Parastichy number and hand (R & L)						
Fibonacci (RH)	−137.508°	1R	2L	3R	5L	8R	13L	21R
Fibonacci (LH)	+137.508°	1L	2R	3L	5R	8L	13R	21L
Lucas (RH)	−99.502°	1R	3L	4R	7L	11R	18L	29R
Lucas (LH)	+99.502°	1L	3R	4L	7R	11L	18R	29L
Bijugate (LH)	+68.754°	2L	4R	6L	10R	16L	26R	42L
Trijugate (RH)	−45.836°	3R	6L	9R	15L	24R	39L	63R
Higher Accessory (LH)	+55.00°	2L	5R	7L	12R	19L	31R	50L

3. Experimental Methods

3.1 *Plant Culture*

Three varieties of *H. annuus* were used in this investigation:

(A) Suncross 52 hybrid seed supplied by Arthur Yates and Co. Pte. Ltd. This seed produces uniform plants with a high oil content provided adequate irrigation is available.

(B) Open pollinated, black, low oil content seed obtained from the local Grain Store.

(C) Large striped "Giant Russian" seed that produces magnificent large heads valued horticulturally, and seeds easily dehulled with kernels suitable for human consumption.

These varieties were grown in sunny positions in both garden beds and plastic containers. The garden beds consisted of sandy loam with well rotten compost and lime added to adjust the pH to between 5.5 and 6.0. The feeding programme consisted of an initial application of an all–purpose plant food that provided fertilizer for the entire growing season. A drip irrigation system was used to water the beds at 3- to 5-day intervals depending on the weather. The black plastic containers used were 45 cm in diameter, and were filled with a commercial brand potting mix that supplied balanced nutrients and trace elements for initial growth of the sunflower seedlings. A liquid fertilizer was applied to aid development. The containers were irrigated at 1- to 3-day intervals using 4 drips per container.

The only disease/pest problem that arose was the presence of white fly, which was controlled with an insecticide applied in the evening. It was noted that the temperature of the potting mix reached 45°C during fine hot weather. This did not appear to be detrimental to the development of the seedlings. During fine weather the flowers were observed to attract bees, which then assisted in pollination.

3.2 *Analysis Procedure*

The first step was to label the plants carefully (date sown, etc.). The observations consisted of recording (a) the divergence angle of the leaves about the stem, with respect to the direction of growth, (b) the direction and characteristic number of parastichies in the flowerhead, and (c) the correlation between (a) and (b) via the bracts and ray flowers.

The analysis of the stem and the buds was necessarily completed at a particular time of growth. Leaves near the base of the stem were opposite (this stage usually consisted of 4 or 5 pairs), then further up the stem, the leaves separated and the divergence angle became stabilized before development of the capitulum.

The stem analysis consisted of firstly determining the direction of the spiral. For purely pragmatic reasons it seemed natural to label this right-handed if the leaves spiralled about the stem in accordance with the curling of the fingers of the right-hand about the thumb. The thumb represents upward growth (see Fig. 2). Secondly, the angular separation

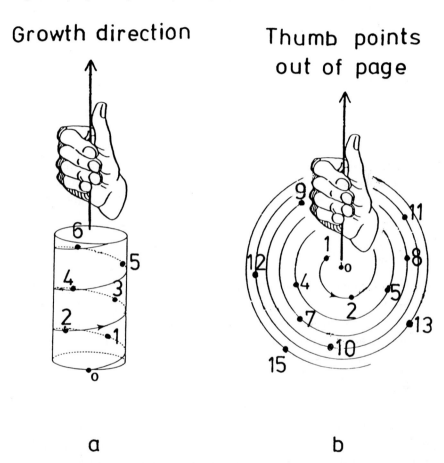

Fig. 2. Illustrating convention for positioning of thumb and fingers for chirality determination of (a) three-dimensional spiral and (b) planar spiral.

of consecutive leaves was determined. If the angle between the leaves was approximately 99.5° then the stem was classified as having a Lucas spiral (from Table 1); if the angle was approximately 137.5°, it was labelled Fibonacci. (The angle measurement was immediately checked against that corresponding to the parastichy numbers counted in the capitula.)

The analysis of the flowers could be satisfactorily completed prior to maturity. For example, capitula with diameters from 5 mm upward yielded conclusive results when sectioned with a razor blade. The procedure involved the counting of the most predominant parastichies, and labelling these with an R or an L, depending on whether the parastichies rotated in a clockwise (R) or anticlockwise (L) direction, reading from the rim inward toward the centre, looking into the flowerhead from the top; see e.g., Fig. 3(a), labelled 55R, 34L.

For most cases the analysis was carried out by eye. Smaller capitula required use of a Wild M420 macroscope with Intralux 5000 illumination. The flowerbuds examined in this manner were repeatedly sectioned (a 1–2 mm layer being removed each time) until the visibility of structure under the protective bracts was optimal. Viewed at an appropriate magnification, the set of two parastichy numbers could then be determined and noted. A Nikon F3 camera was used to photograph unusual subjects.

4. Results

4.1 *Seed Germination and Chirality*

Chirality is first expressed visibly in sunflower when the first two alternate leaves are visible on the stem. To investigate whether the chirality could be detected as the seeds germinated, we carefully removed 11 seeds radially from each of two open pollinated heads of opposite chirality, and inverted these while otherwise maintaining their mutual orientation and sown them in the dark with water and high humidity in upturned nail-brushes, which allowed the seeds to swell while imbibing water and maintaining their orientation. The radicle is highly geotropic and, under the above conditions, emerging vertically from the husk it bends over to develop downward. The horizontal angle of the plant in which this initial bending occurred was noted and the seedling then planted normally to obtain the chirality of the plant when the first alternate leaves appeared.

Lack of any correlation between the angle of the emerging radicle and the plants' chirality for seed from either of the two heads indicated that even if the chirality was predetermined in the seed, it was not expressed in the direction in which the radicle emerged.

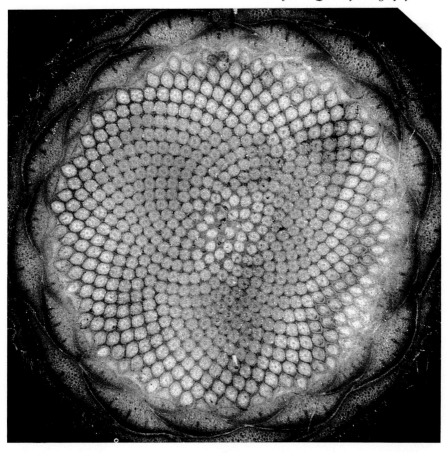

Fig. 3a. *H. annuus* capitula showing floret development for (55R, 34L) Fibonacci phyllotaxis.

4.2 Stem Development

It became apparent that the relative positions of the leaves (ϕ_d) evolved as follows: up to 8–10 leaves appear in pairs, with $\phi_d \simeq 90°$; later, new leaves form singly, finally with establishment of, typically, the Fibonacci ($\phi_d = 137.503°$) or Lucas ($\phi_d = 99.508°$) divergence angle. After approximately 30 days the capitula, with associated bracts, began to develop on the apical shoot. Full inflorescence then developed after approximately 50–60 days. (Further botanical details may be found in

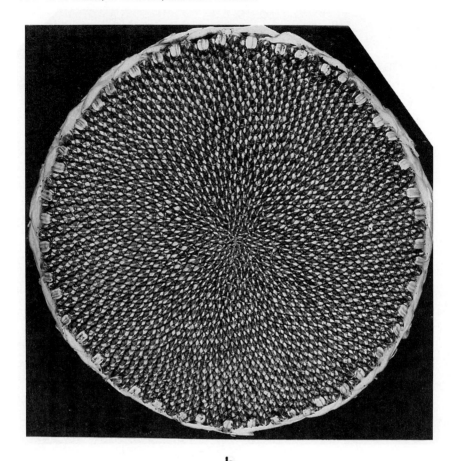

b

Fig. 3b. (76R, 47L) Lucas phyllotaxis in a mature seed packing.

Heiser's *The Sunflower*.)[11] Here we are concerned mainly with the chirality aspects. Thus the divergence angle, once established, was found to be approximately constant (within ≤ 5°, say) from the upper stem, for the leaves, through the bracts on the back of the flowerhead, via the ray flowers and into the final seed packing, after florets were brushed away. Thus the chirality is clearly established for the whole flowerhead, with, for example, the fingers of the right (or left) hand following the sense of rotation of a right (or left) hand screw rotated by ϕ_d in the direction of new growth. Note that $|\phi_d| \leq 180°$, if ambiguity is to be avoided. In the case of a typical single-headed sunflower the spiral flows up the stem, onto the back of the capitula, via the bracts, ray flowers, and then inward from the rim of the capitula toward the centre of the floret (or seed) packing. As shown in Table 2 the whole sunflower/chirality relationship just described was found to hold for 109 out of 111 cases examined. Two exceptions remain unexplained, and they may be due to experimental error; one must be extremely conscious not to befuddle left and right, up and down, in and out, 34 and 21, and writing down correctly what was observed. Alternatively, there may be a small probability that a sunflower capitulum may experience a reversal of chirality, either the bracts with respect to the stem, or capitula with respect to the bracts. Certainly, in a parallel study of a multiheaded sunflower species *H. tuberosus* we noted relatively frequent reversals of chirality, along the main stem or even lateral branches.[12] However, the result for single-headed *H. annuus* is quite clear: the stem, bracts, and capitulum all possess identical chirality, with ≤ 1–2% error, if one follows the sense of rotation of a right- (or left-) handed thread, always taking the direction from old to new primordia. Clearly, it is the chirality of the apical shoot[13] that is definitive in the case examined here.

Table 2. Phyllotactic and Chiral Relationship between Leaves on Stem and Flowerhead

Top Head/Leaves	*H. annuus*		
	A	B	C
Matching	103	4	2
Nonmatching	1	1	–
Total	104	5	2

4.3 Capitulum Development

Scanning electron microscopic results for *H. annuus* varieties have been published by Moncur,[14] Marc and Steer,[15] and Palmer and Steer,[2] where it was clear that primordia, which develop into ray flowers, appear first on an approximately circular rim (Fig. 4). Additional primordia then follow, forming parastichies curving inward toward a notional centre. Macroscopic observations by eye of dissected capitula having diameters in the range 1–20 cm show a remarkable continuity between the disposition of branches (or leaves), bracts (small green leaves that occur on the underside of the developing flowerhead), the ray flowers, and ovaries (seed cells). The whole set of structural elements of one developing capitulum is shown in Fig. 5. The numbers of bracts and ray flowers were often equal to or in close proximity to Fibonacci or Lucas numbers.

Further observations, with the aid of the macroscope (1x–80x), enabled us to visualize the initiation of primordia on the capitulum. Thus Fig. 6 shows the stages in the development of capitula in the range 1.5–3.5 mm diameter. At first a flat platform (new capitulum) forms, covered with a green jelly-like substance (Fig. 6a). This was covered in turn with green leaf-like structures set obviously at divergence angles appropriate for Fibonacci or Lucas phyllotaxis. Within one or two days seed cell nuclei appear (crystallize) starting at the rim (Fig. 6b) then proceeding inward giving the earliest indication of a spiral arrangement. At a later stage the capitulum is almost covered with primordia. Note the continuity between the outer bracts, ray flower nuclei, and seed cells. Continuity of divergence angle may be traced back through the green bracts on the top surface, then to the underside of the developing flowerbud, and, finally to the leaf arrangement on the stem.

Thus it appears that the divergence angle for new primordia is set already by the pre-existing leaf and bract structures associated with the apical shoot (see Williams[13]), which presumably already determines the divergence angle appropriate for Fibonacci, Lucas, or higher accessory phyllotaxis.

5. A Chirality Convention

The following is a development of Fan et al.[8] and Bursill et al.[9], which exposed some ambiguities possible in labelling the chirality of spiral structure.

As shown in Figs. 1a, b the spirals for adjacent parastichy numbers rotate alternatively to the right and to the left. It is not obvious how to assign chirality, without first looking at the (hidden) sense of rotation of the underlying generating spiral. Fig. 1 makes the point, emphasized by both Church[4] and Cook,[5] that assignment of a label left or right is simply a convention, we could equally well reverse the labels. There are two distinct possible structures, related by mirror or inversion symmetry operations, analgous to the relationship between left and right hands.

Fig. 4. Showing top (a) and bottom (b) views of typical *H. annuus* flower.

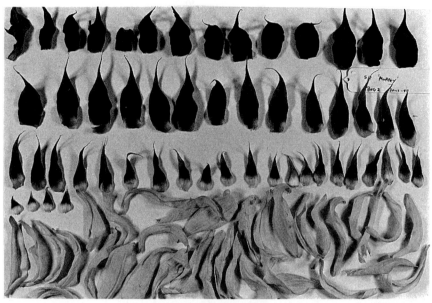

Fig. 5. Complete set of bracts and ray flowers for one sunflower capitulum. In this case 53 bracts occurred prior to ray flower development (note proximity to Fibonacci number). There were 34 ray flowers.

Given that the generating spiral has been located in space, one may attempt to label it by placing it relative to the sense of rotation of a right- or left-handed screw: pragmatically, place the thumb in the direction of movement of a screw, with the fingers curling in the direction of rotation of the screw, then assign the appropriate label left or right. Note that a bolt and nut both have the same hand. This convention works well with three-dimensional spirals, based on screws, cones, or cylinders, for example, but ambiguity arises if we have a planar spiral. Do the fingers follow the outward or inwardly directed spiral; that is, do the fingers curl outward from the centre, or inward from the rim. Furthermore where is the thumb pointing with respect to the planar spiral, up or down?

For planar spiral *lattice* structures, further ambiguity arises since, as shown in Figs. 7a, b, the one set of lattice points may be equally well

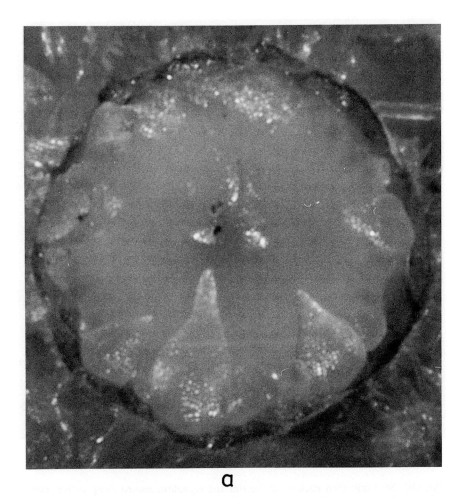

Fig. 6a. Initial development of a flat capitulum. Note a few immature green bracts still maintaining the Fibonacci divergence angle.

generated using divergence angles ϕ_d and $2\pi\text{-}\phi_d$; in fact any of the divergence angles

$$\phi_d = 2\pi/\alpha \pm 2\pi n \qquad (12)$$

when n is an integer, may be applied, provided the amplitude (in Eq. 1) of the radial component is modified appropriately. After some consideration it seems most reasonable to restrict the conventional assignment

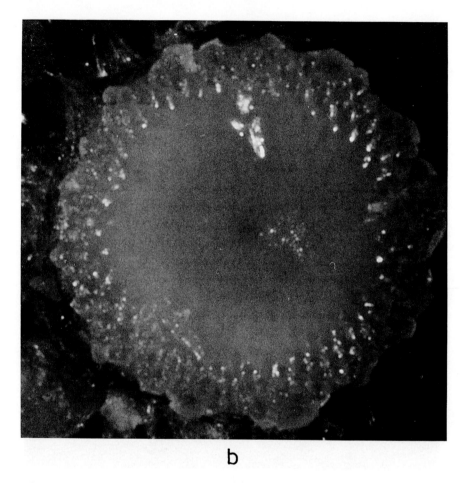

b

Fig. 6b. Seed cells then nucleate at the rim and crystallize inward along visible spirals (bracts removed). Note the continuity between the outer bracts, ray flower nuclei, and seed cells. (Scale: capitullum diameter is 2 mm.)

of ϕ_d as follows:

$$\phi_d^R = \phi_d$$
$$\{R: -\pi \leq \phi_d \leq \pi\} \tag{13}$$

where ϕ_d^R is the reduced divergence angle. Obviously, a negative value for ϕ_d^R implies a left-handed spiral, if $+\phi_d^R$ is taken as right-handed. If the

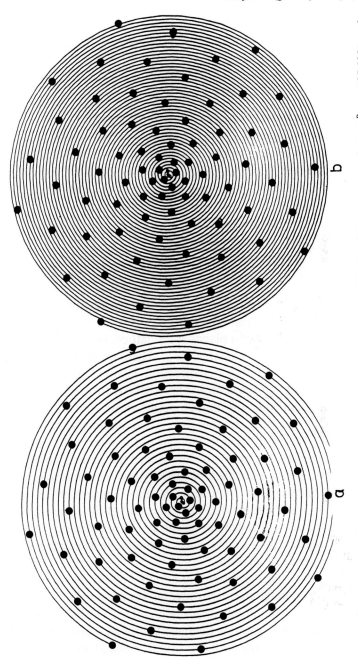

Fig. 7. (a) and (b) show identical spiral lattices threaded by right- or left- handed generating spirals ($\phi_d^R = +137.508°$ and $\phi_d = -222.402°$, respectively).

difference between enantiomorphic pairs is disregarded, then the range of ϕ_d giving distinct spiral structures may be reduced further to

$$\phi_d^S = \phi_d$$
$$\{R: |\phi_d| \leq \pi\} \tag{14}$$

where ϕ_d^S may be called the structural divergence angle.

At a mathematical level, at least in crystallography, chiral objects (i.e., those that do not possess mirror or inversion symmetry) are said to be right- (left-) handed if the basis vectors $\mathfrak{a}_1$, $\mathfrak{a}_2$, $\mathfrak{a}_3$ form a right- (left-) handed set. Converting to polar coordinates the outwardly directed spiral may be described by

$$r = An^k \text{ (power law)} \tag{15}$$

or

$$r = A' K^n \text{ (logarithmic)}, \tag{16}$$

whereas the inwardly directed equivalents are described by Eqs. 1 and 2, including of course the restrictions on ϕ_d given by Eq. 13. In order to complete the description of a planar spiral lattice, we must assign a polar direction (normal to the plane) as $+z$. Then we may say a spiral is right-handed if the thumb ($+z$) direction is up and the fingers curl inward from the rim to the centre. Alternatively, this *same* spiral would be called right-handed if the thumb points down and the fingers curl outward from the centre to the rim. Then obviously, if the thumb direction reverses with respect to the curl of the fingers, the opposite (or left) chirality applies. We have now introduced the elements of ambiguity that must be defined if a satisfactory chirality convention is to be derived. Clearly it is necessary to impose a chirality convention in order to distinguish left- and right-handed structures and at the same time specify the divergence angle of the generating spiral. For applications to naturally occurring structures, such a convention should have associated with it an unambiguous practical algorithm.

It is remarkable that the procedure adopted above, pragmatically and naturally, in describing the experimental observations, satisfied quite appropriately all the requirements just alluded to. Thus we simply traced the direction of growth, rotating the fingers through successive divergence angles $n\phi_d$ ($n = 1, 2, \ldots H$), with the thumb pointed in the direction of growth. For the capitulum, it is natural to look into the flower, as opposed to the growth direction, so that the thumb naturally points upward along the direction of growth. This is appropriate, since one usually looks down, for example, at the right hand, in this same aspect, i.e., with the thumb pointing upward! An individual parastichy may be labelled left- or right-handed if it follows the inward curl of the fingers of the left- (or right-) hand when the upward-directed thumb is placed over the spiral. There is then no ambiguity in writing down the handedness of the visible parastichies (see, for example, Fig. 3). Finally we need only to consult Table 1, which gives effectively the relationships between the value of t (from Eq. 5), the divergence angle, the chirality of the parastichies in each case, and the chirality of the generating spiral. Note that we restrict the conventional assignment of ϕ_d as given by Eq. 13, the positive z direction (thumb) points upward along the direction of growth, and we use the inwardly directed generating spirals (Eqs. 1 and 2).

A practical algorithm for assignment of chirality is as follows:

(a) locate the normal growth direction (i.e., assign a polar axial direction);
(b) identify the predominant parastichy numbers and assign handedness to each of these; and
(c) consult Table 1 to identify the handedness of the generating spiral, within the convention for the reduced divergence angle given by Eq. 13.

In cases where the divergence angles and parastichy numbers are not the common ones shown in Table 1, it is necessary to count the parastichy numbers and deduce a suitable reduced divergence angle and generating spiral, and extend Table 1 if appropriate.

6. Chirality Analysis

6.1 *Occurrence of Matching Phyllotactic Behaviour*

In all but one of the 104 variety A plants, the stem matched the flowerhead with respect to both chirality and divergence angle. The variety B plants also had one nonmatching stem and flowerhead pair, but this was out of a total of only five. The two variety C plants had stems and flowerheads matching in chirality and divergence angle. All results are shown in Table 2. Fig. 3 gives examples of (a) Fibonacci (LH) and (b) Lucas (RH) capitula.

In the course of the study it became apparent that the bracts and ray flowers share the same chirality and divergence angle as both the stem and flowerhead. It would appear as if both the bracts and ray flowers act to continue the chirality and divergence angle of the stem through to the flowerhead.

6.2 Chirality Distribution

As can be seen from Table 3, the study enabled an evaluation of the chirality distribution. For the variety A plants, the phyllotactic behaviour was entirely Fibonacci and the ratio of left- to right-handed chirality was 0.385:0.615. For the variety B plants it was 0.666:0.333; again, the phyllotactic behaviour was entirely Fibonacci. The variety C plants proved an exception, one plant being left-handed Fibonacci whilst the other was left-handed Lucas. When the sample size (104 variety A plants) is taken into account it can be concluded that there is no significant deviation from the binomial distribution expected for a random choice of chirality for these plants.

Table 3. Chirality Distribution (from Analysis of Leaves)

	H. annuus		
	A	B	C
L.H. Lucas	–	–	1 (50%)
R.H. Lucas	–	–	–
L.H. Fibonacci	40 (38%)	2 (40%)	1 (50%)
R.H. Fibonacci	64 (62%)	3 (60%)	–
Total	104	5	2

6.3 Occurrence of Lucas Flowers

The only Lucas flower, that is, a flower which exhibited Lucas numbers of parastichies, was found on one of the cultivar C plants (Fig. 3b).

7. Quasicrystallography of *H. annuus L.*

A variety of mathematical descriptions of the spiral floret and seed arrangements in sunflowers exist (see Jean[1]). Considerable attention has been given to the (apparent) constancy of the divergence angle, and to the preferred packing that results from using an irrational divergence angle ϕ_d derived from a Fibonacci sequence.[9,16,17,18,19] The position of successive primordia has most commonly been described by an equiangular or logarithmic spiral (Eq. 16), which reduces to the parabolic spiral $r_n = A n^{1/2}$ for cells of uniform size.[20,22] Ridley[17] established the more general theoretical model linking the cell areas ($\pi \rho_m^2$) with their radial distances (r_n), where r_n and ρ_n are *any* smooth functions of n, and the cells are located on any surface of revolution.

In order to give a quantitative description of the seed or floret arrangement of sunflowers, based on empirical findings, we analyzed seed cell centres using digitized coordinates obtained from photographs of *H. annuus* and *H. tuberosus*.[23,24] Figs. 3a and b show examples of *H. annuus* for intermediate and mature stages of floret and seed development, respectively. Power law spirals gave a better fit than logarithmic spirals to the *H. annuus* capitula. However, the quality of the fitting was improved significantly by using a combined power law and logarithmic spiral

$$r = A N^k p^{N-1}, \tag{17}$$

which implies that steady (exponential) growth in the outer regions must give way to the accommodation of new seeds (bifurcations) as the developing capitulum closes in on its centre.[24]

The measurements of the angular displacements between successive seed cells and radial distances were both found to fluctuate around the expected Fibonacci (137.5°) or Lucas (99.5°) angles for all six capitula analyzed. Although these oscillations were larger near the centre, they persisted to the outer regions of each flowerhead to a surprising extent ($\pm 5°$) even for the most regular specimens. The same pattern of oscillations was also seen in the plastochrone ratio r_{n+1}/r_n. Figs. 8a and b show typical plots of ϕ_d and r_{n+1}/r_n. Although many theoretical studies have been made concerning sunflower seed packings, none of the previous computer simulations were subjected to precise quantitative comparison with experiment. It is already necessary to formulate theoretical models that allow for fluctuations in both divergence angle and radial functions. Even on the most general terms, one has to expect thermodynamic fluctuations in real systems, as well as nonideal growth conditions (kinetics) that will lead to fluctuations in seed size and position.

An analysis of the most efficient disc packings on a parabolic spiral lattice,[18] which allowed a systematic relaxation of the ideal spiral, showed that the Fibonacci divergence angle (137.508° or $2\pi/1+\tau$) is favoured slightly ahead of the Lucas angle (99.502° or $2\pi/(3+\tau)$) or other higher order series ($2\pi/(t+\tau-1)$). In that study the divergence angle was constrained to a constant value for each packing, but the results emphasized that close-packing considerations should be important for selecting the divergence angle. In subsequent work[19] the constraint on ϕ_d was relaxed, and the discs were allowed to expand according to a radial growth law. The introduction of a close-packing rule then allowed the nature of the disc-packings to be explored, with interesting results. Thus, both Fibonacci and Lucas numbers of visible spirals arose naturally, depending only on the choice of initial conditions. Both the rate of convergence toward an ideal spiral, and chirality, were determined by the placement of the first few discs. Fig. 9 shows the variation of ϕ_d (a) for origin between first and second disc, and (b) with origin slightly off the line joining the first two discs. Thus the divergence angle may converge rapidly onto the Fibonacci angle (Fig. 9a) or show persistent oscillations (Fig. 9b). The effects of further shifts of origin may be quite significant; in fact there are many close-packed spiral-like structures, differing only slightly in packing efficiently.[18,19] The phenomena discovered above, both in the analysis of the experimental data and in the close-packing models, have similarities to some nonlinear problems that concern deterministic routes to chaos.[25]

Clearly, further progress in understanding the crystallography of sunflowers, and hence packings of similar, but not congruent, objects in general, will be dependent on the combination of the application of crystallographic techniques of analysis, based on digitized image data and processing, with the results of theories of phase transitions incorporating nonlinear chaotic phenomena and analysis. The significance of such work, for physicists, is that by understanding such structures on the macroscopic scale, one may be led to new synthetic routes for synthesis of interesting new materials on microscopic scales.

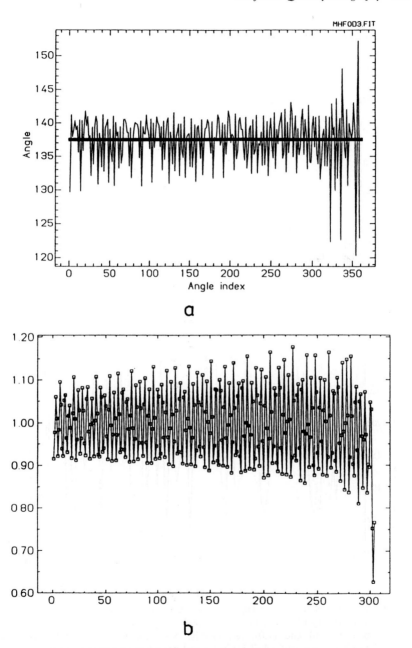

Fig. 8. Typical plots of ϕ_d and r_{n+1}/r_n (plastochrone ratio) for *H. annuus*. Note persistent oscillations in both plots, with deviations increasing as the notional centre of the capitulum is approached.

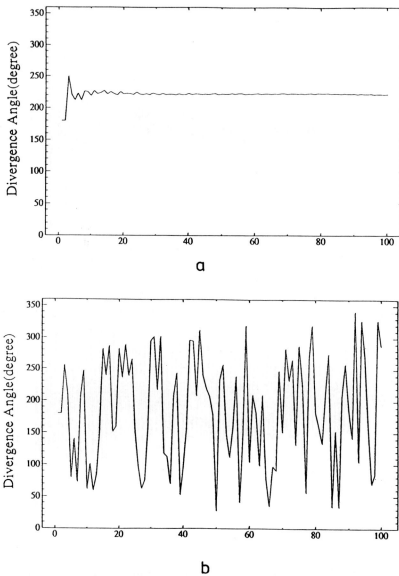

Fig. 9. Variation of divergence angle (ϕ_d) for close-packings of growing discs ($R = 0.5n$) with (a) origin between first and second disc, and (b) with first and second discs at $(-0.05, 0)$ and $(1, 0)$, respectively. Note convergence to Fibonacci angle $\phi_d = 222.5°$ ($= 360° - 137.5°$) in (a) and persistent oscillations in (b).

8. Discussion

The study of three varieties of *H. annuus* has shown that, almost without exception, the characteristic phyllotaxis and chirality of the plant stem will match that of the flowerhead. Of the 111 plants studied and completely analyzed, it was found that only 2 stems had chirality and divergence angle opposite to that of their respective flowers. It may be that these exceptions were incorrectly analyzed. Another explanation could be that the phyllotactic behaviour of these stem/flowerhead combinations represents a genetic or capricious error. The present sample was, however, too small to test for such effects.

It is apparent that both the bracts and ray flowers share the same chirality and divergence angle as the stem and top flowerhead. From the SEM results,[2,15] as well as those above, it is clear that the bracts and ray flowers play a role in transmitting of phyllotactic information from the stem to the flowerhead. Whether this is genetically determined for both the stem and flowerhead together or perhaps separately is not known, but the occurrence of overwhelmingly matching phyllotactic characteristics tends to indicate that a single genetic factor governs the behaviour of both the stem and flowerhead, via control of the apical shoot. The bracts and ray flowers share the same chirality and divergence angle as the leaves and flowerhead, forming a serially developing homology. In addition, the "passing on" of phyllotactic information by the bracts to the primordia formed on the rim of the capitulum was observed and recorded. The chirality expressed by whole plants has been found to be essentially random and not significantly deviated from a binomial distribution. It is not known whether chirality is determined genetically or capriciously. The question remains open as to whether environmental factors can physically determine chirality at various stages of development.

It has become clear from these above results that the highly regimented behaviour of real plants (e.g., *H. annuus*) requires very special setting-up of growth spirals on the sunflower capitulum. Growth outward from a fixed centre with predominantly Fibonacci divergence angles requires a measure of control over initial conditions, which appears most unlikely. The observations depicted in Fig. 6 reveal how the initial growth of spirals on the capitulum may be controlled already by the existing divergence angle of the leaves on the stem and the bracts on the developing flowerbud just prior to capitulum development.

The above-mentioned theoretical results[18,19] lead to an understanding of spiral lattice structures as close-packings appropriate for a given size distribution or growth law. However, they do not describe the kinetics of growth.

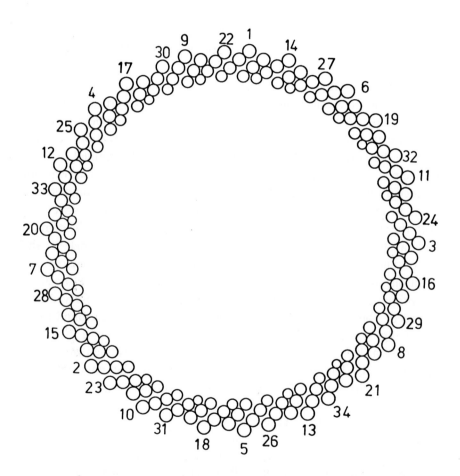

Fig. 10. Schematic model suggesting rim-constrained growth of natural and synthetic spiral structures. Note there are discs in contact at the circular rim, placed in order of decreasing size with angular sequence appropriate for the Fibonacci divergence angle $\phi_d = 137.508°$.

The mechanism by which spiral structures may form naturally, or by a synthetic route for inorganic materials, requires a precise setting of the initial growth centres. This may be achieved most readily in both natural and synthetic structures by a rim-constrained growth, starting with precisely set initial divergence angles. Such a process is suggested schematically in Fig. 10, which shows discs nucleated around a circular rim. There should be a Fibonacci (or Lucas) number of such growth centres. Discs are used here for convenience. In general these simply represent nucleation centres for structural elements of arbitrary shape, provided certain matching rules are satisfied, for example, as implied by Voronoi polygons[20,21] or the lozenge (diamond) shapes of seeds in *H. annuus* or other sunflowers (see Fig. 3). Further cells will then grow in contract with cells growing in from the rim, and the close-packing considerations predict that a spiral structure will evolve, retaining the same mean divergence angle. In reality both the growth law and the divergence angles may tend to develop oscillations about a mean function or a mean value ϕ_d as required to accommodate some fluctuations in the initial conditions and in the positions of new growth nucleation sites.

Following our interest in synthesizing new quasicrystalline structures,[7] it now seems appropriate to attempt to make spiral lattice structures using the principles just discussed.

Acknowledgements

This work has been supported financially by the Australian Research Committee.

References
1. R. V. Jean, *Mathematical Approach to Pattern and Form in Plant Growth*, Wiley-Interscience, New York (1984).
2. J. Marc and J. H. Palmer, Photoperiodic sensitivity of inflorescence initiation and development in sunflower, *Field Crop Research* **4** (1981) 155–164.
3. L. Bravais and A. Bravais, Essai sur la Disposition des Feuilles Curvisereés, *Annals de la Science Naturelle* **7** (1937) 42–110, 193–221, 291–348.
4. A. H. Church, *On the Relation of Phyllotaxis to Mechanical Laws*, Williams and Norgate, London (1904).
5. T. A. Cook, *The Curves of Life*, Dover, New York (1974; reprint of 1914 Edn.).
6. M. Gardner, *The Ambidextrous Universe*, 2nd Edn., Pelican, New York (1982).

7. L. A. Bursill, Peng JuLin, and Fan XuDong, Spiral lattice concepts, *Mod. Phys. Letts.* **B1** (1987) 195–206.
8. Fan XuDong, Peng JuLin, and L. A. Bursill, Algorithm for the determination of spiral lattice chirality, *Inter J. Mod. Physics* **B2** (1988) 121–129.
9. L. A. Bursill, G. W. Ryan, Fan XuDong, J. L. Rouse, Peng JuLin, and Anne Perkins, Basis for synthesis of spiral lattice quasicrystals, *Mod. Phys. Letts* **B3** (1989) 1071–1085.
10. R. F. Williams and E. G. Brittain, A geometrical model of phyllotaxis, *Austral. J. Botany* **32** (1984) 43–72.
11. C. B. Heiser, *The Sunflower*, Norman Press (1976).
12. A. R. Needham, J. L. Rouse, and L. A. Bursill, *Chirality and Phyllotaxis of Helianthus tuberosus: A Multihead Sunflower* (in preparation).
13. R. F. Williams, *The Shoot Apex and Plant Growth*, Cambridge Univ. Press (1975).
14. M. W. Moncur, *Floral Initiation in Field Crops*, CSIRO, Melbourne (1981).
15. J. H. Palmer and B. T. Steer, The generative area as the site of floret initiation in the sunflower capitulum and its integration to predict floret number, *Field Crops Research* **11** (1985) 11–12.
16. I. Adler, A model of contact pressure in phyllotaxis, *J Theor. Biol.* **45** (1974) 1–79. ibid, The consequences of contact pressure in phyllotaxis, *J. Theor. Biol.* **65** (1977) 29–77.
17. J. N. Ridley, Ideal phyllotaxis on general surfaces of revolution, *Math. Bioscience* **79** (1986) 1–24.
18. Fan XuDong, L. A. Bursill, and Peng JuLin, Packing of equal discs on a parabolic spiral lattice, *Mod. Phys. Letts.* **B3** (1989) 119–124.
19. L. A. Bursill and Fan XuDong, Close-packing of growing discs, *Mod. Phys. Letts.* **B2** (1988) 1245–1252.
20. N. Rivier, A botanical quasicrystal, Proc. Inter. Workshop Aperiodic Solids, Les Houches, March 1986; *J. de Physique*, Colloque C3 Suppl. No. 7 (1986) C3-299–310.
21. N. Rivier, R. Occelli, J. Pantaloni, and A. Lissowski, Structure of Bénard convection cells, phyllotaxis and crystallography in cylindrical symmetry, *J. Physiqe* **45** (1984) 49–63.
22. L. A. Bursill and Fan XuDong, Scaling laws for spiral aggregation of contact discs, *Mod. Phys. Letts.* **B3** (1989) 1293–1306.
23. Fan XuDong, L. A. Bursill, and Peng JuLin, Fourier transforms and structural analysis of spiral lattices, *Inter. J. Mod. Phys.* **B2** (1988) 131–146.
24. G. W. Ryan, J. L. Rouse, and L. A. Bursill, Quantitative analysis of sunflower seed packing, *J. Theor. Biol.* **147** (1991) 303–328.
25. H. G. Schuster, *Deterministic Chaos: An Introduction*, Weinheim, Physik-Verlag (1984).

ON THE ORIGINS OF SPIRAL SYMMETRY IN PLANTS

Roger V. Jean

1. Four Keys and Two Remote Mechanisms to Help Unveil the Mystery—a theory of phyllotaxis

The very familiar spiral patterns exhibited, for example, by the scales on a pineapple and on a pine cone, by the florets in the capitulum of the daisy, and by the fruits in the sunflower capitulum are called *phyllotactic patterns*. They are the object of discussion of this paper. These patterns are described by a few parameters, among which are (a) the phyllotaxis expressed by a pair of integers (m, n), and (b) the divergence angle between consecutively born primordia (e.g., scales, florets, leaves, seeds). Here m and n are the numbers of spirals in the two sets defining a pattern. The mystery of phyllotaxis is that m and n are generally consecutive terms of the Fibonacci sequence 1, 1, 2, 3, 5, 8, ... Otherwise stated the divergence angle is generally given by ϕ^{-2} or 137.51°, where $\phi = (\sqrt{5} + 1)/2$. We are not going to describe these patterns in detail. The reason is that we are rather interested in morphogenetical parallelism. The description of all the facets of the phenomenon and an access to all the literature on the subject can be found in the author's articles given in the references.

The *systemic and holistic viewpoint* introduced by the author in phyllotaxis drove him to stress the fact of the presence, in other areas of research, of patterns similar to those found in phyllotaxis (see Ref. 1). For example, "phyllotactic" patterns can be observed in the arrangement of tentacles of jellyfishes. A first key for the understanding of the origins of phyllotaxis is in *morphogenetical parallelism*, in the recent recognition of the importance of such similarities between patterns in various areas. The article will present many of these similarities. According to Meyen,[2] a well-known paleobotanist, such comparisons between various objects or organisms showing similarities of structure or form will bring a revolution in biology, though at first sight it may look senseless and improper.

The *evolutionary and phylogenetic viewpoint* introduced by the author in phyllotaxis[3,4] drove him to look in the plant kindgom for ancestors of the phyllotactic patterns. He was led to stress the importance of Zimmermann's telome theory, according to which all plants derive from primitive plants of the *Rhynia* type, as the result of six elementary morphogenetic processes including overtopping, a process so important in phyllotaxis (see Ref. 5). The *telome theory* is a second key for unlocking the mysteries of phyllotaxis. And considering the similarities or isomorphisms mentioned earlier, a question arises: why stop at the beginning of plant evolution when the origins we are looking for are apparently more remote?

A third key for the understanding of phyllotactic patterns is handed to us by a recent alternative to neo-Darwinism, called *autoevolutionism*. Darwinism and neo-Darwinism generally consider biological patterns as the outcomes of genetic activities. But there are no genes in minerals where similar patterns are found. Autoevolutionism, still a minority opinion not yet accepted by the mainstream, is a new state of mind on evolution that is critical of neo-Darwinism (see Ref. 6). It dares go beyond the Darwinian gene; this apparently allows us to explain better the origins of forms and functions in biology. In autoevolutionism, morphogenetical parallelism is called homology, or isomorphism. Using the tenets of autoevolutionism, a further step can apparently be done regarding the origins of phyllotactic patterns. One of these tenets is that we simply have to go back to the evolutions of minerals and of chemicals to integrate biological evolution into its natural context. In this attempt to address major issues of phyllotaxis and pattern formation, the critique of genocentrism and neo-Darwinism will appear of general significance.

What is the common denominator of the similarities that are going to be presented here? It seems to be the branching processes found at every level of evolution. Among the oldest forms of growth, they arise as the result of simple properties of space. The author has shown[5,7,8] that the skeletons of the phyllotactic patterns are tree-like structures, hierarchies developing into fractals. The same kind of trees is found, for example, in the ramified structures of algae, in the structure of the system of tributaries of rivers, as well as at the chemical and mineral levels, as we will see. Those trees represent the fundamental isomorphism as well as the isofunctionality of universal processes involved. The *properties of space–time* itself, of which branching is a manifestation, appear as a fourth key for understanding phyllotaxis.

With the search for the origins of the phyllotactic patterns, naturally comes the search for the remote mechanisms responsible for the existence of those patterns. As there is a correspondence between matter and energy through Einstein's well-known law, there is a correspondence between form and function, as it is beginning to be more and more recognized. Isomorphism means isofunctionalism. This paper stresses the importance of *gnomonic growth*, and of *branching processes*. Gnomonic growth (formally defined in Sec. 4) produced the beautiful logarithmic spiral on the shell of the very old mollusc called *Nautilus pompilius*, by the constant addition of new material at the open end of the shell, according to the same accretionary law. And the fundamental tenet of the telome theory is that leaves and vascular systems in higher plants are the results of simple branching processes that bring us back to early land plants. But arrangements of leaves is what phyllotaxis is mainly about, and vascular phyllotaxis underlies leaf phyllotaxis. Gnomonic growth and branching processes in phyllotaxis are expressions of an elementary rhythm of growth that is the essence of the regularity in patterns. They give rise, as we will see, to *dendrites* and *fractals*, to buds and trees. Phyllotactic patterns are fundamentally branched structures incorporating a gnomon.

There are three mathematical models throwing light into the mechanisms of gnomonic growth and branching processes in phyllotaxis: van der Linden's model,[9] Jean's ϕ-algorithm,[8] and Jean's minimal entropy model.[3] The first one uses a geometrical gnomon, the second a numerical one, and the third a geometrico-numerical one. They are all concerned with fractals. The first two models do not deal with the most fundamental problem of phyllotaxis: the question of the reasons for the initial configurations of the primordia from which to start to apply the gnomon.

The third model has an answer to the question. Given that almost nothing is known about the intimate processes of evolution, about the properties of space–time responsible for it, and about the ontogenetic mechanisms inside shoot apices, the answer has been inferred in terms of the ultimate effect phyllotactic patterns are considered to achieve, that is, the minimization of the entropy of plants. The model (a) identifies tree-like structures with phyllotactic patterns (Jean's hierarchical representation of phyllotaxis), (b) puts forward an algorithm able to generate those trees (with L-Systems), and (c) proposes a formula for entropy that allows us to compare the various structures and to order them according to increasing entropy costs, given that evolution is concerned with thermodynamics and entropy, a fundamental way to attack all biological phenomena (see Ref. 10).

This model is part of the author's holistic and evolutionary *theory of phyllotaxis*, that is a body of knowledge that obeys the following criteria: (a) it is based on observations that can be put into a coherent body of doctrine and permits the formulation of mathematical laws (see Refs. 3, 8), (b) it makes definite predictions that can be confirmed or disconfirmed (see Ref. 11), (c) it suggests experimentations to biologists and it gives more coherent solutions to phyllotactic problems (see Refs. 4, 12, 13), and (d) it provides a framework in which to organize phyllotactic data (see Refs. 13, 14). The theory stresses the importance of a global approach making use of the results of various sectors of biology, and stressing the importance of comparative morphology not only in biology but throughout nature.

2. Isomorphisms with Phyllotactic Patterns—beyond genetics, inside artifacts

The main types of leaf arrangement are distichy, decussation, whorled, and spiral arrangements. The well-known facts that I want to point out about them is that these patterns can be transformed into one another, either naturally (see, for example, Ref. 15, pp. 115–120) or experimentally (see, for example, Ref. 14). Also, for a specific plant, the pattern adopted is clearly dependent upon the relative sizes of the apical meristem and of the primordia. Consequently, these patterns are not determined by heredity. It follows that the core of the phenomenon of phyllotaxis lies deeper than it is generally thought. This is precisely why it has been bypassed.

The aim of comparative morphology is the discovery of homologies in the organic world, and sometimes only in a specific part of it. We are interested in homologies of phyllotactic patterns, and these do not stop at the artificial frontier between inanimate and animate worlds. Those homologies reveal fundamental mechanisms, beyond the gene. The scaly patterns on the anteater *Manis temminckii* (Fig. 1.3), on the pine cone *Pinus pinea* (Fig. 1.4), on the surface of the shell of the mollusc *Conus milneedwardsi* (Fig. 1.2), and on a crystal of common salt NaCl (Fig. 1.1), show an interesting isomorphism in the arrangement of scales. It follows that research in order to understand the origins of phyllotaxis must not be concentrated on pine cones alone, but on all objects showing similar arrangements of scales. A previous paper[1] shows that the patterns of tentacles on jellyfishes and of bud formation on *Hydra* are strictly comparable to those of leaf primordia and of amino-acid residues in polypeptide chains. One very good illustration of phyllotaxis is the pattern in capituli of Compositae (e.g., daisy, sunflower, chrysanthemum) such as the one shown in Fig. 2.1. The capitulum pattern has a striking resemblance to that on the diatom in Fig. 2.2. In the division of Phycophyta (algae) we find the very small, unicellular, brown diatoms, and the very large brown *Fucus*. The ramified structure of *Fucus* may seem to be very far from the structure of the capitulum, but the next section will show that this is not the case.

A recent article[1] underlines the structural relationship found between phyllotactic patterns and the pattern on the Tobacco mosaic virus, which has been described using phyllotactic parameters. A virus is a colloidal crystal, a term that covers organic, inorganic, as well as biological materials. And the colloidal crystal has an intermediate size between microscopic and macroscopic sizes. Geometrically, there is no difference between the virus, a piece of opal, and a ball of latex; they are all colloidal crystals showing compact packing of spheres. Many studies in phyllotaxis deal with packing of spheres or circles representing primordia. Just as in the colloidal crystals, the phyllotactic arrangements are the results of packing efficiency, as it has often been demonstrated (see, for example, Ref. 18).

According to Rothen and Pieranski,[19] colloidal crystals "represent a very pedagogical model of real crystals, at a different scale." The

328 Roger V. Jean

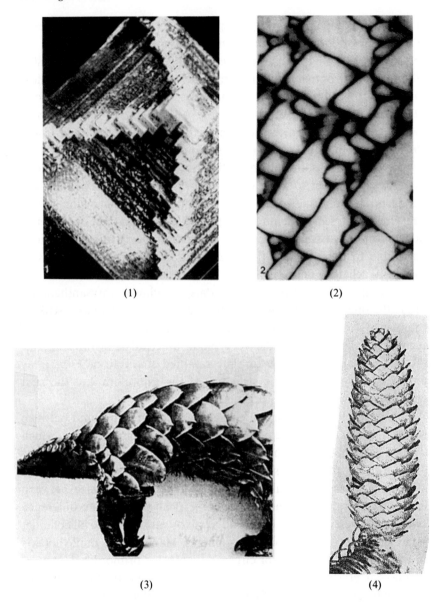

Fig. 1. (1) A crystal of halite, NaCl; (2) surface of the shell of a mollusc (both figures from Ref. 6; reprinted with permission from Elsevier Science Publishers); (3) the pangolin (from Ref 16); (4) a spruce cone, *Picea excelsa*, having (5, 8) phyllotaxis.

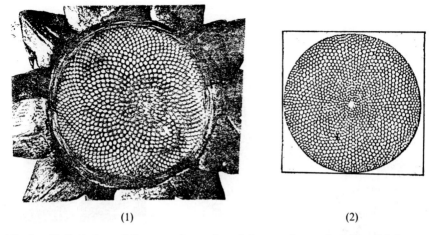

Fig. 2. (1) Capitulum of *Cynara scolymus* shaved down to the ovaries of the disk-florets, showing the phyllotactic system (55, 89) (from Ref. 17); (2) valve-view of *Coscinodiscus*, a Bacillariophyta universally distributed in fresh water, the sea and the soil (from Ref. 15).

crystallographic restriction means that pentagonal symmetry is absent from the world of crystals. But this type of symmetry can be observed in quasi-crystals: the microscopic structure of rapidly cooled metal alloys shows indeed fivefold rotational symmetry. It is also impossible to tile a plane using only shapes that have fivefold symmetry. But the two-dimensional analogue of quasi-crystals, Penrose tilings, have it. It is interesting to remark that ϕ, the golden number, which is so important in phyllotaxis (see, for example, Ref. 1), arises in this tiling: the tiling is made of two types of parallelograms, and in an infinite Penrose tiling the ratio between the numbers of units in these two types is ϕ (see Ref. 20).

In growing crystals there exists two great categories of structures: dendrites and fractals. Fig. 3.1 shows the apical dome of a plant growing its primordia in a phyllotactic arrangement. The dome has a structural similarity with the dendrite in Fig. 3.2, and with the leaf of the common fern in Fig. 3.3. We will see in the next section that phyllotactic patterns are fundamentally fractals; they are thus illustrative of a connection between the two crystalline structures.

The homologies between forms and functions in plants, animals, and minerals that I am putting forward in this article to underline similarities

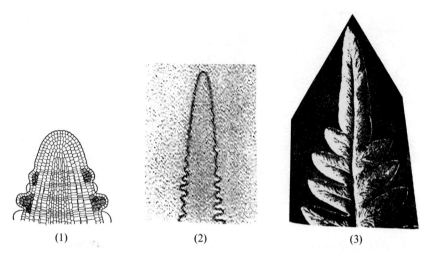

Fig. 3. (1) Longitudinal section of the parabolic shoot apex of *Hippuris vulgaris* showing the leaf primordia (from Ref. 15); (2) crystal of ammonium chloride growing in a dendritic way at a parabolic speed (from Ref. 20); (3) tip of a leaf of *Pteridium aquilinum*, the most widespread fern (from Ref. 21).

with phyllotactic patterns may appear strange to many. It is only that they are far removed from the trail of thoughts generally springing from Darwinism and neo-Darwinism. But "Darwin merely translated the prevailing English political ideas and applied them to nature" (see Ref. 23). For autoevolutionists, "selection and randomness is a morass of concepts, words used to cover ignorance, and they must be banished from evolution." In neo-Darwinism the gene appears to be omnipotent, and what occurred before it is considered irrelevant to evolution. Whatever it may be about these polemical statements, evolution is a phenomenon that started long before the evolution of biological organisms. "Molecular analysis leads to the inescapable conclusion that life has no beginning; it is a process inherent to the structure of the universe" (from Ref. 6). A recent finding is that the elementary particles, the chemical elements, and the minerals have had their autonomous evolutions, which have made their imprints on biological evolution. One of the central theme in Lima-de-Faria's stimulating book is that every biological pattern and every biological function has its predecessors in the mineral, chemical, and elementary particle worlds, so that nothing essentially new arose at the biological level; and though he is a geneticist he professes that forms and functions in biology have not been created by genes and chromosomes.

When there is an isomorphism of structures, beyond a variety of chemical composition, "there is a common chemical denominator that is not always evident. It may be a radical or an atomic or electronic configuration that is at the basis of the isomorphism. Equally important may be the pressure, the temperature and the gas or liquid in which the materials are allowed to expand. The lesson to be learned is that the tendency to consider such similarities as accidents, curiosities or mere analogies derives mainly from the ignorance of the physical and chemical processes underlying them" (from Ref. 6).

Phyllotaxis-like patterns are commonly seen in the global cloud cover pattern, in particular in the hurricane cloud bands. Selvam[24,25] developed a cell dynamical system model for atmospheric flows that predicts the spiral circulation pattern with the golden mean winding number as intrinsic to the quantum-like mechanics of atmospheric flows. And further, such logarithmic spiral patterns are signatures of deterministic chaos indicating long range spatio-temporal correlations (her letter of August 29, 1990). I think that the omnipresent properties of space, in which everything moves, are predominantly important in the structurating and patterning of matter and energy and in the existence of the basic laws of growth that will be considered in the next sections to explain the generation of phyllotactic patterns.

In an article I recently recommended for publication, Friedman[26] compares a mechanism of generation of phyllotactic patterns with features of a model of polymer conformation. She shows interesting analogies between arrangement of plant primordia and molecular patterns, particularly in the case of polymer chains. Some biopolymers, as well as inorganic polymers exhibit a helical backbone conformation similar to the spiral arrangement of primodia around a stem, while others assume nonspiral configurations. A few may display more than one type of symmetry within a single natural sample, recalling plants exhibiting transitions between phyllotactic patterns. The acropetal influence that may drive plant phyllotaxis is shown to have its chemical counterpart: substituants on a backbone tend to maximize their distance from one another due to the repulsive forces between pairs of electrons. Friedman insists more particularly on alternate whorl of three, as observed in oleander. These have an intriguing resemblance to the staggered conformation of the ethane molecule. She says that the eclipsed conformation, corresponding to superposed whorls in plants, is however unfavorable in polymers. This is because of a rotational barrier that prevents the two CH_3 in ethane (C_2H_6) to rotate relative to each other around the C–C bound, from one staggered conformation to another, via the eclipsed state. The author proposes the idea of repulsion between the hydrogen atoms in polymers to explain this barrier, as in models of phyllotaxis that postulate a repulsive influence of primordia on each other, to explain phyllotactic patterns. She concludes that the factors governing the existence of symmetry in plants and in polymers are similar in many respects.

Similarities with phyllotactic patterns are not found in nature only, as Fig. 4 shows. Philosophers, such as Teilhard de Chardin, were right to put forward the idea that man-made designs and forms are extensions of natural ones, and in a sense are still nature. Crystal sections and wallpaper designs obey the same structural rules. The presence of the golden number ϕ and of the golden rectangle (see below) in phyllotactic patterns, is mimicked in the works of musicians (e.g., Bartok), architects (e.g., Le Corbusier), and artists (e.g., Dali), from the Ancient to the Modern Worlds. Creators naturally find molds of the Creator.

3. Isofunctionalism With Phyllotactic Patterns—generation of fractals

Dichotomous branching is probably the most primitive pattern of growth in plants. Palms, which are among the oldest of flowering plants, and which provide the most dramatic examples of phyllotactic spiral patterns, would descend from a dichotomous ancestor. Today, contrary to most of the palm trees, *Hyphaene thebaica* shows a trunk ramifying dichotomously. But the branching process called overtopping, a variation on the theme of dichotomous branching, is probably responsible for the spirality in the leaf pattern on the trunk of palm-trees, and in higher plants in general. An overtopping pattern arises when the doubled branches grow unequally, with the growth of one member exceeding that of the other (see Fig. 5). "Overtopping might have been the means by

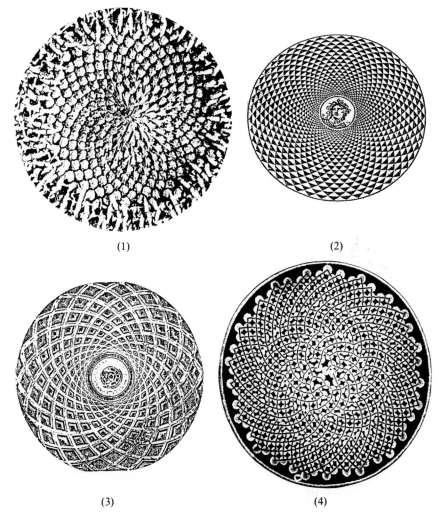

Fig. 4. (1) Head of a Chrysanthemum (21, 34); (2) a Roman mosaic from the Museo nationale in Rome; (3) cupola of a chapel of the Renaissance; (4) a "Fibonacci" carpet by R. F. Williams. The outer zone shows "(34, 55) phyllotaxis," the middle zone "(34, 21)," and the inner zone "(13, 21)."

which morphological differentiation and division of functions was introduced into a system of telomes (branchlets) originally all exactly alike" (from Ref. 15).

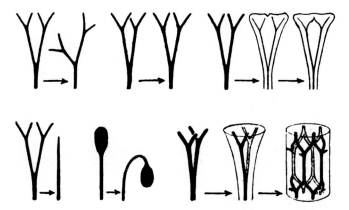

Fig. 5. Elementary processes of the telome theory leading to vascular plants: overtopping, planation, fusion in the leaf, reduction, recurvation, fusion in the axis giving vascular systems (from Ref. 24).

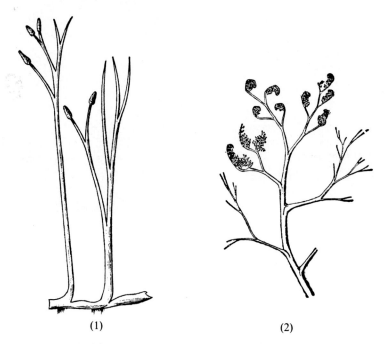

Fig. 6. (1) *Rhynia major*, the ancestral land plant; (2) a primitive frond, from the Devonian, of *Protopteridium hostimense* (both figures from Ref. 15).

This type of branching is among the six elementary processes in Zimmermann's telome theory, which have led to the structure of plants as we see them today. This theory is often said to be the only coherent explanation of the evolution of plants (see Ref. 27, for alternatives). The primordial telomes are the branchlets of the Psilophyta. In the division of Pteridophyta, the Psilophytales are the oldest known land plants. Among them, the most primitive members (Rhyniaceae, see Fig. 6.1) had dichotomous stems, with a single vascular bundle, but entirely lacked leaves and roots. In the same division, the Primofilices show a range of forms that lie almost without discontinuity between early land plants and the morphologically complex plants of today (see Fig. 6.2)

The Psilophytales still resemble quite closely in their external forms the dichotomously branched thalli of many of the more highly organized brown algae (Fig. 7.4). And the latter shows affinity with the Chrysomonadales (Fig. 7.3). The algal ancestor of the Pteridophyta is believed to be found in the Chlorophyceae, the green algae, among which are *Ulothrix* and *Cladophora* (Figs. 7.1, 7.2). That a physical imprint induced the branching process in plants can be imagined when we consider that such a branching phenomenon already exists at the level of elementary inorganic material (Fig. 7.5).

How are these branching processes represented at the level of higher plants? Bolle's work[29] (see also Ref. 7) reveals it. The botanist showed the relevance in phyllotaxis of the principle of overtopping. There is an amazing coincidence between, for example, the vascular skeleton linking the foliar traces and his induction lines for a capitulum of *Cephalaria* (see Fig. 8.4). Bolle thinks that in almost all dicotyledonous plants one of the two cotyledons is the point of departure of a bifurcated induction line, which induces the first pair of leaves. If the two branches of a bifurcated induction line have different lengths and slopes, then there will be a spiral system of leaves if the induction line having a smaller slope stays simple, and the other one bifurcates again, until the system is fully developed along the sequence 3, 5, 8, 13, ... In Bolle's theory the distribution of simple and double nodes is of fundamental importance, and is at the basis of his phyllotactic laws. In Church's theory too, where the emergence of the periodic sequence 2, 1, 2, 1, 2 / 2, 1, 2, 1, 2 / 2, ..., is often mentioned, manifesting a fundamental rhythm in phyllotactic

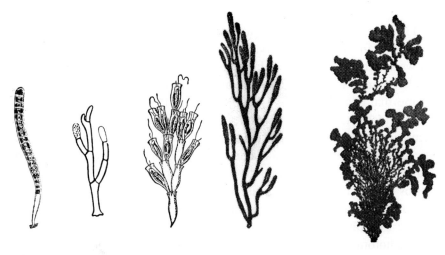

Fig. 7. (1) A young filament of the green alga *Ulotrix*; (2) diagrammatic representation of the green alga *Cladophora*; (3) chrysomonadales, *Dinobryon*; (4) *Fucus bifurcatus*; (5) native copper dendrite (1, 2, and 3 from Ref. 15; 4 and 5 from Ref. 6 reprinted with permission from Elsevier Science Publishers).

patterns, that is a succession of simple and double nodes at every level, as seen in Fig. 9.1. This figure is easily obtainable from Bolle's Fig. 8.4, which is the hierarchy representing the usual phyllotactic pattern expressed by the Fibonacci sequence 1, 1, 2, 3, 5, 8, 13, ..., in Jean's approach. In this figure one can easily read most important phyllotactic parameters, that is the divergence angle and the phyllotaxis (m, n) of the system.

Church[17] insists on the idea that the phyllotactic mechanism is an old function of marine vegetation, that the answer to the problem of phyllotaxis is to be found in the morphogenesis of the brown algae that show the essential features of the phenomenon, and that phyllotaxis as we know it in higher plants is but the amplification of phyto-benthic factors. We must not be surprised then that Fig. 9.1 also represents each of the two branches of *Fucus* in Fig. 8.2, treated with Strahler's method for ordering ramified patterns. Morphologically, vascular systems are hierarchies arising from the primordial telomes, themselves expressing the processes of overtopping and branching found in algae. Experimentally, the phenomenon of sectorial translocation of substances injected in plants reveals their underlying hierarchical organization. These themes

On the Origins of Spiral Symmetry in Plants 337

are developed in my articles,[5,7,18] where is justified my hierarchical representation of phyllotactic patterns. A hierarchy extended indefinitely becomes a fractal. These are fractals of dimension two represented by the exponent two in Jean's allometry-type model.[33] L-systems can be used to generate these fractals.[3,34]

In the network of tributaries of the river in Fig. 8.1, when two tributaries of the same class i, or of different class j and i, $j > i$, meet, the

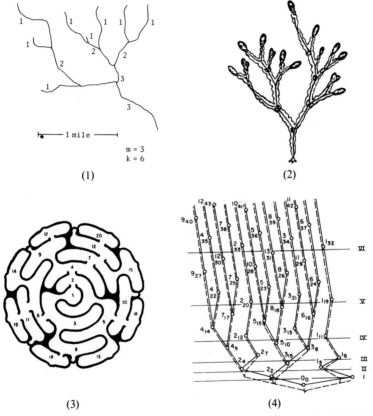

Fig. 8. (1) The tributaries of a stream in South Carolina (from Ref. 30). This is a Fibonacci network with 8 tributaries of class 1; eight being the sixth Fibonacci number, k = 6. (2) *Fucus spiralis*, a brown alga. (3) Representation of a mechanism that generates the main phyllotactic patterns (from Ref. 31). (4) In dotted lines, the links between the foliar traces of the capitulum of *Cephalaria*. The heavy lines are Bolle's induction lines for the spiral system of leaves in the capitulum (from Ref 29).

resulting class is $i+1$ or j respectively, m is the highest class, and k is the order of the number of tributaries of class 1 (no tributaries) in the sequence 1, 1, 2, 3, 5, 8, 13, ... The number of tributaries of class 1 being a Fibonacci number, the river and its tributaries is called a Fibonacci network. Its graph, Fig. 9.2, obtained from the Strahler's topological method, is essentially Fig. 9.1.

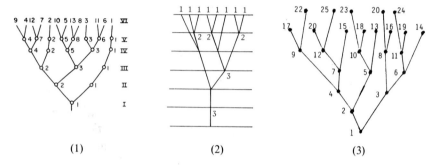

Fig. 9: (1) Phyllotactic hierarchy obtained from the vascular system of Fig. 8.4. Each level contains a Fibonacci number of simple and double nodes. The numbers m and n of nodes in two consecutive levels express the phyllotaxis (m, n) of the pattern. Each level reveals an approximation of the divergence angle. Level V for example represents the fraction 3/8, given that we have to go three times to the left, starting at 1, to pass through the consecutive numbers, that is to meet eight leaves. The fractions obtained from consecutive levels converge toward ϕ^{-2}, the limit divergence of most of the phyllotactic patterns (from Ref. 7). (2) Graph of the stream in Fig. 8.1 (from Ref. 32). (3) The hierarchy obtained from Fig. 8.3 (from Ref. 5).

For Hermant,[31] an architect, the essential geometrical characteristics of the main plant structures, and the main functions of phyllotaxis can be deduced from Fig. 8.3, which can be developed further on by using a two-step algorithm for the generation of consecutive rings. From this figure we easily obtain Fig. 9.3, which shows simple binary ramification. Fig. 8 is thus an illustration of isofunctionality through various substratum, physical, algal, and vascular. This isofunctionality is revealed further on when, following Hermant,[35] we consider the fractal in Fig. 10.1, and reverse the process, which led us to the hierarchical representation of phyllotactic patterns. From such a scheme we can apparently reconstruct buds and flowers. Indeed the hierarchy of scalene triangles is slightly more complex than the hierarchy in Fig. 9.1, but it is essentially the same. When points A and B are brought together in a circular movement, the result (Fig. 10.2) closely resembles the cross-section of a bud with its primordia (Fig 10.4), and the diagram of a flower revealing the relative arrangement of its parts (Fig. 10.3). There is a need for more studies in that direction by which the analogies in Fig. 10 will bear up more convincingly under scrutiny. Hermant's work suggests the fractal nature of phyllotaxis. My allometry-type model, using the plastochrone ratio, shows a fractal dimension 2 in phyllotactic patterns.

4. Mathematical Modelling of Isofunctionalism—gnomon and rhythm

When one looks for the origins of patterns in plants, one obviously deals with the essential mechanisms governing the generation of those patterns. Most of the models in phyllotaxis are oriented toward an ontogenetic perspective, while the problem is phylogenetic. The hypothesis of the diffusion of a phytohormone governing the generation of patterns is common to many models. But these models deal with a situation that is made more complex by subsequent growth processes, and not with the situation that has given rise to those patterns in the first place. The more backward we go regarding the ancestors of those patterns, the more likely we are to find elementary rules governing their formation.

There are three mathematical models dealing with those rules, one by van der Linden,[9] and two by Jean.[3,4,8] The interest of these models is that they do not require the generally used divergence angle and plastochrone ratio. These two parameters are, with the phyllotaxis of a system expressed by a pair of integers, the most important parameters to determine when one considers the practical problem of identifying and describing the patterns (see Ref. 14). But the problem of pattern generation is different from the problem of pattern recognition. The success with which the three models reproduce any type of pattern is indicative of the fact that the other models, using divergence angles and plastochrone ratios and which are much less able to reproduce the patterns, introduce too strong restrictions that prevent the fundamental natural processes involved to be fully operational.

As pointed out by D'Arcy Thompson,[36] we can conceive no simpler law of growth, the simplest of laws which Nature tends to follow, than

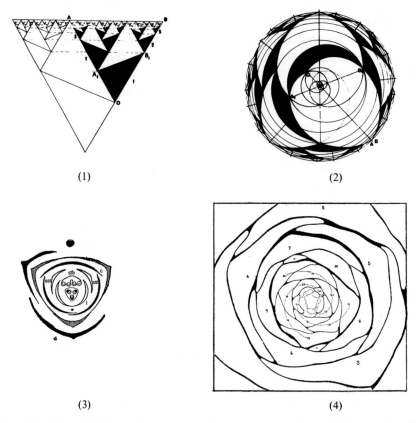

Fig. 10. (1) A hierarchy of congruent scalene black triangles, presenting levels with consecutive Fibonacci numbers, and revealing the architecture of the complexity found in the next diagrams. (2) A topological deformation of the fractal, obtained by bringing together points A and B, in a circular movement. (3) Floral diagram of *Kaempferia ovalifolia*, a Zingiberaceae, showing a (3, 3) pattern. (4) Transverse section of foliage-shoot of *Saxifraga umbrosa*, showing the phyllotactic system (3, 5). (1 and 2 from Ref. 35, 3 from Ref. 15, and 4 from Ref. 17).

that observed in the growth of the shell of the *Nautilus*, which widens and lengthens in the same unvarying proportions. The result is that the shell or the organism grows in size but does not change in shape; it shows constant similarity of form. This is precisely what *symmetry* is literally about: the repetition of the same proportions. This is the way crystals grow, it is called accretionary, or additive, or *gnomonic growth*. It comes from a single growth gradient, and is present at all levels of evolution, just like spirality.

A gnomon can be geometrical or numerical. A geometrical gnomon is an elementary structure that repeats itself, to a near similarity relation, at every step, and that is the building block of the whole. Aristotle defined it as any figure which, being added to any figure whatsoever, leaves the resultant figure similar to itself. This definition includes the case of numbers considered geometrically. A numerical gnomon is a rule or algorithm by which a number in a structure is approximately generated, over and over again. The Ancients defined it as what must be added to a quantity to obtain a similar quantity. Thus $2k + 1$ is a gnomon of k^2, given that $k^2 + 2k + 1 = (k + 1)^2$. What is the importance of gnomonic growth? It is simply that a gnomon is a mathematical expression of rhythm. Because of this rhythm a pattern is worked out: if there is no rhythm there is no pattern.

As an example of gnomon, consider a rectangle with sides of lengths $m - n$ and n. Add a square, the gnomon, with sides of lengths n as illustrated in Fig. 11 with precise values, and then a square with sides of lengths m in the same direction, and so on along the Fibonacci-type sequence $m - n, n, m, m + n, 2m + n, \ldots$. It is easily proved that the ratio of the sides of each rectangle is approximately equal to ϕ, where ϕ is the golden ratio $(\sqrt{5} + 1)/2$, and as we move along the sequence this ratio gets closer to ϕ. It follows, then, that the rectangles become closer to a golden rectangle whose sides are exactly in the ratio $\phi/1$. By linking the vertices x, y, z, w, u, and so on in Fig. 11, in the appropriate way, we begin to generate a logarithmic spiral, the essence of constant similarity form, and the abstract symbol of beauty.

Van der Linden's model is based on the idea of gnomonic growth. This model supposes an initial placement of the primordia based on a geometrical gnomon. Fig. 12 and its caption illustrates what is meant. The gnomon is made of the three circles on the left. It takes in charge the developing process primordium by primordium, in a dynamical way, till the capitulum or pine cone is fully reproduced. The relative arrangement of primordia n and its two nearest primordia p and q must be repeated with primordia $n + 1$ (to be placed), $p + 1$ and $q + 1$ because of the gnomon. But the sizes of other primordia between $p + 1$ and $q + 1$ may

342 Roger V. Jean

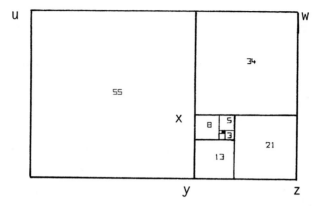

Fig. 11. A numerical gnomon ϕ, based on squares with sides of consecutive lengths 1, 1, 2, 3, 5, 8, 13, 21, 34, 55,... generates quasi-similar rectangles approaching the ratio $\phi/1$.

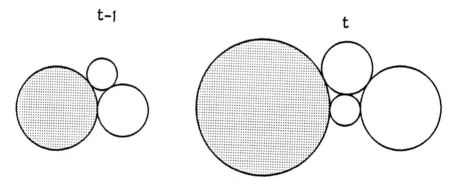

Fig. 12. An initial placement of four primordia. The tinted circle on the left has grown to become the tinted circle on the right, a primordium has been added by growth (the smallest one on the right). But the drawing of the structure at time t is such that if it is turned in some way (giving an angle corresponding to the divergence angle) then it fits onto the drawing of the structure at time t-1, so that we have a geometrical gnomon that allows us to construct the structure at t from that at t-1 (from Ref. 9).

not allow the contact of both of them with $n+1$. That is why the model gives supplementary rules allowing us to choose between possible places for primordium $n+1$ between existing primordia determined by the gnomon. A sigmoid growth function of the circular primordia can be adjusted so that the phyllotaxis can rise to any pair of consecutive

Fibonacci numbers. It follows that the divergence angle between consecutively born primordia can settle around the limit-divergence angle corresponding to the Fibonacci-type sequence we are finally concerned with (see Ref. 5), with any degree of accuracy.

A variant of this model uses the similarity between soap-bubble-like units and primordia, instead of tangent circles. Fig. 13, made with 5,000 such units, illustrates the result of the application of the gnomon: a capitulum (89, 144). Fig. 14 reveals the hidden structure of Fig. 13: lines bifurcating or not, from the centre to the rim (233 lines on the rim, starting to bifurcate to give the next Fibonacci number). Fig. 14 can easily be transformed into a tree-like structure, a hierarchy of simple and double nodes, simply by pulling apart the two lines on each side of the arrow for example, thus reversing the process explained in Fig. 10. The model thus bring to light a unity of structure between two growth processes, gnomonic and branching.

Jean's ϕ-model uses a numerical gnomon in the form of a minimality principle. For every integer m the gnomon determines an integer n, and thus the initial placement of the first m primordia, that is the system must show phyllotaxis $(m - n, n)$. The n is the integer by which the function of the integer k, $\log(m/k\phi)$, is closest to 0. The numerical gnomon also requires that after the initial configuration is set up at time m, a spiral rhythm of gnomonic growth starts. A dynamical process is thus replaced by a deterministic one. This brings the system to grow along the Fibonacci-type sequence starting with the first two terms $m - n$, n. The ratio of consecutive terms gets closer to ϕ as we move along the Fibonacci-type sequence. Geometrically, the gnomon is a square, as Fig. 11 illustrates.

This numerical gnomon respects Bravais and Bravais's well-known biologically plausible assumptions. These assumptions are that the divergence angle of the system must be an irrational number, and that the consecutive neighbors of the vertical axis in the cylindrical lattice of phyllotaxis must alternate on each side of this axis (see Ref. 1). These assumptions are tantamount to saying that the consecutive extensions of $(m - n, n)$ must indefinitely alternate: $(m - n, n)$, (m, n), $(m, m + n)$, $(2m + n, m + n)$,.... It follows that, as in van der Linden's algorithm, the divergence angle corresponding to the sequence can be approximated to any degree of accuracy, as the system grows. Also, for every m,

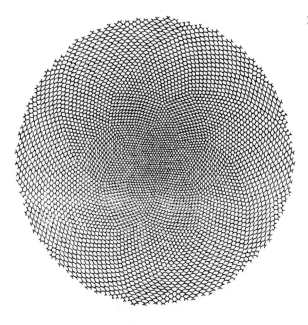

Fig. 13. A capitulum (89, 144) constructed without using divergence angles and plastochrone ratios (from Ref. 97).

Fig. 14. The hidden structure in Fig. 13 (from Ref. 9). It reveals the hierarchy of Bolle's bifurcated and simple induction lines, and the rise of phyllotaxis from the inner to the outer region.

representing the time for the emergence of the first m primordia, there is an initial pattern, and this allows us to order the various patterns according to increasing time. Finally it is meaningful to recall here that L. Bravais and A. Bravais were respectively a botanist and a mathematician-crystallographer (e.g., Bravais lattices), and they were the pioneers who initiated, in the 1830s, the mathematical treatment of phyllotaxis, and more precisely the cylindrical treatment on which most of the models are based.

The initial placement of the primordia determines an interval of values for the divergence angle (see Ref. 5). Once the initial placement or phyllotaxis of the system is determined, the ϕ model and the preceding one have a gnomon that determines every further step of the growth, which becomes deterministic. In both models any spiral pattern along any Fibonacci-type sequence can arise, depending on the initial placement. The common characteristic of these two models is that a rhythm of growth, always the same, starts with the initial configuration. They thus take care of one of the two main aspects of the problem of phyllotaxis, namely the rise of phyllotaxis from $(m-n, n)$, to (m, n), to $(m, m+n)$, and so forth, by which the initial divergences converge to a limit; and among these limits are the usual divergence angles. But they do not address the most important aspect of the problem, that of the initial placement of the primordia.

The question is: by which mechanism or constraint an initial configuration is preferred to another? None of the models, with one exception though, have an answer to this question. They all suppose an initial configuration and they generally ask the computer to try to continue to reproduce it, using rules governing for example the diffusion of a phytohormone or the contact pressures between the primordia. The choice is made in Jean's minimal entropy model[3]: the initial configuration chosen is the one that minimizes the bulk entropy of the plant at time m (this entropy can be calculated from a formula). The first two levels of the phyllotactic hierarchy being thus chosen, all the others are, because of the gnomon by which a double node gives rise to a double node and to a simple node, and a simple node gives rise to a double node. In this simple way, Fibonacci-type hierarchies are generated, level by level.

5. The Whole is in the Part—space and creation

This article deals with morphogenesis, the study of the origins and development of forms and patterns. We have seen that the search for origins in phyllotaxis is concerned with the search for the mechanisms responsible for the patterns observed. Researchers have always tried to reproduce the amazing and complex patterns observed at the ontogenetic level of highly developed plants, but phylogeny reveals simple hidden structures, and fundamental mechanisms at work in phyllotaxis. It is the latter that must be modelized. The key words in this area are *evolution* and *system*, or *phylogenism* and *holism*. In mathematical modelling as well as in nature, the key words are also *bioentropy*, and, most of all, *rhythm*.

Researchers have postulated that behind the untold diversity of plants hides a limited number of basic structures and patterns. Among them are the spirals and the ramifications, the dendrites and the fractals, which abound in nature (see, for example, Refs. 6, 16, 37). The spirals are visible in the structure of galaxies as well as in colonies of *Heliodoma* (invertebrate), in the unfolding leaves of ferns, and in the fronds of *Cycas revoluta*, a primitive seed-bearing plant, as well as on many shells. Related curves, helices, are found in proteins. These curves combine to form elaborate patterns on sunflowers and pine cones, on the trunks of palm-trees made by leaf scars, and of primordia in transverse sections of buds. Ramifications are also very important in phyllotaxis, as we have seen. But ramifications occur in minerals, such as pyrolusite, which forms branching crystals dendrites in sedimentary rocks, as well as in fungi and in Purkinje cells, in the structure of the tributaries of rivers as well as in colonies of *Aglaophenia* (invertebrate), in lightnings, and so on. The basic structures arising in phyllotaxis come from two universal natural processes of growth: rhythmic gnomonic growth and rhythmic branching present at all levels of evolution. In phyllotaxis, models dealing with these have been presented. They are able to generate the phyllotactic patterns from initial configurations. And what sets the initial configuration is the necessity that some rhythm starts sometimes during a growing process, and that entropy be minimized, order be increased, or fixed, and maximum energy be canalized, at that time, to give a chance to that initial pattern to continue to exist and to develop.

On the Origins of Spiral Symmetry in Plants 347

We learned from Einstein about the control exerted by curved space. The paper pleads for the general idea that the patterns and functions we are concerned with in phyllotaxis do not come from the gene: they are properties of space-time (see Fig. 15). This conclusion is in agreement

Fig. 15. The properties of space-time. A pattern around a central cap (about 10 cm in diameter) of an honeycomb wheel of my car. The four caps displayed similar patterns generated by the turbulance, the centrifugal forces, raised by the rotation of the wheels at 120 km/hr over 300 km on an highway covered with a thin layer of melting snow. It offers a fascinating resemblance with the structure of a daisy: it has a capitulum, florets (protuberances) in the capitulum, and petals.

with autoevolutionism according to which the gene only fixes a structural aspect present at the levels of evolution, which preceded biological evolution: the evolution of elementary particles from quarks and leptons, the evolution of chemical elements from hydrogen, the evolution of minerals that produced seven crystallization systems. Genes do not create symmetries, they only choose between those that exist in minerals and quasi-crystals, and that are inherent to the cytoplasmic construction. One of the tenets of autoevolutionism is that order can only arise from order and that no order, no form, and no function is created or lost, rather, it is only transformed by combination. With respect to phyllotaxis, in agreement with the previous idea of autoevolutionists, I suggest that when successful patterns have been discovered by nature, then genes become responsible for the fixation and rapid repetition of these patterns.

In Ref. 4, I showed that distichy and decussation, alternate and superposed whorls, spirodistichy, spiromonostichy, and so on are extreme cases of spiral patterns. From *Ulotrix* to the giant sunflower via *Fucus spiralis*, fractals are developing in time through organic matter, as it already did through inorganic matter, from a single branch to a fully ramified structure. Complex phyllotatic patterns are the end result of a process by which nature reproduces at the biological level the hierarchies present at that level in the form of a chemico-physico-mineral imprint. This follows from the properties of light, gravity, temperature, from those of the carbon atom, from those of the minerals inside the tissues, and so forth.

Phyllotaxis being not a closed system, must not be treated, as it is generally done, as a self-contained phenomenon. The phenomenon cannot be separated anymore from the rest of the world and from evolution. The part is controlled by the totality of the systems. This article is a sketch of the universal framework in which the analysis and modelling of phyllotaxis must take place. Given the actual state of development of physics (see Ref. 38) we can say that phyllotaxis is the tip of an iceberg resulting from the Big Bang. The former phenomenon will be fully understood when the processes initiated by the latter will be clarified. Mathematical modelling can bring light into these processes by working on deductive mechanisms that are able to reconstruct the fascinating patterns, and by imagining inductive models that are able to initiate the patterns. "Our experience hitherto justifies us in believing that nature is the realization of the simplest conceivable mathematical ideas" (A. Einstein).

Hermant's relatively old work regarding "transverse sections of buds" (see Fig. 10) should be developed with the help of the techniques of today. Computer generation of phyllotactic patterns (e.g., mainly the capitulum of the sunflower) is a subject that has been dealt with successfully by many authors (such as Refs. 9 and 8). Others have reproduced with a remarkable realism, colorful shrubs and trees, and flowers (e.g., Ref. 39), using rules of production such as L-systems. These reproductions may be of some help in the determination of the causes and mechanisms that induce the natural things, by the necessity they generate of throwing a new look at nature each time. This possibility handed to us by the development of computer graphic techniques raises the problem of the meaning of these reproductions, with respect to the understanding of the fundamental causes in phyllotaxis, and of the processes of creation itself.

The common denominators between an exact replica of a capitulum generated on the screen of a computer and the capitulum itself are probably the following: they are both natural objects (see the text on Fig. 4), they both move or grow in space-time, they are both a result of some properties of light, and they both proceed from the power of experimental and dynamic thinking. The two creations are in my opinion essentially alike, one mimicking the other, just like a spotlight mimicks the sun. The difference between the two specimens of capituli is at the level of their substances, their radiating media. And before that man will be able to create instantaneouly a tri-dimensional capitulum as we know it, he will have to discover new properties of the primordial substance from which everything is made, of light itself with its various radiations modulating the substance and inducing forms, patterns, colors, and harmony. In the meantime the best we can do is to try to imagine those properties that transcend science as it stands today, and to build holistic models that take into consideration the ultimate effects those patterns supposedly achieve (such as maximization of energy or minimization of entropy), and their functions in the environment (such as mutual services with pollinators, production of food, manifestation of beauty through ergonomy and harmony).

Acknowledgements

The author is thankful to the visiting professor, Dr P. Auger, from Dijon, France, for commenting on an earlier manuscript.
This article was supported by Canada CRSNG Grant A6240.

References
1. R. V. Jean, Nomothetical modelling of spiral symmetry in botany, *Symmetry* **1** (1990) 81–91; and in *Fivefold Symmetry*, I. Hargittai, ed. World Scientific, Singapore (1992).

2. S. V. Meyen, Plant morphology in its nomothetical aspects, *Bot. Rev.* **39** (1973) 205–260.
3. R. V. Jean, Model of pattern generation on plants based on the principle of minimal entropy production, in *Thermodynamics and Patterns Formation in Biology* (pp. 249–264), I. Lamprecht and A.I. Zotin, eds. W. De Gruyter & Co., Berlin, New York (1988a).
4. R. V. Jean, Phyllotactic pattern generation: A conceptual model, *Annals of Botany (London)* **61** (1988b) 293–303.
5. R. V. Jean, A basic theorem on and a fundamental approach to pattern formation on plants, *Math. Biosci.* **79** (1986a) 127–154.
6. A. Lima-de-Faria, *Evolution Without Selection.* Elsevier, New York (1988).
7. R. V. Jean, The hierarchical control of phyllotaxis, *Ann. Bot. (London)* **49** (1982) 747–760.
8. R. V. Jean, A synergetic approach to plant pattern generation, *Math. Biosci.* **98** (1990c) 13–42.
9. F. van der Linden, Creating phyllotaxis, the dislodgement model, *Math. Biosci.* (1990, in press).
10. A. L. Lehninger, *Principles of Biochemistry.* Worth, New York (1982).
11. R. V. Jean, *Model Testing in Phyllotaxis.* Manuscript submitted for publication (1990).
12. R. V. Jean, An interpretation of Fujita's frequency diagrams in phyllotaxis, *Bull. Math. Biol.* **48** (1986b) 77–86.
13. R. V. Jean, Phyllotaxis: a reappraisal, *Can. J. Bot.* **67** (1990a) 3103–3107.
14. R. V. Jean, A mathematical model and a method for the practical assessment of the phyllotactic patterns, *J. Theor. Biol.* **129** (1987) 69–90.
15. D. Denffer, W. Schumacher, K. Magdefrau, and F. Ehrendorfer, *Strasburger's Textbook of Botany.* Longman, London, New York (1971).
16. P. S. Stevens, *Patterns in Nature.* Little, Brown and Co, Boston (1974).
17. A. H. Church, *On the Relation of Phyllotaxis to Mechanical Laws.* William & Norgate, London (1904).
18. J. N. Ridley, Packing efficiency in sunflower heads, *Math. Biosci.* **58** (1982) 129–139.
19. F. Rothen and P. Pieranski, Les cristaux colloidaux, *La Recherche,* **175**(17) (1986) 312–321.
20. D. R. Nelson, Quasicrystals, *Scientific American* **255**(2) (1986) 42–51.
21. Y. Sawada and H. Honjo, Mais d'où vient donc la forme des dentrites?, *La Recherche* **175**(17) (1986) 522–524.
22. F. A. Novak, *The Pictorial Encyclopedia of Plants and Flowers.* J. G. Barton (ed.), Crown Publishers, New York (1966).
23. E. Radl, *History of Biological Theories.* Oxford University Press, Oxford (1930).
24. M. A. Selvam, Deterministic chaos model for self-organized adaptive networks in atmospheric flows, *Proc. IEEE 1989 Nat. Aerospace and Electronics Conference.* Dayton, Ohio, 1145–1152 (1989).
25. M. A. Selvam, A cell dynamical system model for turbulent shear flows in the planetary atmospheric boundary layer, *Proc. Ninth Symp. on Turbulence and Diffusion,* May 1990, Roskilde, Denmark, Am. Meteorological Soc., Boston, Mass., 262–265 (1990).

26. D. Friedman, *Determination of Spiral Symmetry in Plants and Polymers*. See this volume.
27. G. Cusset, The conceptual bases of plant morphology, *Acta Biotheoretica* **31A** (1982) 8–86.
28. W. Zimmermann, Main results of the telome theory, *Paleobotanist* **1** (1953) 456–470.
29. F. Bolle, Theorie der blattstellung, *Verh. Bot. Prov. Brandenb.* **79** (1939) 152–192.
30. W. E. Sharp, An analysis of the laws of stream order for Fibonacci drainage patterns, *Water Resources Research* **7** (1971) 1548–1557.
31. A. Hermant, Structures et formes naturelles, géométrie et architecture des plantes, *Techniques et Architecture* **VI** (1946) #9-10, 421–431.
32. W. E. Sharp, Fibonacci drainage patterns, *Fibonacci Quarterly* **10** (1972) 643–655.
33. R. V. Jean, Allometric relations in plant growth *J. Math. Biol.* **18** (1983) 189–200.
34. R. V. Jean, An L-system approach to nonnegative matrices for the spectral analysis of discrete growth functions of populations, *Math. Biosci.* **55** (1981) 155–168.
35. A. Hermant, Croissance et topologie, architecture des plantes, *Techniques et Architecture* (1947).
36. D. W. Thompson, *On Growth and Form*. Cambridge University Press, Cambridge (1917).
37. T. A. Cook, *The Curves of Life*. Constable & Co., London (1914).
38. S. W. Hawking, *A Brief History of Time. From the Big Bang to Black Holes*. Bantam Book, New York, London (1988).
39. P. Pruzinkiewicz and J. Hanan, *Fractals and Plants*. Collection Lecture Notes in Biomathematics, Springer-Verlag, New York (1989).

GREEN SPIRALS

Robert Dixon

1. Introduction

Little is said in the literature of spiral phyllotaxis about the shape of the spirals in question, because interest centres on the great puzzle of their numbers. In popular accounts there is the occasional crude suggestion that the spirals of, for example, florets on the head of a daisy or sunflower are equiangular in shape, like those of mollusc shells. But we expect the patterns of phyllotaxis to be explained as tendencies toward optimal packing and/or spacing and could, therefore, only give rise to equiangular spiral alignments if the constituent members of the arrangement form a sequence of exponentially increasing size.[1] Of course, the growth of real plant parts deviates considerably from the curve of exponential increase as it slows toward an upper finite limiting size, tracing out a growth curve whose shape we call sigmoidal. Clearly, a close-packing or even-spacing of a spiral sequence of plant parts whose sizes increase sigmoidally must deviate from the equiangular in shape.

We aim in this paper to discuss this question of spiral shape in spiral phyllotaxis. We shall derive some general equations and solve them for some special cases, in a manner suitable for those who may wish to formulate the primary spiral in computer graphic models[2] of spiral phyllotaxis.

However, we begin by recalling that background story, already alluded to, the great puzzle of Fibonacci numbers, which in turn becomes the question of why the golden angle offers an optimal divergence for the purposes of phyllotaxis. Why should plants show a marked tendency toward a particular and common value of divergence, and why should this value be the golden section of a circle?

2. Golden Angle Divergence

Although anticipated by da Vinci and Kepler, it was the 18th century botanist Bonnet who originated the modern study of spiral phyllotaxis with his observations of a curious law of nature in plant forms involving Fibonnaci numbers. In the case of spiral sucessions of single leaves or twigs on a plant stem, the rule might be, for example, that 8 leaves bud per 5 rotations. In the case of composite flowers, such as daisies or sunflowers, or of composite fruits, such as pineapples or pine cones, the apparent arrays of multiple spirals occur in opposing sets of Fibonnaci numbers, such as 3 by 5 for pine cones, 5 by 8 for pineapples, 12 by 21 for daisies, and 34 by 55 for sunflowers. By the 19th century, it was realized that all of these examples are versions of a common pattern. The multiple spirals (visually striking alignments of neighbouring members) arise from a single spiral sequence of buddings from a parent stem; and in all cases the Fibonnaci ratios indicate a tendency toward a special divergence value equal to the golden section of a circle.

$$\phi = 0.6180339\ldots \text{revolution},$$

or

$$360\phi = 222.49\ldots \text{degrees},$$

or

$$2\pi\phi = 3.8832\ldots \text{radians}.$$

The *golden ratio* is defined as the division of a whole into two parts that bear the same proportion to each other as do the larger part to the whole. ϕ is therefore the positive solution of the following quadratic equation:

$$\phi/1 = (1 - \phi)/\phi$$
$$1 - \phi = \phi^2$$
$$\phi = (\sqrt{5} - 1)/2.$$

ϕ is irrational, and its convergents (sequence of increasingly closer rationals) are formed by consecutive pairs of Fibonnaci numbers:

$$0, 1, 1, 2, 3, 5, 8, 13, 21, 34, 55, 89, \ldots$$
$$F_0 = 0,\ F_1 = 1,\ F_{n+2} = F_{n+1} + F_n$$
$F_n/F_{n+1} \to \phi$ as a limit with increasing n.

So the puzzle in spiral phyllotaxis shifts from asking about Fibonacci numbers to asking why golden angle divergence? The puzzle can be set as a purely mathematical question of optimal forms, or as a physical and biological problem of causes. The latter provides the centre of interest in current research, but we shall say no more about it here. Our subject is purely mathematical.

In a marvellous marriage of geometry and number theory, making use of continued fractions and comparing Klien's square lattice diagram for number theory with the skew lattice diagrams of phyllotaxis, Coxeter[4] provides the definite answer to the mathematical question. He shows that ϕ is the only irrational without intermediate convergents (thus independently rediscovering a proposition originally demonstrated and used for the same purpose by DeCandolle in 1881) and that it therefore provides the only value of divergence to strictly fulfill the condition observed in spiral phyllotaxis that Coxeter finds most clearly stated in Tait's study, here called Tait's condition: that the succession of nearest angular neighbours to any given leaf in its following sequence of leaves alternates from side to side.

We can perhaps illustrate the plant geometry and the problem of optimal divergence in the following simple manner. Consider, so to speak, the sun's-eye view of a vertical plant stem (say, a foxglove or a thistle) putting out radial leaves sequentially with regular divergence (Fig. 1). We see the leaves radiating from the stem axis, "trying" to occupy as much of the surrounding space and sunlight as possible with maximum economy. The plant must "try" to put out leaves in all directions and yet delay as long as possible putting out in directions close to those already occupied. Any irrational divergence ensures that no two leaves in the sequence radiate in exactly the same direction. But ϕ is the only irrational divergence for which the succession of nearer and nearer approaches from right and left strictly alternate (Tait's condition). And connected with this fact is the conclusion that of all irrationals, ϕ is the slowest converger.

356 Robert Dixon

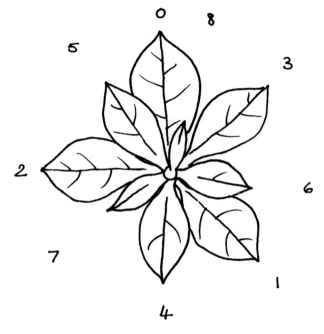

Fig. 1. Succession of leaf directions with golden angle divergence.

The relevance of Tait's condition to the existence of an optimal value of divergence arises in the following way. The member parts making up plant arrangements (packings or spacings) — unlike the member parts of crystals or honeycombs — are not fixed in size but are growing. Each part typically grows with a sigmoidal growth curve and out of step with all the others. At any one time, the plant members in a sequence are all of different sizes and ages and growing at different rates. Therefore, in the course of time, the arrangement of nearest neighbours in a close-packing is continually undergoing change. The divergence of ϕ offers an optimal arrangement throughout the growth of the plant, with the mesh of secondary spirals passing with greatest possible economy of space[3] from one convergent ratio pair to the next. So when looking at daisies, for example, notice not only Fibonacci numbers but also changes of Fibonacci numbers: across the disc of the daisy head from inner (younger) to outer (older) florets.

Golden angle spiral phyllotaxis provides an optimal arrangement for the purposes of close-packing or even-spacing for any sequence of like parts budding and growing on a parent stem in sequence. Buds appear at the growth tip[5] of the stem in succession, while the stem itself continually extends beyond them, leading to a roughly conical-cum-cylindrical arrangement of parts. The pattern may apply just as well to leaves, petals, twigs, florets, fruits, or thorns. For our purposes, the pattern should be thought of as a purely ideal geometry toward which real plants show a strong prevalent tendency.

Golden angle divergence is the chief subject of most spiral phyllotaxis studies and its literature is venerable. Interested readers will find an outline history and bibliography included in Adler.[6] But we shall say no more here about the question of golden angle divergence. Our subject is the shape of the spiral, to which we now attend. The aim is to discuss formulae for the purposes of computer graphic modelling. We shall define the pattern we seek as a spiral succession of points with golden angle divergence, such that the points (representing sequential members of a plant) are spaced equitably/economically.

3. What is a Spiral?

We start by giving an appropriate definition of the word "spiral." In mathematics this term is used mostly to denote plane curves only. However, both the etymology and current wider usage incorporate a broader generality of curves winding through space. For example, the spiral of a spiral staircase winds on a cylinder and is a circular helix, while the concho-spirals of snail shells wind on cones. The spirals in plants can be approximately helical, or concho or other assorted shapes. In view of this, I propose to adopt for present purposes the following definitions:

    i) A *plane spiral* is any curve in the plane traced by a point that continually rotates about a fixed point of the plane, called the pole of the spiral, while continuously receding from (or approaching to) the pole.

In polar coordinates (R, A), taking the pole of the spiral as origin, the above definition is equivalent to $R = f(A)$, for any continuous monotonic function f.

    ii) A *spiral* is any smooth and continuous curve on any surface of revolution that cuts all fundamental generators of the surface obliquely.

Note the following. A "fundamental generator" is the profile curve of the surface of revolution formed by intersection with a plane through the axis of revolution. The "circle of revolution" for any point on the surface of revolution is the path it traces if rotated fully about the axis of revolution, cutting all fundamental generators at right angles. Any spiral that cuts all fundamental generators by the same angle is a special case of spiral called a *loxodrome* of the surface.

We adopt cylindrical coordinates (R, A, D) to determine the spiral path of a moving point P, based in the obvious way upon the axis of revolution (Fig. 2). Our aim will be to formulate any required spiral by giving both R and D as explicit functions of A.

4. Models of Spiral Phyllotaxis

Our specific concern to model plant patterns confines the present interest to fairly tame examples of surfaces of revolution. We shall look in particular at cylinder, cone, and sphere as simple ideals toward which plant forms in spiral phyllotaxis are apt to approach. The arrangement of florets on the heads of daisies and sunflowers approximates to a plane disc, which we can consider as a special case of the cone.

The surface of revolution in our model is to be thought of as that which connects any set of corresponding points, one to each of the plant members in the sequence. Our graphic model is therefore both a snapshot in time of a growing plant and a two-dimensional sample from a fully three-dimensional form.

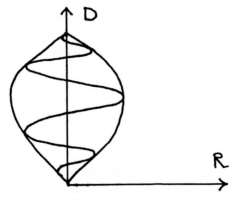

Fig. 2. Cylindrical coordinates for a spiral.

The model of spiral phyllotaxis that we seek is this: to arrange a sequence of points spirally on a surface of revolution such that they are equitably and economically spaced or packed.

In formulating the required spiral, interest centres on the rate at which a spiral "covers" the area of the surface of revolution as it winds. This we measure in terms of the rate of surface area swept out per wind by the circle of revolution at the spiralling point. For the purposes of mathematical exploration it is convenient to distinguish two classes of spiral in this respect:

  i) spirals that advance across the surface of revolution at a constant areal rate we call *grey spirals*;
  ii) spirals whose rate of advance plausibly matches the growth curve of a plant part we call *green spirals*.

Grey spirals are special cases of green spirals, and may be thought of as limiting cases approached in many plants with increasing maturity.

For each of these two classes of spiral, we shall consider in general terms two alternative problems, according to whether we are given the surface of revolution by R as an explicit function of D or as an explicit function of A, which we shall label as options (a) and (b), respectively. Our problem in each case will be to seek a formula that gives D in terms of A. These turn out to be integral equations, which we solve in the particular cases of sphere, cylinder, cone, and disc.

4.1 Grey Spirals

(a) Consider the general case in which the surface of revolution is given as

$$R = f(D),$$

where f is a continuous positive function on an interval of D, which typically starts and ends at zero with one maximum in between.

We wish to find a formula $D = g(A)$. Let H be the cumulative area of surface swept out by the circle of revolution at P, and let h be the increment of area swept out per revolution of the spiral. For grey spirals h is constant, and

$$H = Ah/2\pi + \text{constant}.$$

So

$$dh/dA = h/2\pi,$$

but
$$dH = 2\pi R \sqrt{dR^2 + dD^2},$$

from which we can form the following equation that expresses the general solution:

$$A = k\int f(D) \sqrt{f'(D)^2 + 1}\, dD, \quad \text{where } k = 4\pi^2/h. \tag{1a}$$

To obtain D in terms of A for any given $f(D)$, however, we must first solve the integral to obtain $A = g^{-1}(D)$, and then find the inverse function $D = g(A)$. This we can do to obtain particular solutions for cylinder (Fig. 3), sphere (Fig. 4), and cone. Surprisingly, the solutions for sphere and cylinder of radius r are identical:

$$D = kA, \quad \text{where } k = h/4\pi^2 r. \quad \text{(cylinder and sphere)}$$

This result recalls, incidentally, the fact in cartography that a Peter's Projection, which is cylindrical projection of the Earth's globe, is area-preserving.

The solution for a cone of slant angle a is

$$D = \sqrt{kA}, \quad \text{where } k = h/[2\pi^2 \tan(a)\sec(a)]. \quad \text{(cone)}$$

When a is a right angle, the cone degenerates to the special case of a plane disc (Fig. 5). The solution here is obtained from $H = \pi R^2$, as

$$R = \sqrt{kA}, \quad \text{where } k = h/2\pi^2. \tag{disc}$$

Plane spirals of the form $R = \sqrt{kA}$ are known as Fermat spirals.

(b) Now we consider the general case of a surface of revolution not given explicitly, but via

$$R = j(A).$$

Again $dH/dA = h/2\pi$ and $dH = 2\pi R\sqrt{dR^2 + dD^2}$.

This allows us to form the equation expressing the general solution more directly:

$$D = \int [kj(A)^{-2} - j'(A)^2]\, dA, \quad \text{where } k = h^2/16\pi^4. \tag{1b}$$

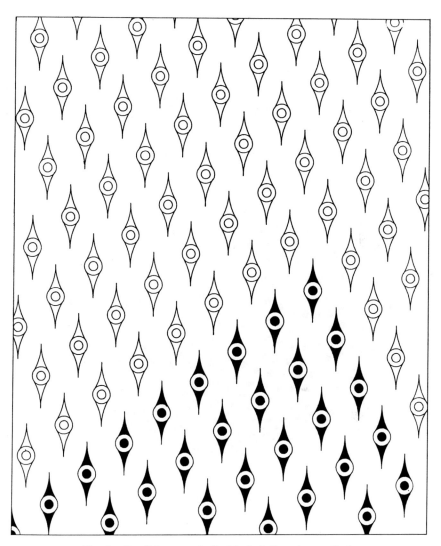

Fig. 3. Fossil fern tree (cylindrical phyllotaxis, unrolled).

Grey spirals allow us to arrange a sequence of points on the surface of revolution so that each point "occupies" the same portion of area of the surface, up to ϕh square units per point. Mature plant forms are inclined to show grey spirals, when the leaves or fruits have all stopped growing and have reached equal size.

Fig. 4. Spherical phyllotaxis (equal members).

4.2 Green Spirals

As mentioned above, the decisive condition in determining golden angle divergence as an optimal solution in spiral phyllotaxis is the fact that the plant as a whole and each of its parts are all the time growing and

Fig. 5. Voronoi daisy (disc phyllotaxis of equal members).

growing non-uniformly. The non-uniformity applies in both space and time: at any one time, different parts will be growing at different rates, while the growth rate of any one part will vary with time. Both these departures from uniform growth are gentle — neighbouring members of a sequence tend to be equal in size and growing at the same rate and young parts tend to grow exponentially — but significant. The general condition is illustrated in Fig. 6. The growth curves of the whole plant and each of its parts are *sigmoidal* in shape (Fig. 7).

Fig. 6. Sunflower (disc phyllotaxis with sigmoidal growth).

Biologists frequently model such curves with the logistic function

$$s = m/(1 + e^{-nt}),$$

which posits a maximum size m and proposes that the rate of growth at any one time is proportional to the difference between maximum size

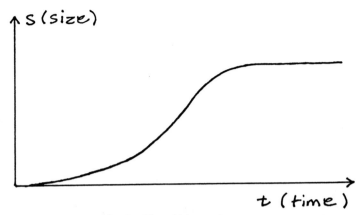

Fig. 7. Sigmoidal growth curve.

and present size. However this is known to be only approximately in fit, and roughly similar shapes of growth curve may be variously modelled, as by, for example,

$$s = arctan(e^{nt})$$

or some other suitable formula.

The leaves or fruits in our spatial sequence begin life in a sequence of staggered times. For simplicity sake we can assume their lives start at regularly staggered intervals of time and proceed through the same pattern of growth at the same rate. The regular intervals of time between the corresponding growth stages of successive leaves or fruits in the sequence are matched by the regular intervals of our angle coordinate A in the spiral succession. Therefore, by a suitable scaling of units, we can replace t by A in any growth formula for s:

$$A_{i+1} = A_i + 2\pi\phi$$
$$s_i = s(A_i),$$

where $s(A)$ is a linear growth function of a plant part.

We are interested in the area occupied by a leaf or fruit, which varies as the square of s. So, to formulate a green spiral we let h, the increment of H per revolution, which was a constant in the formulae for grey spirals, become a variable:

$$h_i = s(A_i)^2.$$

(b) With these considerations of growth functions in mind, it is convenient to consider option (b) before (a). That is to say, we consider the general case where the surface of revolution is defined in terms of A:

$$R = j(A).$$

Formula (1b) applies, but with h as the variable $s(A)^2$, instead of being constant as for grey spirals, giving us:

$$D = \int \left[\frac{s(A)^2}{kj(A)^2} - j'(A)^2 \right] dA, \quad \text{where } k = 16\pi^4. \quad (2\text{bi})$$

However, with some loss of generality, it is natural enough to link the growth of R with the growth of plant parts, $s(A)$. The simplest case being:

$$R = j(A) = c.s(A),$$

where c is constant. Together with $h = s(A)^2$, this converts formula (1b) to

$$D = \int \left[\frac{s(A)^2}{k^2} - c^2 S'(A)^2 \right] dA, \quad \text{where } k = 4\pi^2 c. \quad (2\text{bii})$$

(a) Now we consider the general case of a surface of revolution given as a function of D:

$$R = f(D).$$

Again we can let $h = s(A)^2$, so that formula (1a) becomes:

$$\int S(A)^2 dA = 4\pi \int f(D) \sqrt{f'(D)^2 + 1}.dD. \quad (2\text{ai})$$

Alternatively, let us suppose that h, the increment of H per revolution, varies as the square of R. That is,

$$dH/dA = pR^2,$$

for some constant p. This condition will give us a spiral that is a loxodrome of the surface of revolution:

$$dH = 2\pi R \sqrt{dR^2 + dD^2}.$$

This provides us with an implicit general solution:

$$A = \frac{2\pi}{P} \int \frac{1}{f(D)} \sqrt{f'(D)^2 + 1}.dD. \quad (2\text{aii})$$

This can be integrated and transposed easily enough for the particular cases of sphere and cone. For a sphere of radius r we obtain:

$$D = 2r/(1 + e^{-Ap/\pi}). \qquad \text{(sphere)}$$

Note that the loxodrome of a sphere is a logistic function of A. For a cone we get a concho spiral:

$$D = e^{Ap \cdot \sin(a)/2\pi}, \qquad \text{(cone)}$$

and for a disc we get an equiangular spiral:

$$R = e^{Ap/2\pi}. \qquad \text{(disc)}$$

4.3 Numerical Approximations

For other specific surfaces of revolution, not considered here, the integrals in our formulae for green and grey spirals may prove intractable. In such cases we can devise approximation formulae from (1b), by putting $dA = 2\pi\phi$ and averaging R. Let the ith point, P_i, have cylindrical coordinates (R_i, A_i, D_i), with $A_{i+1} = A_i + 2\pi\phi$; and let the increment of surface area swept by the circle of revolution from P_i to P_{i+1} be h_{i+1}. In the case where we are given $R = j(A)$, formula (1b) converts to

$$D_{i+1} = D_i + \sqrt{\left[\frac{\phi h_{i+1}}{R_{i+1} + R_i}\right]^2 - [R_{i+1} - R_i]^2}. \qquad (3b)$$

In the case where R is not known in terms of A but is given in terms of D, we take $R_{i+1} = R_i$, and the above approximation degenerates to

$$D_{i+1} = D_i + \frac{\phi h_{i+1}}{2\pi R_i}. \qquad (3a)$$

Finally, we note that each plant member may be allotted a region of the surface of revolution up to an areal value of ϕh, per point in the spiral phyllotaxis.

Glossary

Phyllotaxis. Literally, "leaf-arrangement", but applying to any multiple plant members. *Spiral* phyllotaxis, the arrangement of a set of members that bud sequentially in a spiral succession about their parent stem.

Primary spiral. The spiral connecting the multiple parts in sequence; also called the "fundamental" or "genetic" spiral.

Secondary spirals. The apparent multiple spirals generated by the single primary spiral arising in a close-packing spiral phyllotaxis, spirals of touching neighbours; also called "parastichies." The members of any secondary spiral occur at regular intervals of the primary sequence and so form a spiral whose shape is a linear transformation of the shape of the primary spiral.

Divergence. The angle about the stem axis between consecutive members in the primary spiral sequence.

Convergents of an irrational magnitude, the sequence of rationals with increasing denominator that approach the true value from above and below, being of two mutually exclusive categories: (i) the *principal* or *ordinary* convergents; and (ii) the *intermediate* convergents. The distinction and connection between principal and intermediate convergents is illustrated by Klein's diagram[4] and yields a numerical law that the reader may like to spot in the following example, showing the convergents of π, with intermediate convergents in brackets: 3/1, 22/7, (25/8, 47/15, 69/22,..., 311/99), 333/106, 355/113,...

References
1. R. Dixon, The mathematics and computer graphics of spirals in plants, *Leonardo* **XVI-2** (1983) 86–90
2. R. Dixon, *Mathographics*. Basil Blackwell (1987).
3. R. Dixon, Spiral phyllotaxis, *Computers Math. Applic.* **17** (1989) No. 4–6, pp. 535–538; and in *Symmetry 2: Unifying Human Understanding*, I. Hargittai, ed. Pergamon Press, Oxford (1989).
4. H. S. M. Coxeter, The role of intermediate convergents in Tait's explanation for phyllotaxis, *J. Algebra.* **20** (1972) 167–175. Also see Coxter, *Introduction to Geometry.* Wiley (1961).
5. R. O. Erickson, The geometry of phyllotaxis, in *The Growth and Functioning of Leaves*, J. E. Dale and F. I. Milthorpe, eds. Cambridge Univ. Press (1983). Includes electron scanning micrographs demonstrating the pattern of approximate golden angle divergence emerging strongly on a tiny scale at a very early stage on budding on a stem's growth tip.
6. I. Adler, A model of contact pressure in phyllotaxis, *J. Theor. Biol.* **45** (1974) 1–79.

THE FORM, FUNCTION, AND SYNTHESIS OF THE MOLLUSCAN SHELL

Michael Cortie

1. Introduction

The shells of nearly all molluscs, including limpets and bivalves, are characterized by an elegant and symmetric design, based on a spirally-coiled cone. It has been known since 1638 that the spiral form of such shells exhibited the property of self-similarity during growth, implying also that the projection of any particular generating spiral of the shell onto a plane perpendicular to the axis of symmetry produced an equiangular, or logarithmic, spiral.[1-4] The radius vectors of a spiral of this type intersect the curve at a characteristic, and constant, angle. The planar spirals can be described by the formula[1,2]:

$$r = r_0 \cdot e^{\Theta \cdot \cot(\alpha)} \quad (1)$$

where α is the constant angle between any radius vector and the spiral. (The other symbols are defined in the list of symbols in the Appendix.)

A particular property of the equiangular spiral is that it preserves its shape and form as it grows. Thus a large shell is, in terms of linear dimensions, a scaled-up version of itself at an earlier stage in its life. The extreme forms of the equiangular spiral are the straight line and the circle,[4] corresponding to α values of 0° and 90°, respectively.

An excellent review of the equiangular spiral in living organisms is available in the now classic book, *On Growth and Form*.[2] In the present work, the nature of the spirals in the shells of molluscs is examined more closely, and a mathematical model capable of generating realistic mollusc shells is presented. Finally, the model is used to examine the issue of the evolution of limpet-shaped shells.

2. The Spiral Form in the Mollusca

2.1 *Why the Spirals in Mollusc Shells are Equiangular*

The existence of equiangular spirals on the surface of molluscs shells has been variously explained as evidence of a life energy field or a life force, as largely irrelevant, as necessary to guarantee optimum strength, as the result of the shell growing along energy-efficient trajectories driven by so-called clockspring spirals, or as the inevitable result of isometric, self-similar growth.[2,5-8] One of the oldest explanations, due to Moseley,[1] may be the most satisfactory: "Now the law of the logarithmic spiral ... is the only one according to which the Mollusk can wind its spiral dwelling in an uniform direction through the space round its axis, in respect to that axis." Thus the equiangular spirals and self-similar growth of most molluscs may be understood in the context of modern biology as being the result of a genetically determined, and constant, inclination of the aperture with respect to the axis of symmetry.

2.2 *Determination of the Value of* α

Various methods for the determination of the α value of the spirals of mollusc shells have been described elsewhere.[1-3,8,9] In general, a series of measurements are made of the rate of expansion of the spiral, and the angle, α, is determined by averaging or regression analysis. The spiral formed by the suture line (the line that is formed by the intersection of two adjacent whorls) is frequently used, but if the shell is self-similar, then any of its other generating spirals is suitable.[2] An alternative parameter, W, is sometimes used to describe the curvature of these spirals,[4,8] but W and α are related[2,10] and

$$W = e^{2\pi \cdot \cot(\alpha)}. \tag{2}$$

Although W and α are both measures of the rate of expansion of the equiangular spiral, α has been used here since it covers the range of spirals possible in molluscs more conveniently than W. Table 1 compares the W and α values for some typical mollusc shells.

Table 1. α and W values for some typical molluscan spirals

Genus	α (°)	W
Patella	5	1.55×10^{31}
Dentalium	30	53252
Arca	55	81.4
Nautilus	80	3.03
Turritella	88.2	1.22

Fig. 1 shows the spiral that is present in the shell of the mollusc *Haliotis spadicea*. The r-Θ data were obtained by measuring the positions of the little holes present in the shell using a low-power microscope equipped with an *x-y* stage. The equiangular spiral takes the form of a straight line on the coordinate system used in this figure. It is evident that the data are well described by the superimposed equiangular spiral.

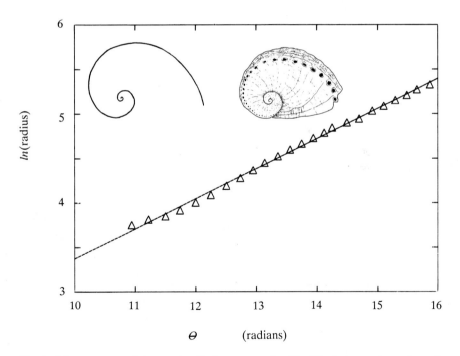

Fig. 1   Measurements of the angular displacement and radii of the holes on the surface of the shell of the abalone *Haliotis spadicea*. The best-fitting equiangular spiral is shown.

2.3 Distribution and Variation of α Values

Table 1 has shown that the values of α vary from about 5° for flattened limpet-like shells to close to 90° for such tightly coiled shells as *Turritella*. Fig. 2 shows measured r-Θ data for a range of mollusc shells. The best-fitting equiangular spirals are shown superimposed on each set of data. The radius in each case has been scaled so that it has the same magnitude at an arbitrary starting value of Θ. The α values shown range from 23° (the limpet *Hipponix*) to 88° (a *Turritella* shell). Although the magnitude of α is not correlated to the evolutionary process in molluscs,[2,8] it is related to some extent to the taxonomic position. Thus the Pelecypoda (bivalve molluscs) are all found to possess values of α between 30° and 60°, whereas the shelled Cephalopoda (also molluscs) have values between 80° and 85°.

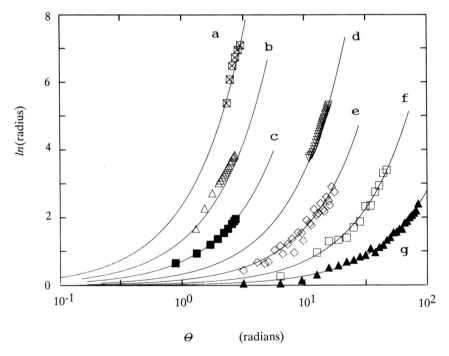

Fig. 2 Measurements of the spirals of a range of molluscs with superimposed best-fitting equiangular spirals: (a) *Hipponix*, (b) *Cardium*, (c) *Arca*, (d) *Haliotis*, (e) *Janthina*, (f) *Thatcheria*, (g) *Turritella*.

The shells within a genus, and to a lesser extent a family, are generally associated with a characteristic value of α.[1,2] Table 2, for example, shows data for five families of gastropods, and it can be seen that there is a common range of α values in each family. More detailed information of this type has been published by Raup[11] for fossil *Cephalopoda*, and in this case families also share similar spiral shapes. Furthermore, the α value has been found to exhibit even less variation at the species level.[1,2] Data published[12] for the fresh water snail *Bellamya jousseaume*, for instance, shows that the α values of 60 specimens collected at a variety of locations had a mean of 84.4° and a standard deviation of 0.8°. As another example, analysis of the average dimensions of over one thousand fossilized specimens of the landsnail *Poecilozonites bermudensis*[13] reveals that despite significant changes in the habitat and shell shape of this snail over the past 300,000 years, their α values remain unchanged at 87.5°, with a standard deviation of about 0.3°.

The consistency of the α values of many types of mollusc permits the use of this parameter, or an equivalent variation of it, as an aid in the resolution of problems of molluscan systematics.[12,14,15] The parameter has also proved to be of use in the study of shape changes during ontogeny,[16-18] as well as in the study or morphological variation as the result of either environmental[13,19] or geographical[20] factors.

Although α spans the full range of values within the shelled Mollusca, most species lie at the upper end of the range. Fig. 3 shows a histogram of α values for the 581 species of shelled mollusc meriting text entries in a recent regional book on marine molluscs.[21] This information was obtained by assigning each family of shells to the appropriate α interval and then coupling this information with the number of species per family to produce the histogram. Exact α values were not measured for every

Table 2. A sample of the variation of α at the family level in the Gastropoda

Family	Mean α of sample (°)	Standard deviation (°)	Size of sample
Haliotidae	72.9	2.5	7
Naticidae	81.6	1.6	6
Trochidae	84.7	1.3	6
Conidae	87.2	1.3	11
Turritellidae	88.3	0.2	6

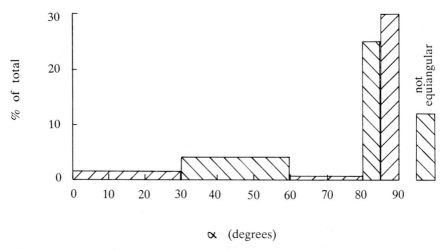

Fig. 3 Distribution of α values in a representative sample of modern marine molluscs.

one of the bivalve families and many were placed in the 30° to 60° interval without an attempt being made to further differentiate them. It is evident that most species of mollusc possess shells with α values in excess of 80°. A few limpet-shaped gastropods have shells with α values less than 30°. Only bivalves and the scaphopods have shells with α values between 30° and 60°. About 12% of the species are equipped with shells that do not appear to be based on an equiangular spiral.

3. Models for Molluscan Shell Shape

3.1 Previous Models

Mathematical models for the shape of mollusc shells were first developed, but apparently never applied,[2] in the nineteenth century by Mosely[1,9] and Blake.[22] The idea of modelling shells in three dimensions was reexamined by Raup[3,8,11,23] in the 1960s. Raup's model allowed for a circular aperture, any value of horizontal expansion rate, W, and a variation in shell shape from planispiral to helicoid. While it was a very useful, and influential, tool for theoretical morphology, the model was not able to simulate the shapes of real molluscs.[24] Løvtrup and von Sydow[10,25] have independently developed an equivalent model to that of Raup that is similar in scope and that has equivalent parameters.

Illert[5-7] has published a model for molluscan shells in which the aperture can be inclined in one direction. In principle, the model of Illert can accommodate an aperture of any analytical form, but only the equations for a circular aperture were provided. The model of Illert also allowed sinusoidal collabral corrugations. Savazzi[26] modelled the shape of both gastropod and bivalve shells with approximate difference equations. In this approach, the shape of the aperture is given by experimental data and is not limited to circular shapes. Savazzi allowed the apertural plane to be tilted such that it no longer intersected the axis of coiling (a different type of inclination to that permitted by Blake[22] and Illert[5-7]). Pickover[27] has simulated the shape of seashells using interpenetrating spheres.

In the present work, a model is presented that has an elliptical, or any other shaped, aperture which can be independently tilted in three directions and which can be positioned anywhere with respect to the axis of the shell. This model effectively contains, as subsets, the models of Blake,[22] Raup,[8] Løvtrup,[10,25] Illert,[5-7] and Savazzi.[26] In addition, the model allows for surface ornamentation in the form of collabral ridges, bumps, tubercles, and spines, as well as for nonisometric changes in the shell's growth. A discussion of the parameters of this model have been published elsewhere.[28]

3.2 A Sixteen-Parameter Model for the Mollusc Shell Shape

A sixteen-parameter model has been developed[28] that can simulate the shapes of many mollusc shells. The notation of Moseley,[1,9] Thompson,[2] Løvtrup,[10] and Illert[5] was used where possible.

The formula for a planar, equiangular spiral was given in Eq. (1). The spirals of most mollusc shells, however, are not planar. Consider a point in space, denoted g in Fig. 4, located within the aperture of a shell. Let the value of Θ at this position be zero. The spiral path traced out by point g during ontogeny is given, for a cylindrical coordinate system, by

$$r = A.\sin(\beta).e^{\Theta.\cot(\alpha)} \tag{3}$$

$$z = -A.\cos(\beta).e^{\Theta.\cot(\alpha)}, \tag{4}$$

where A and β are defined in Fig. 4 and in the Appendix. A surface can be constructed around the spiral traced out by point g by adding a suitable parametric equation to Eqs. (3) and (4):

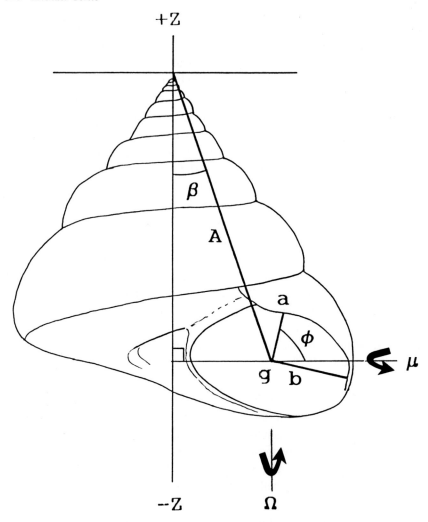

Fig. 4. A hypothetical shell showing the location, and geometric significance, of the several of the model's parameters. The three methods of rotating the aperture are shown as the angles ϕ, μ, and Ω.

$$r = A. \sin(\beta) + R. \cos(s) \qquad (5)$$

$$z = -A. \cos(\beta) + R. \sin(s) . \qquad (6)$$

See the list of symbols in the Appendix for a definition of R and s. If R is a constant, then the aperture is circular. For the present model, however, R is an expression that generates an elliptical aperture and, if required, a shell surface with nodules.

The shape of the aperture, R, is given by:

$$R = R(\Theta, s) = R_e + k, \qquad (7)$$

where R_e, an ellipse, is given by:

$$R_e = [a^{-2}.\cos^2(s) + b^{-2}.\sin^2(s)]^{-0.5}. \qquad (8)$$

Surface sculpture, k, of any type can be superimposed on top of the ellipse, R_e. A bell-shaped hump that can vary periodically in altitude with both Θ and s was adopted for the present model. The amplitude of the hump is given by

$$k = L.e^{-[2(s-P)/W_1]^2}.e^{-[2.g(\Theta)/W_2]^2}. \qquad (9)$$

The function, $g(\Theta)$, yields a number that varies periodically as Θ increases and was expressed as:

$$g(\Theta) = 360/N*(\Theta*N/360 - \text{round}(\Theta/N/360)). \qquad (10)$$

The function "round" rounds up or down according to the normal rules.

Finally, the plane of the aperture is tilted through the angles ϕ, Ω and μ shown in Fig. 4. After converting to Cartesian coordinates, the model becomes:

$$x = D.[A.\sin(\beta).\cos(\theta) + R.\cos(s + \phi).\cos(\Theta + \Omega)$$
$$- R.\sin(\mu).\sin(s + \phi).\sin(\Theta)].e^{\Theta.\cot(\alpha)} \qquad (11)$$

$$y = [-A.\sin(\beta).\sin(\theta) - R.\cos(s + \phi).\sin(\Theta + \Omega)$$
$$- R.\sin(\mu).\sin(\Theta + \Omega).\cos(\Theta)].e^{\Theta.\cot(\alpha)} \qquad (12)$$

$$z = [-A.\cos(\beta) + R.\sin(s + \phi).\cos(\mu)].e^{\Theta.\cot(\alpha)}. \qquad (13)$$

The model can be extended to accommodate shells with complex, or nonelliptical aperture shapes by alteration of Eq. (7) to Eq. (10). Nonisometric growth in shells can be introduced by means of the differential growth ratio, K,[13,17] or by way of allowing the parameter α to be a function of Θ.[16,28] Other types of ontogenetic change in mollusc shells can be simulated by making other parameters, such as those controlling the aperture shape, functions of Θ.

3.3 Modelling Real Shells

A selection of simulated shells, generated using parameters determined from real specimens, is presented in Fig. 5. Each of these shells is a recognizable depiction of the intended genus, although not necessarily of any particular species. The parameters necessary to generate many of these shells are available elsewhere.[28] The ten shells shown each have approximately elliptical apertures. Obviously, if the shell possesses a pronounced anterior canal, then the elliptical model for aperture shape is unsuitable and it would be necessary to substitute some other expression.[28]

The model is also applicable to the limpet-shaped shells found in families such as the Patellidae, Acmaeidae, and Fissurellidae, as well as to the shells of the nautiloid cephalopods, the pelecypods (bivalves), and to scaphopods (tusk shells). Simulated examples of these shells are shown in Fig. 6.

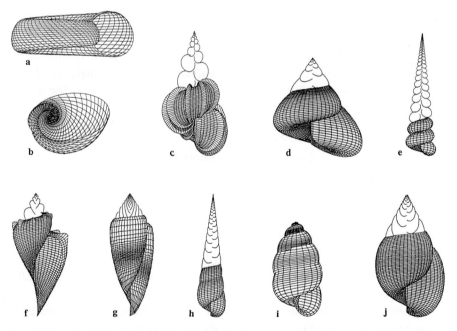

Fig. 5. A variety of gastropod shells simulated using the model: (a) *Planorbis*, (b) *Haliotis* (c) *Epitonium*, (d) *Oxystele*, (e) *Turritella*, (f) *Lyria*, (g) *Conus*, (h) *Terebra*, (i) *Gulella*, (j) *Achatina*.

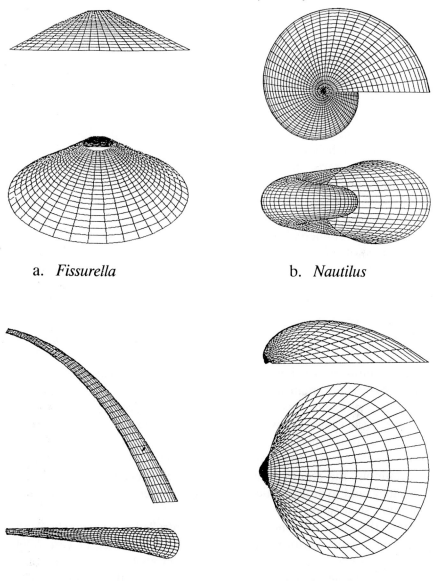

Fig. 6. Application of the model to molluscs other than tightly-coiled gastropods: (a) fissurellid limpet, (b) shelled cephalopod, (c) scaphopod shell, (d) one valve of a bivalve shell.

4. Discussion

4.1 The Biological Significance of the α Parameter

It is clear from Fig. 3 that α values in real molluscs are not evenly distributed. Although the suggestion is sometimes made that mollusc shells have not assumed all possible shapes because the evolutionary process has not yet had enough time to generate all permutations, it is more generally believed that α is determined by function and not by chance.[11,16] Clearly the α value is an important parameter in the determination of shell shape, since it controls the rate of increase in size of the shell. The shape and size, in turn, have an immediate bearing on such important considerations as the ratio of internal volume to apertural area, the ratio of internal volume to mass of calcareous material required, or the ratio of internal volume to hydrodynamic drag. In this view, the mode of life, design of soft parts and chemistry of the environment, should determine the optimum value of α in a species.

Ammonite shell shape has been quite carefully studied in order to determine which factors had affected the living animal. It has been shown[11,23] that these shells, with the exception of the hetermorph ammonites, had tightly wound shells of a shape consistent with an efficient ratio of calcium carbonate shell to internal volume. The efficiency was optimum for α values of more than 85°, but then there were other problems such as a reduced aperture size relative to internal volume. At α values below 79° there was a rapid fall off in "carbonate efficiency," thus explaining the absence of shells with such spirals. Later, Ward[29] showed that the required orientation of the shell in water, its streamlining, the required length of body chamber to achieve neutral bouyancy, and hydrodynamic stability also played a role in determining optimum shell shape.

Since the shape of a mollusc shell is controlled most strongly by the function it must perform, it is possible that molluscs of completely different habitat and taxonomic position can evolve shells of a similar shape. A striking example of this is *Helix aspersa*, the pulmonate land snail that plagues suburban gardens around the world, and *Janthina janthina*, the violet, floating shell that is washed ashore on temperate coasts worldwide after strong winds. *Helix aspersa* has to use the least amount of shell building material possible because calcium is not usually abundant in terrestial soils. *Janthina* is immersed in an environment containing adequate dissolved calcium but has to minimize the mass of its shell as far as possible, because it floats suspended from a raft of bubbles. Both animals, however, still require their shells for protection, at least against their less-powerful predators. Table 3 compares the shapes of these two shells. It can be seen that the α values and other shape parameters of the two species are very similar.

Table 3. Comparing the shell shapes of *Helix aspersa* and *Janthina janthina*, two molluscs that need to have light-weight shells.

Species	95% confidence interval for α (°)	Height/diameter ratio	Shell mass/internal volume (g/cm^3)
Helix aspersa	79.7–82.1	0.94	0.15–0.20
Janthina janthina	79.4–81.4	0.90	0.23–0.27

4.2 Non-Isometric Growth in Mollusc Shells

Although the shells of most living molluscs are based on the equiangular spiral, there are several genera, for example the Cypraeidae, where the growth of the shell is clearly not self-similar. The significance and occurrence of such deviations have been discussed elsewhere.[7,8,13,16–19,30] In general, if a deviation from an equiangular spiral form occurs, then it is believed that it represents an attempt by the growing mollusc to compensate for the increased internal volume of the shell relative to its shell surface area[8,11,16] or shell aperture area[17,18] as the animal grows. This follows from the principle that, as the linear dimensions of the shell increase, the internal volume increases in proportion to the cube of a linear dimension, but any given area only as the square of a linear dimension.[1,31]

Outside of the Mollusca it is interesting to note that the shells of the Brachiopods, though often displaying systematic deviations from self-similarity, are nevertheless to a good first approximation also based on the equiangular spiral.[16] The factor driving ontogenetic change in this phylum also appears to be a need to accommodate the rapid increase in internal volume relative to shell area as the shell grows. This is accomplished by changing the value of α from about 30° to near 80° as the animal matures. Correspondingly, their spirals change from open to relatively closed during growth.[16]

4.3 The Evolution of Limpets

Limpet shells possess low values of α (less than 30°) and are ideally suited for life on exposed, wave-washed rocks due to their intrinsic high ratio of aperture area to internal volume. This, combined in many instances with a very low profile, has led to limpet-shaped shells proliferating in marine and freshwater environments. Although best known from the Archaeogastropoda order of marine molluscs (examples being *Patella* and *Helcion*), limpet-shaped shells have evolved independently in the Meseogastropoda (*Calyptraea* for example) and also amongst the Pulmonates (as in the freshwater limpets *Burnupia* and *Ancylus*, or in the marine pulmonate *Siphonaria*).

The ancestor of each of these limpet shells is believed[32] to have been a (different) tightly coiled gastropod with an α value in excess of 80°. This conclusion has been drawn from the close similarity of the body parts of the limpets to those of other living gastropods in the appropriate families. The product of the evolutionary process is a flattened shell with an α value of less than 30°. There are, however, very few gastropod shells (fossil or living) with α values lying in the range 30° and 60°. (Although the Bellerophont molluscs of the Palezoic are sometimes cited as possessing high rates of whorl expansion,[24] examination of these fossils reveals that their α values are generally not less than 80°.) It is thus clearly a challenge to explain the route by which the shells of normally coiled gastropods evolved into limpet shapes.

The process of shape change that transforms a tightly coiled gastropod shell to a limpet-shaped shell is not mentioned or discussed in a recent text on the subject of molluscan evolution,[32] nor is it mentioned in a standard text on molluscs.[33] Nevertheless, an attempt has been made here to simulate this evolutionary shape change using the mathematical model presented earlier. The ancestral shell is assumed to be similar to *Calliostoma*, a modern member of the *Trochidae*, and the final form is assumed to be similar to *Patella*, the archetypical limpet. The parameters required to simulate these two shells with the model were measured, and the parameters of the hypothetical intermediate forms of shell estimated by linear interpolation. The resultant, hypothetical route through the molluscan morphospace is shown in Fig. 7 and Table 4. It can be seen that the trochid-like shell gradually uncoils and changes from helicoid to planispiral, while the plane of the aperture twists through the angles

Table 4. Parameter values of the shells in the hypothetical evolutionary series.

	Calliostoma	B	C	D	E	F	Helcion	Patella
α	83	73	63	53	43	33	23	3
β	20	20	30	40	50	70	89	89
ϕ	-50	-40	-30	-20	-10	1	1	1
μ	40	40	30	20	10	5	1	1
Ω	-10	-10	-20	-25	-30	-35	-40	-80
s_{min}	-160	-180	-180	-180	-180	-180	-180	-180
s_{max}	175	180	180	180	180	180	180	180
A	9	9	9	9	9	9	9	5
a	8	8	8	8	8	8	8	8
b	7	7	7	7	7	7	7	7

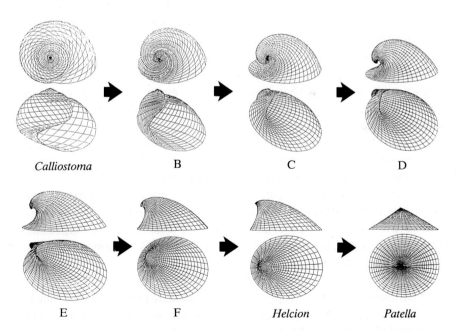

Fig. 7. Hypothetical phylogenetic sequence showing how a limpet might have evolved from a trochacean gastropod.

described by ϕ and Ω. The final change in the series is a reduction in the profile of the shell (to reduce wave drag).

There are, however, no known molluscs, fossil or living, with the shell shapes denoted B through to F. This may be due to the fact that evolution through these forms was rapid and the living animals few in number, with the result that corresponding fossils are so rare that they have not yet been found. Alternatively, the evolution of the molluscs may have taken place along some other path through the molluscan morphospace.

5. Conclusions

The shells of virtually all living molluscs have a shape based on the equiangular spiral, although this may be somewhat masked in adult shells by the effect of ontogenetic change or the environment. The range of possible equiangular angles is 0° to 90°, and living and fossil molluscs are spread unevenly across this range. The uneven distribution of α values in real shells is not thought to be due to chance, and is rather believed to be the result of the optimizing of shell shape for varying lifestyles and environments.

The value of the equiangular angle is relatively constant within a given species and, to a somewhat lesser extent, a given family. There is no correlation between the value of this angle and evolutionary position, although some orders of molluscs do have a relatively characteristic range of values. Deviation from a purely equiangular spiral is the result of a growing animal attempting to compensate for the different rate of change of its body volume and its shell or aperture area.

A mathematical model for mollusc shell shape is described that effectively includes previously published models as subsets. The model is able to simulate the shape of most mollusc shells, including limpet-like forms, scaphopods, bivalves, marine gastropods, and the nonequiangular shells of some pulmonates.

Limpet shells are believed to be closely related to other, more conventionally coiled gastropods, but it is unclear how their shell shape evolved. A hypothetical phylogenetic sequence of shell shapes has been generated using the model. Unfortunately, there are no known living or fossil molluscs with shells in the central portion of this sequence.

Acknowledgement

The author wishes to acknowledge the useful comments of Dr. R. Scholes, Dr. R. Kilburn, and those of the anonymous referee.

Appendix

List of Symbols

Angular parameters

α	: equiangular angle of spiral
β	: angle between Z-axis and line from aperture local origin to XYZ origin.
ϕ	: tilt of ellipse major axis from horizontal plane
Ω	: amount of azimuthal rotation of aperture
μ	: amount of "leaning over" of aperture
s_{min}	: angle at which aperture-generating curve begins
s_{max}	: angle at which aperture-generating curve ends
p	: position of nodule in terms of the angle, s
W_1	: width of nodule in s-direction
W_2	: width of nodule in Θ-direction

Linear dimensions

A	: distance from main origin to local origin of aperture at $\Theta = 0$.
a	: major radius of ellipse at $\Theta = 0$
b	: minor radius of ellipse at $\Theta = 0$
R	: length of aperture-generating vector
L	: height of nodule at $\Theta = 0$

Other

x, y, z	: Cartesian coordinates
r, Θ	: polar coordinates
r_0	: radius of spiral at $\Theta = 0$
N	: number of nodules per whorl
D	: sense of coiling, 1 = dextral, -1 = sinistral
k	: amplitude of the surface ornamentation
K	: differential growth ratio

References

1. H. Moseley, On the geometrical forms of turbinated and discoid shells, *Roy. Soc. Lon. Philos. Trans. Pt. I* (1838) 351–370.

2. D'A. W. Thompson, *On Growth and Form*, 2nd Edn. Cambridge University Press, Cambridge (1942).
3. D. M. Raup, The geometry of coiling in gastropos, *Proc. Nat. Acad. Sci.* **47** (1961) 602–609.
4. H. Weyl, *Symmetry*. Princeton University Press (1966).
5. C. Illert, The mathematics of gnomonic seashells, *Mathematical Biosciences* **63**, (1983) 21–56.
6. C. Illert, Formulation and solution of the classical seashell problem. I. Seashell geometry, *Nuovo Cimento D* **9(7)** (1987) 791–813.
7. C. Illert, Formulation and solution of the classical seashell problem. II. Tubular three-dimensional seashell surfaces, *Nuovo Cimento D* **11(5)** (1989) 761–779.
8. D. M. Raup, Geometric analysis of shell coiling: General problems, *J. of Paleontology* **41(1)** (1966) 1178–1190.
9. H. Moseley, On conchyliometry, *Philos. Mag.* **21** (1842) 300–305.
10. S. Løvtrup and B. von Sydow, D'Arcy Thompson's theorems and the shape of the molluscan shell, *Bull. of Mathematical Biology* **36** (1974) 567–575.
11. D. M. Raup, Geometric analysis of shell coiling: Coiling in ammonoids, *J. of Paleontology* **41(1)** (1967) 43–65.
12. P. H. Joubert, *Taksonomiese aspekte van die genus Bellamya jousseaume 1886 in oosterlike Suid-Afrika*. MSc Thesis, Potchefstroom University for C.H.E., South Africa (1978).
13. S. J. Gould, An evolutionary microcosm: Pleistocene and recent history of the land snail P. (*Poecilozonites*) in Bermuda, *Bull. Mus. Comp. Zool.* **138(7)** (1969) 407–532.
14. T. C. Foin, Systematic implications of larval shell parameters for the genus *Cypraea*(Gastropoda: Mesogastropoda), *J. Moll. Stud.* **48** (1982) 44–54.
15. A. J. Kohn and A. C. Riggs, Morphometry of the *Conus* shell, *Systematic Zoology* **24** (1975) 346–359.
16. G. R. McGhee, Shell form in the biconvex articulate Brachiopoda: A geometric analysis, *Paleobiology* **6(1)** (1980) 57–76.
17. H. E. Andrews, Morphometrics and functional morphology of *Turritella mortoni*, *J. of Paleontology* **48(6)** (1974) 1126–1140.
18. R. A. D. Cameron, Functional aspects of shell geometry in some British land snails, *Biological J. of the Linnean Society* **16** (1981) 157–167.
19. S. U. K. Ekaratne and D. J. Crisp, A geometric analysis of growth in gastropod shells, with particular reference to turbinate forms, *J. Mar. Biol. Ass. U.K.* **63** (1983) 777–797.
20. R. R. Graus, Latitudinal trends in the shell characteristics of marine gastropods, *Lethaia* **7** (1974) 303–314.
21. R. Kilburn and E. Rippey, *Sea Shells of Southern Africa*. Macmillan South Africa, Johannesburg (1982).
22. J. F. Blake, On the measurement of curves formed by cephalopods and other molluscs, *Philos. Mag. Ser.* **5(6)** (1878) 241–263.
23. D. M. Raup and J. A. Chamberlain, Equations for volume and center of gravity in ammonoid shells, *J. of Paleontology* **41(3)** (1967) 566–574.

24. C. S. Hickman, Gastropod morphology and function. In *Mollusks*, T. W. Broadhead, ed. University of Tennessee Studies in Geology **13** (1985) 138–156.
25. S. Løvtrup and M. Løvtrup, The morphogenesis of molluscan shells: A mathematical account using biological parameters, *J. of Morphology* **197** (1988) 53–62.
26. E. Savazzi, Shellgen: A BASIC program for the modeling of molluscan shell ontogeny and morphogenesis, *Computers and Geosciences* **11(5)** (1985) 521–530.
27. C. A. Pickover, A short recipe for seashell synthesis, *IEEE Computer Graphics & Applications* (Nov. 1989) 8–11.
28. M. B. Cortie, Models for mollusc shell shape, *S. Afr. J. Sci.* **85(7)** (1989) 454–460
29. P. Ward, Functional morphology of Cretaceous helically-coiled ammonite shells, *Paleobiology* **5(4)** (1979) 415–422.
30. G. A. Goodfriend, Variation in land-snail shell form and size and its causes: A review, *Syst. Zool.* **35(2)** (1986) 204–223.
31. M. B. Cortie, The effect of geometric similarity on gill size in the Mollusca, *S. Afr. J. Sci.* **85(10)** (1989) 621.
32. E. R. Truemann and M. R. Clarke, *The Molluscs. Vol. 10: Evolution*. Academic Press, Orlando, Florida (1985).
33. J. E. Morton, *Molluscs*, 4th Edn. Hutchinson University Library, London (1967).

ISOMETRIC SYSTEMS IN ISOTROPIC SPACE: AN ARTIST'S PERSONAL STATEMENT ON SPIRAL AND OTHER MAP PROJECTIONS

Agnes Denes

Isometric Systems In Isotropic Space: Map Projections, like most of my work, involves distortions and perspective, probability and space relations, transformations and interactions of phenomena.

In essence mathematical forms are projected over fluid space to create distortions of our globe into three polyhedra: The Pyramid, The Cube and The Dodecahedron; The Doughnut (*tangent torus*), The Egg (*sinusoidal ovoid*) and The Snail (*helical toroid*). Additional forms are The Lemon (*prolate ovoid*), The Hot Dog and The Geoid (see, for example, Ref. 1).

Map Projections creates sculptural form in celestial space and presents analytical propositions in visual form. It is a tantalizing game if one learns to read between coordinates and doesn't mind making sport of the human predicament. *Map Projections* are "sculptured reality," projecting a dynamic world of rapidly changing concepts and measures, where the appearances of things, facts, and events are assumed manifestation of reality, and distortions are the norm. Today's knowledge must be reassessed to cope with the new concepts of probability and catastrophe theories, curved space, and black holes.

In *Map Projections*, the anatomy of form is studied, vectors are built, earth measurements and scale factors rearranged. Longitude and latitude lines are unraveled and continents are allowed to drift. Gravity has been tampered with, earth mass altered, polar tension released. The north pole is forced to meet the south, or they are pulled apart. The boundaries have been transgressed. Art searches for the logic of matter, a glimpse at the formation of form, knowledge gained and abandoned, the game won or lost. And the game is all there is.

Reference
1. A. Denes, Notes on a visual philosophy. *Comp. and Maths. with Appl.* **12B** (1986) Nos 3/4, 835–848, and in *Symmetry: Unifying Human Understanding*, I. Hargittai, ed. Pergamon Press, New York (1986).
2. A. Denes, *Isometric Systems in Isotropic Space: Map Projections*, Visual Studies Workshop Press, Rochester, New York (1979).
3. *Cartes et Figures de la Terre*. Musee National d'Art Moderne, Centre Georges Pompidou, Paris. Cover of catalog: "Map Projections — The Snail, Helical Toroid" — and other map projections by Denes pp. 236–239 (text and illustrations)(1980).

Isometric Systems in Isotropic Space 391

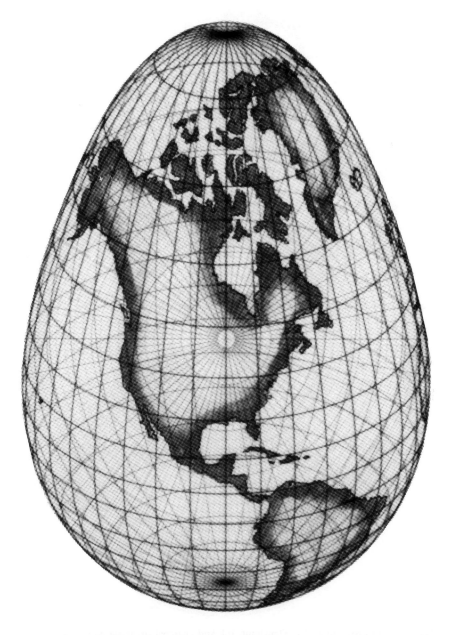

Fig. 1. Map Projections: The Egg. Ink and charcoal on graph paper and mylar.
© Copyright, Agnes Denes, 1974.

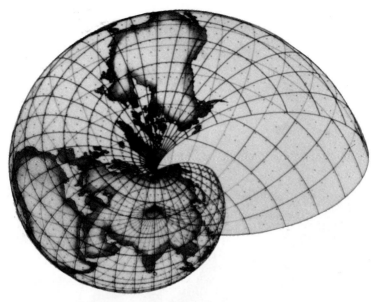

Fig. 2. Map Projections: The Snail. Ink and charcoal on graph paper and mylar (24" × 30"). © Copyright, Agnes Denes, 1974.

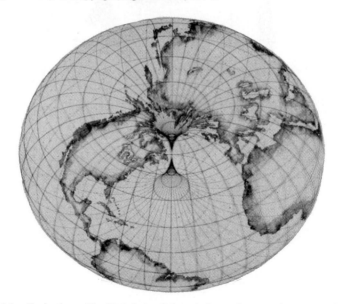

Fig. 3. Map Projections: The Doughnut. Ink and charcoal on graph paper and mylar. © Copyright, Agnes Denes, 1976.

Isometric Systems in Isotropic Space 393

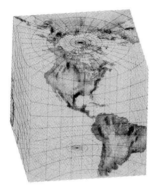
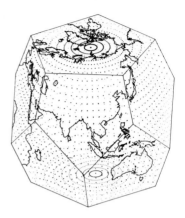
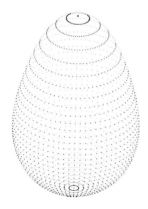
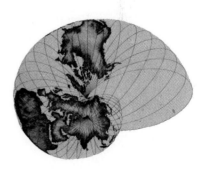

Fig. 4. Map Projections: The Cube, The Dodecahderon, The Egg, The Snail. © Copyright, Agnes Denes, 1974–76.

HELMET

Eleanor Kent

I began with an idea of making portraits of a favorite spiral shell with some eggs and corals, but as I worked the portrait became the record of the transformation of discrete objects into various unpredicted images. Each monoprint evolved from the one before it, and the rearrangement of the shells and eggs obliterated the previous image on the glass. The eggs broke and spread, and the heat of the Xerox began to cook them.

The nearly instant feedback of finished prints accelerated my critical judgment and contributed to a sense of urgency about which compositional moves to take next. I saw each print right after shifting the objects so I was able to decide how to move objects to make the next print. I began to work faster and to follow ideas that came as I was making images.

At several points I set the photocopier to make 3 or 4 prints and moved the objects between each three-color cycle using the spiral shell as the principal feature in the compositions. At these times the process was similar to giving a performance or ritual on top of the Xerox machine. I did not know how the prints would look until the machine stopped. The process was similar to painting on a canvas, where the various stages of the picture were saved as it progressed. The images made were recorded more quickly than a series of still photographs of the process, but more slowly than a movie recording the action in real time.

"*Helmet*" was made midway in the series. The cluster of objects became the profile of a warrior with a spiral decorated helmet. Subsequent images transform the spiral shell into parts of abstract compositions that make allusions to microscopic organic origins and outer space.

Fig. 1 is an example from a series of 28 different xerographic images made during a session of composing and moving a chambered nautilis, seashells, white corals, and eggs on top of the glass platen of a Xerox 6500 color copier. It is 11 inches high by 8.5 inches wide, printed on rag paper. The figure was produced in 1981.

Fig. 1. "Helmet," a xerographic image produced using seashells, corals, and eggs. Monoprint on rag paper.

SPINNING DESCARTES INTO BLAKE: SPIRALS, VORTICES, AND THE DYNAMICS OF DEVIATION

Kevin L. Cope

> Even as I scanned that impressive form and sloshed the Madeira around a bit (setting up an interesting spiraling motion among the floating M. Domestica), I could not help thinking how unmethodical we humanists were.
>
> Joel Gold, *The Wayward Professor*

1. Turbulence and Eighteenth-Century Studies

In the Zigler Museum of Jennings, Louisiana, a banner proclaims an exhibition on "The Eighteenth Century: The Age of Reason." Placards exhort visitors to regard the exhibited pieces as evidences of a neoclassical preoccupation with rationality, mathematics, and symmetry. Yet the works displayed are anything but orderly. Tornadoes whip up the skirts of baroque heroines; lightning dazzles the portrait of a leering gentleman; strange serpents torque around the *Laöcoon*. If this is "order," then it is the order of the tête-à-tête, the rococo love-seat that, facing two directions at once, winds lovers into a coil. It is an order that, like spiral forms, suggests disorder, that *feels* chaotic.

A cultural eddy like the Zigler might seem an odd place to begin a study of spirals in the eighteenth century. This odd strategy is justified by a precept guiding most inquiries into complexity, whether in thermodynamics or literary discourse: that minor influences may yield major consequences. Our contemporary sensitivity to the colossal results of tiny events is just as evident in Faller's study of particular prisoners in eighteenth century jails[1] or Backscheider's investigation of hustling hacks[2] as it is in Plotnick's work on geological strata[3] or Lorenz's paradigmatic work on meteorology. To borrow terms from one eighteenth-century poet, William Blake, the Zigler exhibit "condenses" and "conglobes" an array of confusions concerning the character of early modern intellectual history. Attention to the tiny details of the period will show that the Enlightenment was less enamored with rational order than with the construction of and feeling for it. As Donald Greene once pointed out, Augustans loved "exuberance" more than rationality.[4] It should not be surprising that, in so turbulent a period, spirals and spiral forms should outnumber their less dynamic counterparts.

1.1 *Concentric vs. Vortectical Form*

Literary critics and intellectual historians must accept less precise nomenclature than might their colleagues in the sciences. The term "eighteenth-century studies," for example, has come to designate the study of the wide historical and cultural field between Descartes and Wordsworth. Sloppy as it may seem, this free-wheeling division of history has more merit than most heuristic devices. This "eighteenth century" begins with Descartes' fundamental critique of scholasticism; it ends, symmetrically, with Coleridge's critique of Descartes. From the ruminations of the Cambridge Platonists on the relevance of Cartesianism to British culture, it proceeds to Young's, Smart's, and Cowper's visionary renovation of Cartesian constructs. Consistently an era of "vortecticalism," this "eighteenth century" was troubled and stirred by Descartes' exploration of spiral, helical, and whirlpool forms.

Words like "spiral" or "vortex" must also admit of some historically justified imprecision. In the eighteenth century, few scientific institutions could command respect, let alone enforce terminological consistency. John Dryden and King Charles II, for example, contemptuously withheld their dues from the Royal Society. Unorganized, intellectuals used terms like "spiral" and "vortex" erratically. Blake, for one, relegates all curving forms to one voracious category, "the vortex." This imprecision can be beneficial, for it can support an enriched understanding of the matrix of meanings attached to these terms. Researchers working during the eighteenth century responded to whirlpools in a qualitative as well as quantitative way, with a passion that only now is regaining credibility among philosophers of science. Their language and their theories carried ideological, religious, artistic, and moral as well as technical meaning.

In this essay, I shall listen to the many resonances of vortecticalism reverberating in assorted eighteenth-century materials. I shall suggest that our new scientific interest in instability, dynamics, chaos, and alternative presentations of order can enable literary critics to see past literary periods in a way that was unavailable to their predecessors, much as supercomputing has enabled scientists to see structures that eluded earlier researchers. I shall argue, for one, that most works of the period evidence "vortectical" or "spiral" or "dynamic" rather than "circular" or "concentric" form. Intellectual historians have long struggled to fit the often irregular productions of Enlightenment poets and artists into simplified, geometric schema—to argue, for example, that erotic narratives like those of the naughty Earl of Rochester can be rhetorically schematized as simple sets of concentric circles (see below). I shall respond that such schematizing projects need complication rather than simplification. In the case of Rochester, for example, critics need to consider the experience of the poem over time. They need to remember that, far from evidencing simple, static, and circular structure, works like Rochester's take on a spiral form, a form not unlike our own two-dimensional projections of the decay of periodic systems into less energetic conditions.

The defense of my thesis will constitute only part of my project. I would be guilty of hypocrisy were I to argue for research into complexity while oversimplifying my topic. I shall therefore also try to publicize the wonderful variety and vitality of spiral structures in eighteenth-century culture.

2. Unstable Foundations: René Descartes and the Theory of Vortices

University courses on Descartes focus on his epistemological speculations, but his most influential contribution to cultural history was his natural philosophy, as articulated in *The World* (1629)[5] and *Principles of Philosophy* (1647).[6] Cartesian metaphysics may have been the favorite exercise of specialists, but Cartesian science stirred the popular imagination. It established the cultural medium in which geniuses like Newton or Leibniz could germinate. Scholastics like Thomas Aquinas had characterized the world politically rather than spatially, as an extended city with no particular physical form. Descartes rendered the universe as a sphere, as a big container defined by indefinitely extended radii. Newton could never have represented the universe as "God's sensorium"—as a vast, mostly empty ampitheatre—were it not for Descartes' rendering of the universe as the immensely open location of space. Alexander Pope's God could never "extend throughout extent" without the Cartesian popularization of "extension."

Like nature, Descartes abhorred vacuums. Immense, his universe is immensely full. In such a universe, a philosopher must find some way to explain density. If there are no empty spaces in the Cartesian plenum, how can some atoms be packed more tightly than others? How can air be more extenuated than iron in a universe where there can be no extenuation? Anticipating such queries, Descartes devised his theory of vortices. The universe, Descartes claims, is pock-marked with whirlpools that suck extension into tight yet dynamic knots. Variations in the strengths of vortices yield variations in the densities of substances.

Descartes' vortex theory takes many surprising turns. For one, it dares to explain all motion, from the planetary to the sub-atomic. The planets rotate because they are vortices; they orbit the sun owing to the powerful influence of the solar vortex (indeed, Descartes fears that this "great" vortex might "swallow" its satellites (Figs. 1 and 2)). On a smaller scale, or-

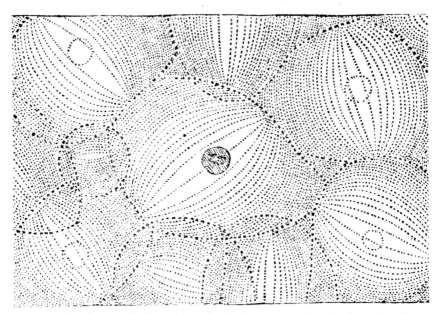

Fig. 1. Early stage in the evolution of the planetary and solar vortices. From Gabriel Daniel, *A Voyage to the World of Cartesius*. LSU Hill Memorial Library.

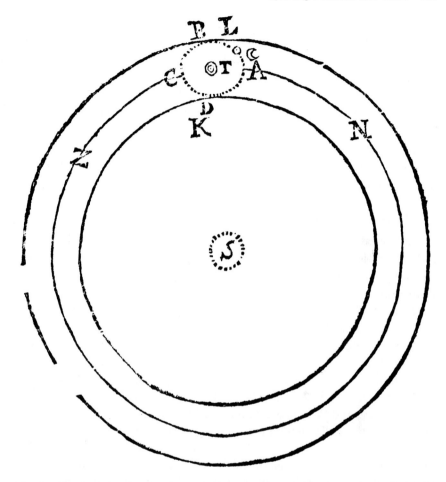

Fig. 2. Relation of the little vortex of the earth to the great vortex of the sun. From Gabriel Daniel, *A Voyage to the World of Cartesius*. LSU Hill Memorial Library.

dinary objects propagate vortices affecting other objects. Cartesian motion, although mathematically linear, is implicitly curvilinear, a vortectical interplay of centrifugal and centripetal forces.

Qualitatively, Cartesian motion is more evocative than Newtonian motion. Newton may analyze forces, but Descartes relishes the mutual influence of dynamic objects. Careful to distinguish "moving" from "being moved," he portrays a universe in which most everything is being affected by some other vortex. Scholars eager to appoint Descartes as the founder of the "age of reason" have neglected the dynamic, unstable quality of his universe. Every object or system is a vortex, a whirlpool so energetic that it might well fly apart. The universe began in motion: "they [the planets] moved individually and separately about their own centres" and then "moved together in groups" (Ref. 6, p. 257). Descartes' nineteenth- and twentieth-century commentators, especially those concerned with the foundations of determinism, may have fixated on Descartes' flirtations with mechanism, but more recent and generous critics, like W. J. Bate, warn that Descartes' first interpreters advanced more liberal exegeses of their master's philosophy.[7]

2.1. Early English Responses to Descartes' Vortex Theory

Controversy was the dominant rhetorical mode of the Restoration and eighteenth century. No philosopher escaped the bruises of peer review. Descartes sustained a savaging, even by the standards of the age. Conversely, those who found something evocative in Descartes' Disneyesque vision of the cosmos as whirlpool offered more than ordinary defenses of their champion.

Among those most offended by Descartes' vortectical mechanicism were the Cambridge Platonists. J. E. Saveson has refuted the notion that all members of this loose group held the same unfavorable view of Descartes,[8] but it is safe to affirm that Cudworth, More, Culverwell, Smith, Sterry, and Whichcote all regarded Descartes' multiscaler vortices as epitomes of intellectual totalitarianism. In providing a comprehensive, deterministic model of the universe, vortecticalism swept away the divine presence. To the chagrin of the vitalist Platonists, it reduced the spontaneity, or "plasticity," of the world. Nothing surprising, warm, or fuzzy could happen in a world spun by hurricanes. Ralph Cudworth, confident author of *The True Intellectual System of the Universe* (1678), waxes livid.

> [Cartesians] suppose heaven and earth, plants and animals, and all things whatsoever in this orderly compages of the world, to have resulted merely from a certain quantity of motion, or agitation, at first impressed upon the matter, and determined to vortex ... it is prodigiously strange, that these Atheists should, in this their ignorance and sottishness, be justified by any professed Theists and Christians of later times, who, atomizing in their physiology also, would fain persuade us in like manner, that this whole mundane system, together with plants and animals, was derived merely from the necessity and unguided motion of the small particles of matter, at first turned round in

a vortex, or else jumbled all together in a chaos ... God in the meantime standing by, only as an idle spectator of this lusus atomorum, this "sportful dance of atoms," and of the various results thereof. (Ref. 9, Vol. ii, pp. 34, 612)

Cudworth's hysterical outburst (which lasts for hundreds of pages) betrays a sense of wonder as well as outrage. To use an Augustan term, he senses the "sublimity" of Descartes' appalling theory. Susceptible to the apocalyptic fears that beset the first vorticians, Cudworth rhapsodizes over the gradual involvement of the earth in the solar vortex, the decay of the earth's rotation, the eventual alignment of the earth's axis with the solar vortex, and a host of other magnificently entropic, vortectical processes (see Fig. 3).

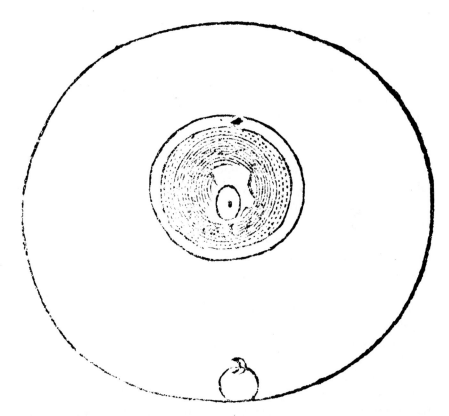

Fig. 3. Irregular, interfering vortices of the earth and moon, and their effects on tidal motion. From Gabriel Daniel, *A Voyage to the World of Cartesius.* LSU Hill Memorial Library.

Cudworth's simultaneous revulsion with and attraction to Cartesian physics points up his character as a reconciler. Cudworth, after all, had preached sermons on brotherly love to the English House of Commons during a bloody civil war. Other writers were less temperate. Joseph Glanvill, a propagandist for the Royal Society, cared little for technical analyses but loved apocalyptic raptures.

> At length when these *greedy flames* shall have devoured what ever was *combustible*, and converted into a *smoak* and *vapour* all grosser *concretions*, the great *orb* of *fire* that the *Cartesian Philosophy* supposeth to constitute the *centre* of this *Globe*, shall perfectly have recovered its pristine *nature*, and so following the *laws* of its *proper motion*, shall fly away out this *vortex*, and become a *wandring comet*, till it settle in some other. (Ref. 10, pp. 179–180)

Henry More, whose principle subject, witchcraft and demonology, nominated him as even more of a crackpot than Glanvill (whose principle subject was reincarnation), decided to call for moderation. Pointing out all the technical errors in Glanvill's interpretation of Descartes—not the least of which is his overlooking the irreversibility of the entropic progress from a "Sun" to a "Terrela" and finally to a planet—he graciously concludes that "this looks like an heedless mistake of this ingenious Writer, who, though he speak the language of *Cartesius*, seems not to have recalled to mind his *Principles*" (Ref. 11, p. 141). Glanvill, however, had committed himself to the heavens of inspiration, where his soul, along with those of "*Cartes, Gassendus, Galileo, Tycho, Harvey, More, Digby*," would, overleaping the likes of More, go "skipping into *Vortexes* beyond their Light and Influence; and with an easie twinkle of an Intellectual Eye look into the Centre, which is obscur'd from the upper Luminaries" (Ref. 12, pp. 240–241).

In scoffing that Glanvill uses the words but not the ideas of Descartes, More proves himself clairvoyant. Writers were already treating the Cartesian vortex as more of a literary motif than a serious theory. Post-Glanvillians would care little whether their use of vortectical themes could be authorized by Descartes' writings. More himself wasn't above making extravagant use of vortices. "*This* Vorticall *Motion being the cause of the generation of all things* . . . and it may be for this cause also the *Pythagoreans* called the *Decad*, that is the World, Generation," as did Moses and all the other famous sages of antiquity (Ref. 13, p. 103, cited in Ref. 14, p. 159). Indeed, as the pious Cudworth feared and the heretical Glanvill hoped, the vortex was becoming a powerful instrument of propaganda in the challenging of a whole range of orthodoxies.

3. Varieties of Vortecticalism: Explicit and Implicit Evidences

Vortecticalism could, then, influence a wide range of genres. Descartes had the bad luck to publish his curious ideas only a few years before the great age of neo-classical satire. The comical roasting of the vortex theory soon proved more popular than the theory itself. Among the most entertaining of the many early send-ups of Descartes is Gabriel Daniel's remarkable *A Voyage to the World of Cartesius* (1692), a work quickly translated into English from the original French, doubtless as a British slap at the great Gallic vortician.[15] Daniel opens with the thesis that Descartes, like Elvis Presley or Bruce Lee, has not really died. Having taken "his usual Dose of Snuff . . . his [Descartes's] soul leaves his Body in the Bed" (pp. 25–6). Along with Father Mersenne, a scholastic theologian famed for having compiled 800 pages of undoubted information by way of refuting skepticism, Daniel says "yes" to drugs and tries the same expedient. Trans-snuffing his soul out of his body, he and Mersenne leave the corporeal world for the "World of Cartesius." A first-rate satirist, Daniel rivals Swift or Voltaire:

> While *M. Descartes* was busied in disclosing to me all his Mysteries, *Father Mersennus* and the *old Gentleman* were diverting themselves, by Vaulting from *Vortex* to *Vortex*, and were but very ordinary Company, to *Aristotle's Plenipotentiaries*, who star'd confusedly, and were exceedingly out of Countenance, and who now, and then joyned them, now and then came to us; comprehending not a Syllable all the time, in that Gallimauphry of *Vortexes*, of the *first, second, and third Element*, of *ragged* and *branched Parts* . . . (p. 221)

Daniel educates his reader about vortices, achieving a neo-classical melding of destructive satire with constructive didactic.

> I saw them [extended substances] more distinctly than your most clarify'd *Cartesians* do your *Chamfer'd* Parts of Matter, wreath'd in the shape of little Skrews, by the Struggle they have to squeeze betwixt the *Balls* of the *Second Element*, or to constitute a little *Vortex*, round the *Loadstone*, and to cause that wonderful affinity that is found betwixt that Stone and the Poles of the Earth, and with it and Iron. (p. 205)

As his characters move between real and theoretical universes—as his tone wavers between pedagoguery and scoffing—so Gilbert straddles the boundary between the philosophical analysis and the literary transformation of the vortex theory.

Not many Augustan satirists considered Descartes worthy of such careful study. Most exploited the materials of vorteclicalism with few pedagogical intentions. Jonathan Swift, for example, deploys vortecticalism in many diverse ways. In his *A Tale of A Tub*, Descartes joins the company of deluded philosophical madmen. "*Cartesius* reckoned to see before he died, the Sentiments of all Philosophers, like so many lesser Stars in his *Romantick* System, rapt and drawn within his own *Vortex* . . . for which the Narrowness of our Mother-Tongue has not yet assigned any other Name, besides that of *Madness* or *Phrenzy*" (Ref. 16, p. 105). In the "Preface" to the same work, he describes the cultural "tub" as "hollow, and dry, and empty, and noisy," and, like philosophical schemes, "given to rotation." Philosophy becomes a rotating structure, vortecticalism a rhetorical *topos*. Swift's mock-georgic "A Description of a City Shower"[17] portrays whirlpools of "Sweepings from Butchers Stalls, Dung, Guts and Blood,/Drown'd Puppies, stinking Sprats, all drench'd in Mud,/Dead Cats and Turnep-Tops." Swift incorporates the vortex into the structure of his poem, indeed makes it a visible part of ugly urban experience. However contemptuous Swift may have been of poor René, the Cartesian vortex has become the unstable foundation of his turbulent poetics.

More subtle early uses of the Cartesian world-view are less "uses" than perceptions. Chaoticians know well that the order within "chaos" is perceived from outside the chaotic system, through the collective graphic representation of millions of discrete events. Something similar can be said about our critical perception of the literary and philosophical "collectors" of the seventeenth and eighteenth centuries. Writers of this era never succumbed to our contemporary yearning for literary coherence. Anthologies comprised of diverse remarks on unrelated topics by disagreeing authors drew no less applause than did coherent philosophical projects.

Readers of the period, however, were weaned on religious literature. Inclined to assume that everything served some purpose, they tended to see providentially directed form emerging from the most disorderly texts. Indeed, many texts were not texts at all, but reorganized transcriptions, by faithful hearers, of charismatic leaders' edicts. One Cambridge Platonist, Benjamin Whichcote, wrote very little but talked a great deal. Whichcote spent most of his useful hours in the pulpit, declaiming maxim after maxim and aphorism after aphorism. His students, resistant to their master's love of disorder, arranged his utterances into his *Moral and Religious Aphorism*,[18] a series of twelve "centuries" comprised of 100 maxims each.

Those few critics who have bothered with the eccentric Whichcote have declared his writings the epitome of circular, concentric form. His centuries are said to be ordered like self-surrounding circles. Century XII allegedly surrounds, balances, and stabilizes century I. Yet Whichcote's aphorisms form up into cycles as well as circles. Experienced in time, they appear as a series of evocations. They repeat and echo one another, but in no easily intuited order. An answer to a maxim in the middle of Century II may appear at the beginning of Century V. It might, then, be prudent to rename the "centuries" as "spirals," as rhetorical circles that appear serially, rolling into one another like a spiraling line.

Many other authors of the period—George Herbert, Robert Herrick, Richard Crashaw, Thomas Traherne—adopted versions of the "century" plan. These authors offer voluminous but scattered writings, literary manifolds that tend to aggregate into concentric form but that achieve this shape only after a long period of writing and reading. Readers can't discern the form of Traherne's *Centuries* at a single glance, as they might be able to do when looking at highly schematic works like Sidney's or Spenser's. In his voluminous *Characteristics*, Lord Shaftesbury announces that his self-consciously miscellaneous work emerges from the variation of its pace, that his enormous array of perspicacious remarks recalls the path of a horse that sometimes trots and sometimes runs but that will, eventually and irregularly, complete its course (Ref. 19, Vol. II, pp. 172, 207, 216). The Marquis of Halifax, for another example, organizes his maxim collections by topic, from the most general to the most particular. Halifax's anticlimacticizing works wane less and less energetic, spiraling, over time, into smaller and smaller topics. His centuries grow shorter and shorter, the radius of his interests diminishing moment by moment.

3.1 *Time, Spirals, and Literary Form*

A term like "century" suggests authorial interest in the temporal extension and historical experience of works. Halifax, Shaftesbury, and their peers write on a colossal scale—hundreds of pages, thousands of aphorisms, dozens of chapters. Their works attack writers who, in the language of the period, represent "scholasticism"—who cling to a belief in a complete, timeless, and bounded system of an Aristotelian world. Halifax and his peers force readers to experience their writing over a long period of time, often exhausting them into an anti- or nonclimactic conclusion. I have compared these fatiguing projects to the two-dimensional projection of decaying periodic processes—to a spiral—on the grounds that these works extend, complicate, and distort the experience of a balanced, concentric, and periodic literary structure over an unexpectedly long interval of time.

The consummate examples of the transformation of synchronic, concentric structures into diachronic, decaying processes are the satires of John Wilmot, Earl of Rochester (1647–1680). An adventurer and a courtier, Rochester was also a man of letters. Although temperamentally inclined to such practical jokes as rowing a dinghy into the middle of a raging sea-battle or peddling patent medicine at county fairs, he seldom failed to obey the neo-classical rules that he was challenging, even when penning mock-epic poetry on outrageous topics like "Signior Dildo."[21]

Rochester's most brilliant critic and editor, David Vieth, has offered the most enticing—and most revisable—attempt to interpret Rochester as an author of rational, "concentric" pieces.[22] Vieth's topics are Rochester's "social" satires, heroic acts of excoriation enumerating all the evils in society. "By the next post such stories I will tell/As joined with these, shall to a volume swell" warns his spokeswoman, "Artemisia." Rochester's reader is stunned, dunned, and numbed by an ever-expanding lists of abuses. Vieth observes that Rochester's works "exhibit a centrifugal effect that leaves a gap at its center and renders its outer boundaries indefinite and unstable" (Ref. 22, p. 129). Arguing against his own theory, Vieth subsequently affirms that poems like *Artemisia to Chloe* "employ 'concentric structure' " in which "the physical center of a literary work receives significantly heavy emphasis, while passage on either side, balanced against each other with the center as a fulcrum, are brought into specially close relationships," a device he attributes to classical "Greek culture" (Ref. 22, p. 139). Vieth supplements his remarks with a diagram (Fig. 4).

Rochester, alas, confounds the otherwise excellent Vieth. The "physical" center of the poem occurs at line 132. Only the conceptual center

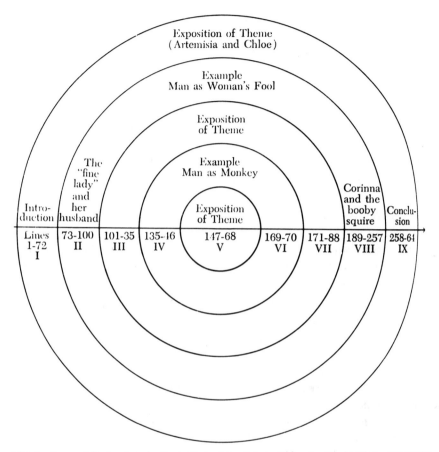

Fig. 4 Concentric structure in Rochester's *Artemisia to Chloe*, by David Vieth. LSU Hill Memorial Library.

occurs where Vieth places it. Curiously, the same topic is discussed at both "centers." Rochester divides his theme from itself, preserving the illusion of concentric structure but establishing several foci for his discursive circles. His reader experiences the center of his poem in two or more places at two or more times. A more appropriate diagram might be drawn with the well-known child's toy, a "spirograph," letting the drawing wheel roll around the geared bar, progressively sweeping out dislocated circles—spirals—around ever-changing centers.

In Rochester's greatest satire, *A Satyr Against Reason and Mankind*, a similar spiral progress occurs. The first 122 lines examine the life of a libertine—"Hunger calls out, my reason bids me eat." The next 51 lines, in counterpoint, explore a Hobbesian, Machiavellian view of the world—"For fear he arms, and is of arms afraid,/By fear to fear successively betrayed." A third segment, lines 174—221, later appended to the original poem, characterizes the ideal religious life. The long part one and the short part two asymmetrically balance one another; parts one and two *ensemble* enter into a de-centered balance with the concluding, shortest, and least energetic segment. The effect, once again, is of a set of circles in some but scarcely a concentric arrangement. Like the revolution of differently sized gears, the scheme of the poem is complicated, elongated, and made dynamical by time.

Rochester was a libertine and a rake, but works by "conservative" authors evidence the same kind of spiral structure. Experts like Martin Battestin have lauded the concentric structure of Henry Fielding's novels.[23] Fielding, critics argue, sets the first and the last chapters of his novels in Edenic country estates, then, in the intervening books, discusses the wicked city. Yet no critics have trifled with the enormous inequity in what Vieth might call the "physical" size of the chapters in these novels, nor have they explored the complex relations between text size, reading time, and the apprehension of literary structure. Were the irregular books of Fielding's novels compared against the equally-sized cantos of Dante's or Spenser's epics, the dynamic, spiraling structure of his books would become apparent.

Tension between rectilinear and dynamic interpretations of history is not uncommon during the era. An illustrator of the great fire of London (1666) portrays an unchanging, rectilinear city smothering under the turbulent, spiraling smoke far off in the background (Fig. 5).

This evocative but perplexing interplay between historical progress, literary convention, and spiral forms also unsettled scientific literature. Stephen Jay Gould has chronicled the rising popularity, during the eighteenth century, of literature concerned with "deep time," time vastly exceeding the 6,000 years allocated to the world by holy writ.[25] Robert Hooke, for example, undertook a painstaking investigation into "Stone[s] of the Common Nautilus-shape," fossil shells insinuating that England had been inundated for more than 40 days and 40 nights (Ref. 26, p. 287; see

Fig. 5. Anonymous illustration of the Great Fire of London, 1666. London Society of Antiquaries.

Fig. 6. Spiral shells. From Robert Hooke's *A Discourse of Earthquakes*. LSU Hill Memorial Library.

Figs. 6, 7). Scrupulously attentive to detail—"the 22nd [stone] was somewhat like the 12th, but of a smaller Spiral" (p. 283)—Hooke found himself drawn headlong into a polemic against the literal truth of Scripture.

Despite his technomania, Hooke is as much a man of letters as Rochester is. For Rochester, the poetic city emerges from minute attention to its cumulative irregularities; for Hooke, the colossal form of history emerges from the details of tiny, spiraling fragments of irregular "Nautil-shells."

3.2 Urbanity and Spiral Structure

The emergence of spiraling literary forms registered an increasing concern with history—with the influence of living, moving, breathing, and sometimes decrepit readers on the construction of literary experiences. This sensitivity to history coincided with a sharpening of interest in city life. Involved in convoluted pathways, often shaped like irregular circles, and full of entropic processes, eighteenth-century cities enact the vortectical world view.

The London landscape of the poet John Gay is more than an occasion for mock-heroic *jeux d' esprit*. Gay's dynamic city relentlessly intrudes on his consciousness, ceaselessly nominating itself as a candidate for poetic representation. "The *Dust-man's* Cart offends thy Cloaths and Eyes,/ When through the Street a Cloud of Ashes flies" (Ref. 27, p. 144). Coupled with his lavish use of emotionally charged words like "offends," Gay's unstinting association of transitive verbs with seemingly inanimate subjects makes his city seem to live, breathe, intend, and support a host of psychological processes normally reserved for conscious beings.

Gay, moreover, is wholeheartedly vortectical. His heart a helix, he perceives the world as a spiral. The poet of the periphery, he never walks down the center of the street, preferring to wander along alleyways—to shimmy toward his destination, the center of the city. He teaches his reader "when to assert the wall," when to get to the side, how to circulate around "winding Alleys" and "perplexing Lanes," and how, in writing, to "tread in Paths to ancient Bards unknown" (I:4–10, 19–20). His dominant motif throughout *Trivia* is that of labyrinthinity. Eighteenth-century labyrinths were almost always laid out on some variation of a spiral plan. Finding the center of the maze involved circulating around a series of rings, gradually stepping inward at each

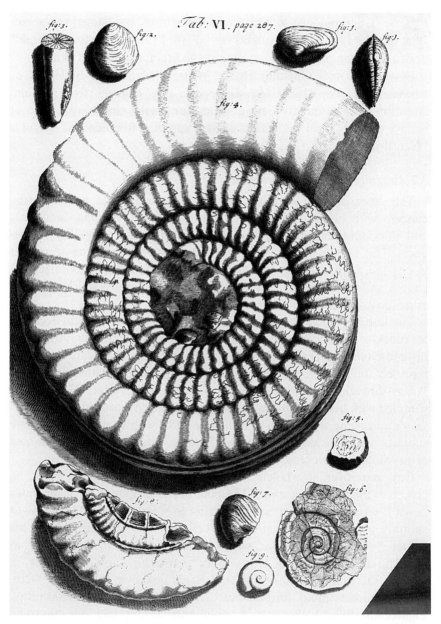

Fig. 7. Spiral shells. From Robert Hooke's *A Discourse of Earthquakes*. LSU Hill Memorial Library.

opportunity. The ever-complicating Gay offers not one labyrinth, not two contrasted labyrinths, but three labyrinths. Outlining the twofold "Labyrinth of *Crete*," at once the classical precedent and the fact of the degraded maze of modern London, he superimposes on both mazes a third maze: a simplified, radiative, and ambient orb made up of shining pathways effused by visionary London streetlights (Ref. 27, pp. 145–146, 164). "Still keep the publick Streets, where oily Rays/Shot from the Crystal Lamp, o'erspread the Ways." Gay's last, visionary labyrinth, the shining pathways shed by a crystalline lamp, is nothing but a periphery, a ray defined by its motion from the city and its encounter with center-seeking viewers. And a turbulent encounter it is: like pencils drawing spirals, Gay's circulating but centripetal citizens are always on the roll: "Shops open, Coaches roll, Carts shake the Ground" (Ref. 27, p. 144).

Gay's urban pastoral is literally whirled around a pencil point. *Trivia*, a point of intersection from which travelers can take any of three roads, is all about regulated indeterminacy, about sweeping to and from the peripheries of the city to and from its infinitesimal center. Both pessimistic and optimistic, *Trivia* comically recounts the dissipation of urbanity into nothingness and the regurgitation of nothingness into urbanity. A poem on urban rather than national history, *Trivia* spins the cultural record around a point on a map.

By no means is Gay the only poet of the urban vortex. The Reverend Thomas Maurice, the mock-heroic poet of schoolmasters and whipped boys, transposed the urban comedy to Oxford, scourging smokers and calumniating "that detested weed, *Virginia* hight,/Which the sage *Don*, in spiral clouds exhales."[29] Maurice's mentor, John Philips, the poet of cider and wine, had already surfed the smoky waves, attacking Londonians addicted to "*Mundungus*, ill-perfuming Scent" (Ref. 30, p. 4). John Gay, for that matter, proved his flexibility by attending to an array of spiral-related forms, from circles and fans to radii. His poem, *The Fan*, a forerunner of the spiraling *Trivia*, imitates an oriental fan, spreading out from a trivial point. Alexander Pope dishes up the most elaborate response to the exciting spiralizing of history. His famous dissertation on entropy, *The Dunciad*, describes a curvilinear progress toward the black hole at the center of culture.

418 Kevin L. Cope

> In vain, in vain, — the all-composing Hour
> Resistless falls: The Muse obeys the Pow'r.
> She [Dulness] comes! she comes! the sable Throne behold
> Of *Night* Primaeval, and of *Chaos* old!
> Before her, *Fancy*'s gilded clouds decay,
> And all its varying Rain-bows die away.
> *Wit* shoots in vain its momentary fires,
> The meteor drops, and in a flash expires.
> As one by one, at dread Medea's strain,
> The sick'ning stars fade off th' ethereal plain;
> As Argus' eyes by Hermes' wand oprest,
> Clos'd one by one to everlasting rest;
> Thus at her felt approach, and secret might,
> *Art* after *Art* goes out, and all is Night. (Ref. 28, p. 799)

One collapsing circle after another, one blinking eye after one decaying rainbow, Pope's urban apocalypse follows a spiral curve into torpidity.

4. Visualizing the Vortex of City and Country Life

Belle-lettristic literature had not cornered the market on vortices. Mapmakers and other art-topographers attempted to represent historical scenes through techniques reminiscent of their literary counterparts'. Projecting the whirl of human history onto two-dimensional planes, they mapped festival scenes onto fan leaves—leaves that could be spread out slowly, over time, that could, as part of the apparel of ladies at festivals, become part of history itself. Thomas Loggon, a dwarf who specialized in fan-painting, portrays the bustle of St. Bartholomew's fair by leading the eye around a fan (Fig. 8; Ref. 31, p. 68).

Loggon's peculiar genre registers broader changes in popular, illustrative cartography. As late as 1751, chartmakers were portraying city squares (sensibly!) *as* squares. Rectilinearity ruled cartography; topographers took a stable, elevated, and even Olympian viewpoint (Fig. 9). By the revolutionary year of 1789, however, city squares were being viewed from ground level, as though the illustrator participated in the observed scene. Dynamic, these vitality-infused representations round the corners off squares, rendering the street as a spiral curving into the distance (Fig. 10).

The most inventive of city illustrators was William Hogarth. Hogarth's numerous "progress" series—"The Rake's Progress," "The Harlot's Progress"—capitalize on the vortectical, serial presentation of human history. Heroic painters like Raphael or Rubens might try to capture an

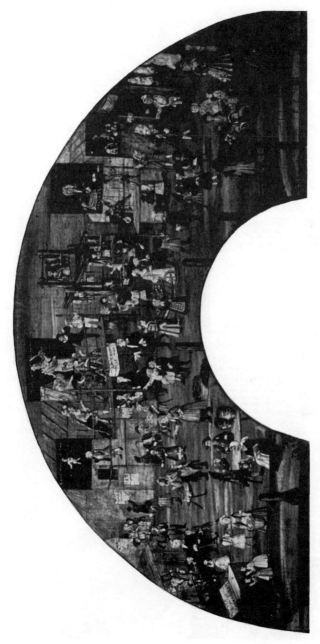

Fig. 8. Thomas Loggon, fan-leaf illustration of St. Bartholomew's Fair, 1740. British Museum.

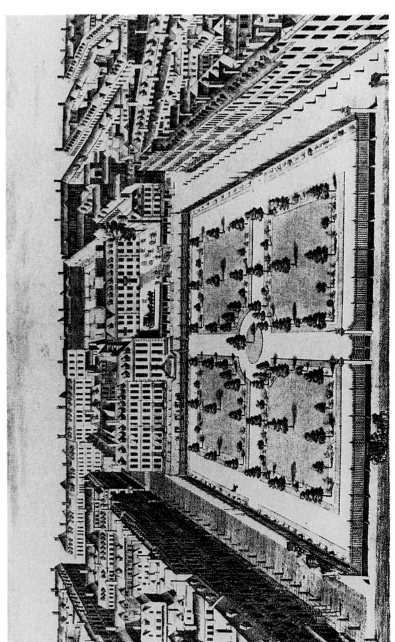

Fig. 9. Leicester Square, 1751. LSU Hill Memorial Library.

Fig. 10. Grosvenor Square, 1789. British Museum.

entire epic in a single canvas, but Hogarth tells his story sequentially, in slides presented seriatim. He exposes the timeless, still-life representation of culture to the stress of historical, dynamical, and entropic processes. When Hogarth maps a city, his perspective is at eye level, his chart full of activity (Fig. 11). Hogarth's picture follows a "serpentine line" (see below). In this engraving, the eye travels over the coach, down to the vacant—trivial?—center, up to the balcony, and then down into the open background. It sweeps out a helix.

4.1 *Hogarth and Serpentinism*

Hogarth raised vortecticalism into a discipline. His *The Analysis of Beauty*[32] proclaims that art should do more than imitate appearances. It should capture the process, the life, beneath the image. All appearances, Hogarth postulates, must be represented with lines. Lines fall into two categories: straight and curved. Curved lines are subdivided into waving and "serpentine," helical lines. Straight lines may represent only

Fig. 11. William Hogarth, "The Industrious 'Prentice, Lord Mayor of London" (1747). LSU Anglo-American Art Museum.

inanimate objects; waving lines represent living "beauty"; serpentine lines capture "grace" and "life." Wavy hair, for example, is touted as the most beautiful of human ornaments (p. 28). "Forms of most grace have the least of the straight line in them" (p. 38). Hogarth waxes enthusiastic:

> The eye hath this sort of enjoyment in winding walks and serpentine rivers, and all sorts of objects, whose forms, as we shall see hereafter, are composed generally of what, I call the *waving* and *serpentine* lines.
>
> Intricacy in form, therefore, I shall define to be that peculiarity in lines, which compose it, that *leads the eye a wanton kind of chace,* and from the pleasure that gives the mind, intitles it to the name of beautiful. (p. 25)

The interdisciplinary Hogarth selects reading as one of the most beautiful of experiences. The scanning of printed lines take the eye on a "wanton kind of chace," tracing out serpentine curves! As this example suggests, Hogarth is fond of temporally extended processes in which serpentine lines unexpectedly emerge from straight lines. The frontispiece to *The Analysis of Beauty* features a coiling serpent ready to burst out of a

Euclidean pyramid (Fig. 12). Hogarth represents the serpentine line itself in ways that show it to be something more than a mere line. He brings out its three-dimensional quality. In one illustration (Fig. 13, panel 26), an undulating serpent coils around an lifeless conical solid.

Top and center in Hogarth's plate, the serpentine line presides over the chaoticum of curves, lines, and events that comprise the subtending portrait of art. This implied presidency of the serpentine may have been predictive. The great theorist of landscape gardening, William Shenstone, lauds the introduction of waving walkways into previously formal, rectangular gardens; Shenstone's gardening disciple, President Thomas Jefferson, installed a serpentine wall at the University of Virginia.

4.2 *The Serpentine Household and Its Environs*

Serpentinism quickly slithered out of the carrel and studio and into daily life. Set in a labyrinth deep beneath the evil Manfred's castle,

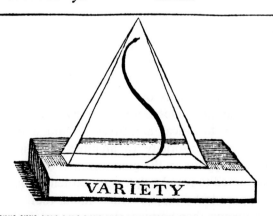

Fig. 12. Frontispiece, Hogarth's *The Analysis of Beauty*. LSU Middleton Library.

Fig. 13. Plate I from Hogarth's *The Analysis of Beauty*. LSU Middleton Library.

Horace Walpole's gothic thriller, *The Castle of Otranto*, domesticates the maze. Every torturer ought to have a spiral-form dungeon in the home! Most buildings erected during the eighteenth century reflected the neo-classical architectural mode. As Diderot and d'Alembert point out in their encyclopedia, the capitals on neo-classical columns sport flashy ornaments, usually spiral volutes. Some columns imitated Trajan's pillar, wrapping continuous historical narratives around a pole by way of imparting motion and historicity to architecture (Fig. 14). Notable examples include the pillars in the great rotunda of Vaux-hall, a music and dance hall annexed to popular "pleasure garden" (Fig. 15). The Vaux Hall rotunda happens to be capped with a radial roof.

Even more impressive than vortectical columns are the late eighteenth-century "panoramas" of Robert Barker and his followers (see Ref. 34). Comparable to Cinemax movies, these enormously popular paintings

Fig. 14. Trajan's Column, c. 112-113 A. D. LSU Hill Memorial Library.

Fig. 15. Music Room at Vaux Hall Gardens, c. 1752. LSU Middleton Library.

offered spectators a 360° view of city and country scenes. Panoramas were temporally as well as physically cyclical. The famed panorama of the countryside around Dixton manor depicts the complete cycle of haymaking (Fig. 16).

Although the panoramic display was circular, its effect was serpentine. Panoramas bludgeoned their less-than-cultivated viewers into the realization that, despite their "all-embracingness," they could only be experienced over time, as the viewer turned to inspect each panel. When a viewer returned to the original first segment, that segment had moved forward in time. Revolving, viewers swept out spiral trajectories along the axis of time. Moreover, panorama buildings were sometimes too small to contain the entire panorama. Visitors to these inadequate facilities could only apprehend the whole scene by looking at the published series of engravings of the panels, as they might look at the plates in one of Hogarth's series (Ref. 34, p. 62). The panorama thus brought spiral form full circle. Rather than leading viewers to perceive

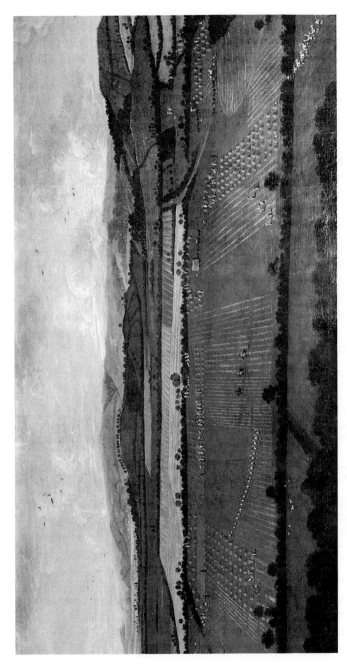

Fig. 16. The Countryside Around Dixton Manor, c. 1725–1735. Cheltenham Museum and Galleries.

spiral form in a work, as happened in Whichcote's aphorisms, panoramas looked down on viewers, silently watching them describe moving circles in time.

5. Enlarging the Cosmic Vortex

As the eighteenth century progressed, ambitious writers aspired to uncover vortices throughout all the physical and spiritual universe. Descartes, sad to say, had treated vortices as ordinary mechanical phenomena. No mystery attached to them. Undeterred, brooding writers in the later eighteenth century continued to load these curious "springs of sense" with moral, psychological, and aesthetical significance. They tried to "animate" the vortex, to exploit its potential for the expression of feeling and to affiliate the whirlwind of nature with the tornado of human thought.

The most lavish animator of vortices was Edward Young, the author of the chilling *The Complaint: Or, Night Thoughts (1744)*.[35] For the relentlessly exclamatory and indefatigably depressed Young, nature whirls into ascending circles of creation, spiraling far beyond the mundanities of physics.

> Orb above orb ascending without end!
> Circle in circle, without end, enclosed!
> Wheel, within wheel: EZEKIEL, like to thine!
> Like thine, it seems a vision or a dream;
> Though seen, we labor to believe it true!
> What involution! what extent! what swarms
> Of worlds, that laugh at earth! immensely great!
> Immensely distant from each other's spheres!
> What, then, the wondrous space through which they roll
> At once ingulfs all human thought;
> 'Tis comprehension's absolute defeat.
>   Nor think thou seest a wild disorder here:
> Through this illustrious chaos to the sight,
> Arrangement neat, and chastest order, reign . . .
> Confusion unconfus'd! (Ref. 35, p. 283)

In declaring "comprehension's absolute defeat," Young puns. He defeats the circumscribing circle, but he allows the vortex to comprehend nature in a larger, poly-dimensional way. "Now fancy glows,/Fired in the vortex of Almighty power" (Ref. 35, p. 298; cited in Ref. 36). Thomas Taylor, an eighteenth-century neo-platonic visionary, likewise suggests that those who "lapse into the whirls of generation" will eventually spiral up to the Milky Way, where, living on a diet of milk (Ref. 37, p. 315–316), they themselves will have a chance to churn up a few vortices!

James Thomson, author of *The Seasons* (1746),[38] worked on a smaller scale than the "immense" Young, yet, varying between the sub-atomic and the circumpolar, he delineates a helical world-view rivaling his complaining contemporary's. Explicating "frost" as labyrinthine "myriads of little salts, or hooked, or shaped/Like double wedges, and diffused immense/Through water, earth, and ether" (Ref. 38, p. 212), he builds additively, algorithmically, and vortectically until he whirls into orbit.

> The full ethereal round,
> Infinite worlds disclosing to the view,
> Shines out intensely keen, and, all one cope
> Of starry glitter, glows from pole to pole.
> From pole to pole the rigid influence falls
> Through the still night incessant, heavy, strong,
> And seizes nature fast. It freezes on,
> Till morn, late-rising o'er the drooping world,
> Lifts her pale eye unjoyous. (Ref. 38, pp. 212–213)

Thomson, a self-styled Newtonian, Copernican, and progressive, seems to follow a conventional trajectory, circling round the earth. Yet his circles are complicated by time, by the fact that, in his self-consciously seasonal, fashionably Newtonian universe, the earth sweeps through space. Thomson, like Young, ends up describing a sequence of periodically displaced circles.

The aggregative, the calendrical, and the domestic come together in the wild maxims of Christopher Smart, a deranged cyclothymic who suffered from a compulsion to pray aloud in public places. Roughly half of Smart's maxims begin with "let," the remainder beginning with "for." Smart had intended to write a concentric, antiphonal work in which members of two choruses chanted the "let" and the responsorial "for" clauses in neat alternation (Ref. 39, p. 1166). In the entropy of his madness, Smith's work devolved into a furious outpouring of both "let" and "for" clauses. His *Jubilate Agno* emerges as a work *aspiring* toward circular, concentric form—as a progressive, temporally extended spiraling *toward* order. "Let" and "for" clauses, moreover, seem to anticipate something. "Let" ... what? "for" ... what? Both sets of clauses are evocative; neither can be resolved by the other. Smart's cat endlessly wreathes around himself, defining this divinely mad cycle of creation.

> For I will consider my cat Jeoffry.
> For he is the servant of the Living God duly and daily serving him.
> For at the first glance of the glory of God in the East he worships in his way.
> For is this done by wreathing his body seven times round with elegant quickness ...
> For this he performs in ten degrees.
> For first he looks upon his fore-paws to see if they are clean.
> For secondly he kicks up behind to clear away there ...
> For he has the subtlety and hissing of a serpent, which in goodness he suppresses. (Ref. 40, pp. 87–88)

Smart's two constant values, seven and ten, are both "complete," mystical numbers, one the number of days in the creation, the other the conventional base of our numbering system. Yet neither is a factor of the other, nor do the two stand in any convenient relationship. Units of seven and ten wind around and on top of one another, gyrating like "Jeoffry," who takes on a serpentine form and character. Smart's work is a process of involution, a wrenching of symmetry out of asymmetry. The consummate poet of deviation, he elicits the ordering circle of God from the ongoing process of irregular behavior.

6. Conclusion: Helices in the Home

Christopher Smart epitomizes the eighteenth-century habit of locating the normative in the aberrant particular. As the period wound down, writers interested in the grand, the cosmological, and the Cartesian turned to idiosyncracy and introspection as tools for disclosing the vortectical habit of the world. The later eighteenth century domesticated the helix, rediscovering its roots in the individual genius.

This reclamatory process is hastened by another irregularly representative poet, William Cowper. A fundamentalist and a paranoid, Cowper believed himself to be the one man guilty of the unpardonable sin. Passing his life in an isolated search for expiation, he heard voices from the heavens while he sought peace in the practice of gardening. Cowper took a particular relish in nurturing vines—cucumbers, melons, ivy, jasmine—that grow by branching and rotating. Beginning with his "agglomerated pile" of dung, *The Task* (1785),[41] Cowper's epic on sofas, devotes large passages to gardening in general and cucumbers in particular. "Cautious he [the gardener] pinches from the second stalk/A pimple, that portends a future sprout,/And interdicts its growth. Thence straight succeed/The branches, sturdy to his utmost wish,/Prolific all, and harbingers of more." Branching, cutting, and turning, Cowper's vine resembles "the spiry myrtle with unwithering leaf" (Ref. 41, pp. 225–227). Cowper's poem branches and turns in much the same way, beginning with a davenport but ultimately winding through the whole universe. Unlike Thomson's, Young's, and Taylor's, Cowper's universe occupies little space. Like that most famous of domesticated spirals, the slinky, it extends a long way yet fits in the palm of the hand.

The undisputed king of later eighteenth-century vorticians is William Blake, the self-appointed prophet of Felpham, England. Although he offers hundreds of commentaries on spirals, helices, and vortices, he condenses his natural history of the whirlpool into a few tranquil lines.

> The nature of infinity is this: That every thing has its
> Own Vortex: and when once a traveller thro Eternity
> Has passed that Vortex, he perceives it roll backward behind
> His path, into a globe itself infolding, like a sun:
> Or like a moon, or like a starry majesty
> While he keeps onwards in his wondrous journey on the earth
> Or like a human form, a friend with whom he livd benevolent.
> As the eye of man views both the east & west encompassing
> Its vortex: and the north & south with all their starry host:
> Also the rising sun & setting moon he views surrounding
> His corn-fields and his valleys of five hundred acres square.
> Thus is the earth one infinite plane, and not as apparent
> To the weak traveller confin'd beneath the moony shade.
> Thus is the heaven a vortex passed already and the earth
> A vortex not yet passed by the traveller thro' Eternity. (Ref. 42, p. 233)

For Blake, vortices do more than explain the world. They explain explanation, down to its psychological roots. The entire vortectical history of perception transpires on Blake's 500-acre plot. Descartes, after all, had already suggested that vision results from the vortex of an object impinging on the vortex of the eye. Extrapolating from Descartes, Blake presents a world in which everything is whirling into perception, in which the most intimate of all systems, the mind, "infolds" itself into the universe. Blake's visionary creator-poet, Los, "bended/His Ear in a spiral circle outward," proceeding from the whorl of the ear to the whirl of creation (Ref. 42, p. 311). Blake's illuminated texts, too, involve both readers and characters in spirals spreading from body to text to nature, from arm to plant to verse (Fig. 17).

For Urizen, Blake's personification of determinism, the encounter with the vortex reenacts his own mental and physical fall from visionary to technician.

> But Urizen said: "Can I not leave this world of Cumbrous wheels, Circle o'er Circle . . .?
> When I bend downward, bending my head downward into the deep,
> 'Tis upward all which way soever I my course begin;
> But when a Vortex, form'd on high by labour & sorrow & care
> And weariness, begins on all my limbs, then sleep revives
> My wearied spirits; waking then 'tis downward all which way
> Soever I my spirits turn, no end I find of all. (Ref 43, p. 317)

"Creating many a vortex" from "Chaos to chaos," "surrounded by a shadowy vortex" (Ref. 43, pp. 316, 355), Urizen proceeds down the curve of a never-ending but always descending helix—a helix that, ironically, physically realizes his metaphysical disorder, his mad lust for geometry.

Like Gay, Hogarth, and other vorticians, Blake associates vortices with the turbulent motion of history. His post-Edenic serpent, "Orc," presides over a "cycle" of destructive oppression and entropic revolution, a repetitive but progressive cycle culminating in the crowning of a coiling snake (Fig. 18). Yet even in the historically expansive Orc cycle, the domestic triumphs over the cosmological. At the center of one revolutionary gyre Blake pivots a constricted human body (Fig. 19). In a sketch for *The Four Zoas*, Blake shows Ezekiel's visionary eyes of history spinning into a spiraling stream from a ring held by an individual (Fig. 20). It is only natural that the greatest history of all, Dante's *Divine Comedy*, should, for Blake, feature a guide borne by a spiral-wheeled car (Fig. 21).

Vortices, as forms, cannot be separated from the people who conceive, perceive, feel, and act upon them. "And all the Sciences were fix'd & the Vortexes began to operate/On all the sons of men, & every human soul terrified/At the turning of the wheels of heaven shrunk away inward, with'ring away" (Ref. 43, p. 317). The heroic duty of the poet is the subduing of the vortex, the controlling of philosophical systems and the joyous navigation of the whirlpool of history. One can either place oneself in the position of Adam and Eve, on the receiving end of a vortex

very one
ound Albions knees.
of thunders round.
roll'd far and wide

nful Hill
to death
hath studied the arts
nds are his abhorrence.
sed !
us bosom !

ion
ds !
ields !

n watry chariots.
rocession
t aloud as they
t Albions House

d on Man :
ready to burst :
s. & the immortal mansion
the deeps :
an endless curse.
f Moral Justice.

Fig. 17. Detail from William Blake, *Jerusalem*, Plate 36. LSU Middleton Library.

Fig. 18. William Blake, *Europe*, Plate 10. LSU Middleton Library.

expelling one from paradise (Fig. 22), or one can anticipate Milton, standing at the vertex of the vortex and turning it out and forward, using it to spin the world into "visionary forms dramatic" (Fig. 23).

Fig 19. William Blake, *America*, Plate 5. British Museum.

Fig. 20. Marginal sketch from William Blake's *The Four Zoas*. Courtesy Nelson Hilton.

Fig. 21. William Blake, *Illustrations to Dante*, "Beatrice Addressing Dante from the Car." Tate Gallery.

Fig. 22. William Blake, *The Expulsion.* Courtesy J. Springer Borck.

Fig. 23. William Blake, *Milton*, Plate 1. British Museum.

For better or worse, the vortex drives creation, if only the creation of chaotic systems. Taken from two perspectives on the vortex, these two pictures are worth more than two thousand books. If there is any lesson to be learned from Blake, it is that the long spin from Whichcote to Wordsworth is likely to cycle back into a familiar, if unstable, domestic space: the immediate, evervortectical, and inveterately personal process of vision.

References
1. L. Faller, *Turned to Account: The Forms and Functions of Criminal Biography in Late-Seventeenth and Early Eighteenth-Century England.* Cambridge University Press, Cambridge (1987).
2. P. Backscheider, *Daniel Defoe: Ambition and Innovation.* University Press of Kentucky, Lexington (1986).
3. R. E. Plotnick, A fractal model for the distribution of stratigraphic hiatuses, *J. of Geology* **94** (1986) 885–890.
4. D. Greene, *The Age of Exuberance.* Random House, New York (1970).
5. R. Descartes, tr. S. Mahoney, *Le Monde, ou Traité de la lumière.* Abaris Books, New York (1979).
6. J. Cottingham, R. Stoothoff, and D. Murdoch, trs., *The Philosophical Writings of Descartes.* Cambridge University Press, Cambridge (1985).
7. W. J. Bate, *From Classic to Romantic: Premises of Taste in Eighteenth-Century England.* Harper, New York (1961).
8. J. E. Saveson, Differing reactions to Descartes among the Cambridge Platonists, *J. of the History of Ideas* **21** (1960) 560–567.
9. R. Cudworth, ed. J. Harrison, *The True Intellectual System of the Universe.* Tegg, London (1845).
10. J. Glanvill, *Lux Orientalis, Or An Enquiry.* Cambridge (1662).
11. H. More, *Annotations upon ... Lux Orientalis.* London (1682).
12. J. Glanvill, *The Vanity of Dogmatizing.* London (1661).
13. H. More, *A Collection of Several Philosophical Writings.* London (1662).
14. D. Hirst, *Hidden Riches: Traditionalism from the Renaissance to Blake.* Eyre & Spottiswoode, London (1964).
15. G. Daniel, tr. T. Taylor, *A Voyage to the World of Cartesius.* London (1694).
16. J. Swift, ed. H. Davis, *A Tale of a Tub.* Blackwell, Oxford (1957).
17. J. Swift, *A Description of a City Shower.* London (1710).
18. B. Whichcote, ed. W. R. Inge, *Moral and Religious Aphorisms.* Elkin Mathews & Marrot, London (1930).
19. A. Cooper, Lord Shaftesbury, ed. J. M. Robertson, *Characteristics.* Peter Smith, Gloucester, Massachusetts (1963).
20. G. Savile, Marquis of Halifax, ed. W. Ralegh, *Complete Works.* Clarendon, Oxford, (1912).
21. K. L. Cope, *Criteria of Certainty.* University Press of Kentucky, Lexington (in press).

22. D. Vieth, Toward an anti-Aristotelian poetic: Rochester's *Satyr Against Mankind* and *Artemisia to Chloe*, with notes on Swift's *Tale of a Tub* and *Gulliver's Travels. Language and Style* **5** (1972) 123–145.
23. M. Battesting, *The Moral Basis of Fielding's Art*. Wesleyan University Press, Middletown, Connecticut (1959).
24. J. Wilmot, Earl of Rochester, ed. D. Vieth, *The Complete Works of John Wilmot, Earl of Rochester*. Yale University Press, New Haven (1968).
25. S. J. Gould, *Time's Arrow, Time's Cycle: Myth and Metaphor in the Discovery of Geological Time*. Harvard University Press, Cambridge (1987).
26. R. Hooke, *The Posthumous Works of Robert Hooke*. London (1705).
27. J. Gay, ed. V. Dearing, *John Gay: Poetry and Prose, Vol I*. Clarendon Press, Oxford (1974).
28. A. Pope, ed. J. Butt, *The Poems of Alexander Pope*. Yale University Press, New Haven (1963).
29. Rev. T. Maurice, *The Oxonian: A Poem in Imitation of the Splendid Shilling*. Oxford (1678).
30. J. Philips, ed. M. G. Lloyd-Thomas, *The Poems of John Philips*. Blackwell, Oxford (1927).
31. I. Brown, *London: An Illustrated History*. Studio Vista, London (1965).
32. W. Hogarth, *The Analysis of Beauty*. London (1753).
33. W. Shenstone, *Unconnected Thoughts on Gardening*. London (1764).
34. R. Hyde, *Panoramania! The Art and Entertainment of the "All-Embracing" View*. Trefoil Publications, London (1988).
35. E. Young, *The Complaint: Or, Night Thoughts*. Andrus, Hartford (1843).
36. N. Hilton, *Literal Imagination: Blake's Vision of Words*. University of California Press, Berkeley and Los Angeles (1983).
37. K. Raine and G. M. Harper, *Thomas Taylor the Platonist: Selected Writings*. Boloingen, Princeton (1969).
38. J. Thomason, ed. J. L. Robertson, *The Complete Poetical Works of James Thomson*. Oxford University Press, London (1908).
39. G. Tillotson, P. Fussell Jr., and M. Waingrow, *Eighteenth-Century Literature*. Harcourt Brace Jovanovich, New York (1969).
40. C. Smart, ed. K. Williamson, *The Poetical Works of Christopher Smart, Vol. I*. Clarendon, Oxford (1980).
41. W. Cowper, ed. W. Benham, *The Poetical Works of William Cowper*. MacMillan, London (1921).
42. W. Blake, ed. D. V. Erdman, *The Illuminated Blake*. Oxford University Press, London (1975).
43. W. Blake, ed. G. Keynes, *Blake: Complete Writings*. Oxford University Press, London (1966).

INDEX

A

aborigines 2
active media 165
alga 336
Archimedean spiral 90
art 123, 389, 395
Art Noveau ix
astronomy 95
attractors 76
auditory never 18
autocatalytic reactions 224

B

bacteria 84
Belousov-Zhabotinski reaction 165, 180, 223
bifurcations 73, 201
biology 73, 221
biochemistry 78, 224
Blake, W. 1
Blake, W. 399
bones 13, 14
bromate oscillator 229
broccoli 260
Bursill, L. 295
buttercup 42

C

cardiac tissue 73
cellular automata 170
chaos 73, 165

chemistry 224
chirality 281, 301
chrysanthemums 333
Coke can 196
Cook, T. 83
Cope, K. 399
Cortie, M. 369
Couette, M. 187
crystals 328

D

deer 13
Denes, A. 389
Descartes, R. 17, 399
Dictyostelium discoideum 222
Dixon, R. 353
DNA 83
double spirals 159
dynamical spirals 73

E

Eckert, E. 63
ecology 221
eggs 41, 391
electromagnetic theory 281
Elmegreen, B. 95
entropy 85, 346
environment 1
ethane 266
evolution vi, 324

F

Fali culture 5
feedback 224
Fibonacci 298, 316, 323, 333, 355

flowers 308, 329
fluids 187
fossils 361
fractals vii, 107, 129, 135, 259, 332
Friedman, D. 251

G

galaxies 73, 95
genetics 326
glucose 78
glycolytic oscillator 78
gnomon 339
golden spiral 47, 323, 354
golden ratio 47, 323, 341, 354
Great Wave 11
green spirals 353
grey spirals 359
growth 365
growth filaments 84, 336
gyrant bifurcations 215

H

handedness 282, 298
Hargittai, I. x
heart tissue 73, 75
Helmet 395
Holden, A. 73
horns v
hysteresis 215

I

inflorescence 42

J

Jean, R. 323
Julia sets 129, 135

K

Kappraff, J. 1
Kent, E. 395
Klein, D. 83
Koros, E. 221

L

labyrinth 417
Lakhtakia, A. 281
limit cycles 76
literature 399
Loeb, A. 47
logarithmic spiral 48

M

Mandelbrot set 140
map projections 389
Markus, M. 165
Michelitsch, M. 129
molecular orbitals 265
molluscs 369
multiarmed spirals 242
musical instruments viii
myth 1

N

nature 1
Needham, A. 295

nerves 77
Newman, R. 123

O

orbitals 265
oregonator model 230
oscillations 73, 221, 317

P

pace makers 234
painting 125
parastichies 295
path curves 35
Penrose lattices 168
period doubling 80
phase model 232
Philip, A. 135
photocopiers 395
phylotaxis 251, 295, 323, 353
Pickover, C. vii, x, 134
pine cone 42, 328
plants 42, 251, 295, 323, 353
polymers 84, 251
projective geometry 27
Pythagorean spirals 63

Q

quasi-periodicity 76
quasicrystallography 295

R

random spirals 83
random walks 84

Rossler, O. 129
rotating spirals 244
Rouse, J. 295

S

Schwenk, T. 6
scrolling walks 87
scroll waves 244
seashells 18, 73, 325, 369, 416
Seitz, W. 83
self-avoiding walks 90
self-similarity 107, 369
self-similar spirals 111
self-organization 182
snails 18
staircases ix
Starry Night 19
Stewart, I. 187
strawberry flower 252
sunflowers 295, 364
symmetry breaking 193

T

Taylor apparatus 188
telome theory 324
Thompson, D. v, 83, 259
time 409
trigger waves 234
universality 85

V

van der Pol oscillator 76
van Gogh, V. 17
Varney, W. 47
vortices 12, 15, 187, 213, 399

walks 84
water 6, 19, 187, 337
water animals 8
waves 8, 11, 19, 102, 166, 221, 232
Wicks, K. 107
Wizard of Oz vi

Y

Young diagram 88

Z

Zeeman catastrophe machine 199